AF356648

Advances in Sustainable River Management

Advances in Sustainable River Management: Reconciling Conflicting Interests under Climate Extremes

Editors

Andrzej Wałega
Alban Kuriqi

MDPI • Basel • Beijing • Wuhan • Barcelona • Belgrade • Manchester • Tokyo • Cluj • Tianjin

Editors
Andrzej Wałega
Department of Sanitary
Engineering and Water
Management
University of Agriculture
in Krakow
Krakow
Poland

Alban Kuriqi
CERIS–Civil Engineering
Research and Innovation
for Sustainability
Instituto Superior Técnico
University of Lisbon
Lisbon
Portugal

Editorial Office
MDPI
St. Alban-Anlage 66
4052 Basel, Switzerland

This is a reprint of articles from the Special Issue published online in the open access journal *Sustainability* (ISSN 2071-1050) (available at: www.mdpi.com/journal/sustainability/special_issues/river_management).

For citation purposes, cite each article independently as indicated on the article page online and as indicated below:

LastName, A.A.; LastName, B.B.; LastName, C.C. Article Title. *Journal Name* **Year**, *Volume Number*, Page Range.

ISBN 978-3-0365-2897-7 (Hbk)
ISBN 978-3-0365-2896-0 (PDF)

Contents

About the Editors

Andrzej Wałega

Andrzej Wałega is a Doctor in the Department of Sanitary Engineering and Water Management at the University of Agriculture in Krakow, Poland, where he teaches hydrology, modeling of hydrological processes, and water management in urban catchments. He holds a Ph.D., a M.S., and a M.Eng. in Environmental Engineering from the University of Agriculture in Krakow. He has been working as a researcher and lecturer since 2005. Currently, he is the President of the Association of Polish Hydrologists. He was an expert in a project study of hydraulic modeling against floods—in its first stage—in preparation for the implementation of EU directive No 2007/60EC, and he was also pivotal in the second stage of this study by supporting the competence and readiness of Georgian institutions in which the Polish Center of International Aid was a leader. He is a Guest Editor in *Sustainability*, *Atmosphere*, *Water* (MDPI), Intelligent Automation and Soft Computing, and a Member of the Editorial Board of *Sustainability*. He has co-authored over 190 journal and conference papers, books and book chapters, and technical guides related to hydrology and water management.

Alban Kuriqi

Alban Kuriqi is a researcher at CERIS–Civil Engineering Research and Innovation for Sustainability, Instituto Superior Técnico, University of Lisbon. He holds a Ph.D. in Civil Engineering focusing on River Restoration and Management from Instituto Superior Técnico, the University of Lisbon. He obtained a five-year integrated diploma M.Sc. Degree in Hydrotechnical Engineering from the Polytechnic University of Tirana. He also holds an M.Sc. degree in Civil Engineering, majoring in Water Resources and Hydraulics from Epoka University and Grenoble Institute of Technology. He has been awarded several scientific awards and fellowship grants. He has co-organized and been a member of five international conferences. He has published over 30 peer-reviewed papers, 3 book chapters and presented over 20 communications at international conferences. He acts as a reviewer for over 90 international journals (i.e., ISI journals), two international conferences, and Marie Skłodowska-Curie's research proposal grants. He also develops activity as an editor for four international journals (i.e., ISI journals).

Preface to "Advances in Sustainable River Management: Reconciling Conflicting Interests under Climate Extremes"

Because different human pressures influence on freshwater ecosystems, there is still a very important discussion about the consequences of these activities mainly for rivers and ways to decrease risk of degradations of river ecosystems. Assurance of optimal living conditions for aquatic organisms is one of the most important principles related to sustainable water management. Conservation and wellbeing of freshwater ecosystems are closely linked to the preservation of the natural hydrological regime. Therefore, the main objective of this book is to contribute in understanding and provide science-based knowledge, new ideas/approaches and solutions in sustainable river management, to improve water management policies and practices following different environmental requirements aspects. In this book are present nine chapters where the following topics were presented: an advanced hydro-economic water allocation and management model (WAMM) to help resolve water shortages and disputes among river basin units under severe drought conditions [1], a possibility of using bottom sediments from dam reservoirs in earth structures was considered [2], using advanced machine learning based on the Artificial Neural Network, optimized with a Genetic Algorithm for seasonal groundwater table depth prediction [3], comparison of different hydrological method to assessment of e-flow [4], modeling flood zones based on advantage 2D hydraulic model Flood2D [5], effects of stream flow changes on cruise tourism [6], sensitivity of simple rainfall-runoff model on errors of flood prediction [7], some problems with human pressures on the Thames ecosystem [8] and role of geosynthetics in civil engineering, mainly in dyke construction [9].

The book can be useful for water managers autorities, civil engineers, researchers and students to show current knowledge best management practices in river ecosystems. The editors are thankful to all authors who submitted very interesting papers, reviewers for very constructive opinions about the submitted papers and editors for support during the publication process.

References

1. Jeong, G.; Kang, D. Hydro-Economic Water Allocation Model for Water Supply Risk Analysis: A Case Study of Namhan River Basin, South Korea. Sustainability 2021, 13, 6005.

2. Koś, K.; Gruchot, A.; Zawisza, E. Bottom Sediments from a Dam Reservoir as a Core in Embankments—Filtration and Stability: A Case Study. Sustainability 2021, 13, 1221.

3. Pandey, K.; Kumar, S.; Malik, A.; Kuriqi, A. Artificial Neural Network Optimized with a Genetic Algorithm for Seasonal Groundwater Table Depth Prediction in Uttar Pradesh, India. Sustainability 2020, 12, 8932.

4. Suwal, N.; Kuriqi, A.; Huang, X.; Delgado, J.; Młyński, D.; Walega, A. Environmental Flows Assessment in Nepal: The Case of Kaligandaki River. Sustainability 2020, 12, 8766.

5. Petroselli, A.; Florek, J.; Młyński, D.; Ksiażek, L.; Wałega, A. New Insights on Flood Mapping Procedure: Two Case Studies in Poland. Sustainability 2020, 12, 8454.

6. Yao, Y.; Mallik, A. Stream Flow Changes and the Sustainability of Cruise Tourism on the LRiver, China. Sustainability 2020, 12, 7711.

7. Młyński, D. Analysis of Problems Related to the Calculation of Flood Frequency Using Rainfall-Runoff Models: A Case Study in Poland. Sustainability 2020, 12, 7187.

8. Richardson, M.; Soloviev, M. The Thames: Arresting Ecosystem Decline and Building Back Better. Sustainability 2021, 13, 6045.

9. Rimoldi, P.; Shamrock, J.; Kawalec, J.; Touze, N. Sustainable Use of Geosynthetics in Dykes. Sustainability 2021, 13, 4445.

Andrzej Wałega, Alban Kuriqi

Editors

 *sustainability*

Editorial

Preface to "Advances in Sustainable River Management: Reconciling Conflicting Interests under Climate Extremes"

Andrzej Wałęga [1,*] and Alban Kuriqi [2,*]

[1] Department of Sanitary Engineering and Water Management, University of Agriculture in Krakow, St. Mickiewicza 24/28, 30-059 Krakow, Poland
[2] CERIS, Instituto Superior Técnico, Universidade de Lisboa, Av. Rovisco Pais 1, 1049-001 Lisbon, Portugal
* Correspondence: andrzej.walega@urk.edu.pl (A.W.); albankuriqi@gmail.com (A.K.)

Citation: Wałęga, A.; Kuriqi, A. Preface to "Advances in Sustainable River Management: Reconciling Conflicting Interests under Climate Extremes". *Sustainability* **2021**, *13*, 10087. https://doi.org/10.3390/su131810087

Received: 11 August 2021
Accepted: 2 September 2021
Published: 9 September 2021

Publisher's Note: MDPI stays neutral with regard to jurisdictional claims in published maps and institutional affiliations.

Safeguarding optimal living conditions for aquatic organisms is one of the most important principles of sustainable water management. The conservation and wellbeing of freshwater ecosystems are closely linked to the preservation of the natural hydrological regime. On the other hand, human activities often alter the natural hydrologic regime and habitat conditions for aquatic ecosystems by substantially affecting several essential life-stages of aquatic organisms, such as migration and spawning of fish macroinvertebrates and other aquatic species. Ever-increasing water exploitation, mainly for water supply, irrigation, and renewable energy, has degraded freshwater ecosystems, notably rivers. Thus, planned or existing hydraulic structures, such as hydropower plants, dams, and water intakes, among others, may adversely affect aquatic organisms' living conditions.

Further, the climate extremes and water scarcity exacerbated by climate change induce additional stress in freshwater ecosystems and may stimulate conflicts among water users. Therefore, ensuring optimal living conditions for aquatic organisms is one of the most critical sustainable development goals to stop biodiversity decline.

Additionally, we are aware that water is needed for several vital human activities, where agricultural and industrial seems to be primary water consumers; in a situation in which the world has observed more frequent droughts and moments of water scarcity, water systems management requires the most advanced approaches and tools for rigorously addressing all dimensions involved in the sustainability of its development.

This book presents nine chapters featuring some of the main lines of research around sustainable river management, emphasizing international experiences across several countries. These chapters represent a collection of articles published in a Special Issue entitled "Advances in Sustainable River Management: Reconciling Conflicting Interests under Climate Extremes," published by *Sustainability* (MDPI) in 2020 and 2021. The editors of this book would like to acknowledge the excellent guidance and efforts from the editorial team at MDPI and the quality of the experience and research presented by the 24 authors who have contributed to this Special Issue's academic and technical success.

This book covers a wide range of literature reviews and original research interventions. Moreover, it presents studies about practical problems with the use of rainfall-runoff models to simulate floods, a new approach to the design of flood zone areas that can help with the assessment risk of flood disaster, as well as discussions about more reasonable hydrological methods for the assessment of e-flows, a crucial issue regarding optimal assurance conditions in aquatic organisms. This book also presents modern tools to support water management and the prediction of water resources. It explores the use of sediment and geotextiles in water engineering, human pressure on the ecological status of rivers, and the importance of proper water management when using rivers.

We hope that this collection of papers will be of use to academics and practitioners in helping to provide sustainable water management strategies under different natural and artificial conditions.

Funding: This research received no external funding.

Conflicts of Interest: The authors declare no conflict of interest.

 *sustainability*

Review

The Thames: Arresting Ecosystem Decline and Building Back Better

Martin Richardson * and Mikhail Soloviev *

Department of Biological Sciences, Royal Holloway University of London, Egham, Surrey TW20 0EX, UK
* Correspondence: Martin.Richardson.2014@live.rhul.ac.uk (M.R.); Mikhail.Soloviev@rhul.ac.uk (M.S.)

Abstract: The Thames is an iconic river of cultural and historical importance. A cyclical process of deterioration during the last two centuries, followed by technology-driven restorations, including two major sanitation projects with a third currently underway, has produced detrimental effects on the Thames ecosystem. This paper overviews the river ecology, pollution and other anthropogenic pressures, which lead to biodiversity loss and the proliferation of non-native, pollution-tolerant species. This article further reviews past and current management, sampling and assessments trends and provides an objective overview of remediation, restoration and monitoring needs, practices and research gaps. Here, we argue that restoration work, if maladapted, can be ineffective in improving resilience or have unexpected side effects that make matters worse rather than better. We explain the need for a broader view of river restoration and management including consideration of species transplants in achieving overall sustainability against a backdrop of accelerating change in the Anthropocene.

Keywords: the Thames; ecosystem; pollution; biodiversity; non-native species; ecosystem management; remediation; restoration; monitoring

 check for updates

Citation: Richardson, M.; Soloviev, M. The Thames: Arresting Ecosystem Decline and Building Back Better. *Sustainability* **2021**, *13*, 6045. https://doi.org/10.3390/su13116045

Academic Editors: Andrzej Wałęga and Alban Kuriqi

Received: 16 April 2021
Accepted: 24 May 2021
Published: 27 May 2021

Publisher's Note: MDPI stays neutral with regard to jurisdictional claims in published maps and institutional affiliations.

1. Introduction

The Thames catchment covers an area of over 16,000 km^2. The Thames system is composed of several connected subsystems which can be divided into 'natural' or geographically described, and manmade, engineered elements, each of which has cobenefits and trade-offs beyond their initial purpose (Figure 1). The river has a low gradient and is well-mixed and generally shallow. Above Richmond, the river has been transformed into a series of lacustrine stretches connected by locks and weirs (Figure 2, Supplementary Figure S1). The dominant factor governing species distributions in the tidal Thames is the increasing salinity from Teddington down towards the estuary although flow regime, brine from desalination plants, groundwater inputs, warm water from sewers and power stations, chemical pollution and low oxygen saturation, each of which can create temporary and localised effects, also play a part. The Teddington weir delineates the tidal river; saline conditions intrude as far as Teddington during times of low flow. Water is released from the Richmond lock during high flow events in the upper catchment, which can alter salinity and flow downriver. The Thames barrier at Woolwich, in association with other elements of the Thames flood protection system (Figure 3, Supplementary Figure S1) [1], provides flood control, particularly for the low-lying areas of London. The barrier is only closed for short durations when flow control is needed, allowing for fish passage most of the time. The barrier was completed in 1984 and designed to withstand a projected annual sea-level rise of 6 mm to 8 mm, which appeared more than adequate at the time. However, analysis of the compounding effects of accelerating sea-level rise, extreme rainfall events and storm surges is raising the possibility of it being overwhelmed sooner [2].

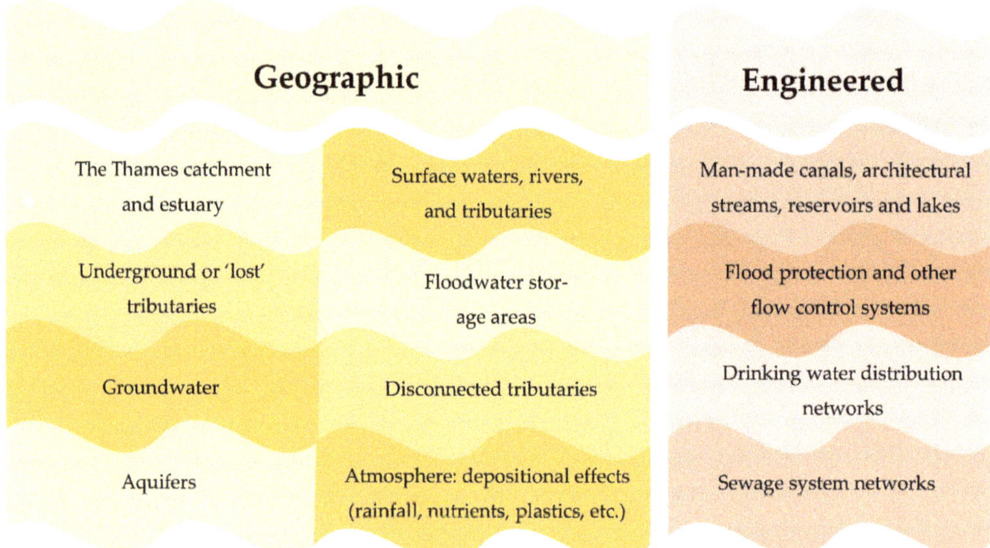

Figure 1. Natural and engineered components of the Thames catchment system.

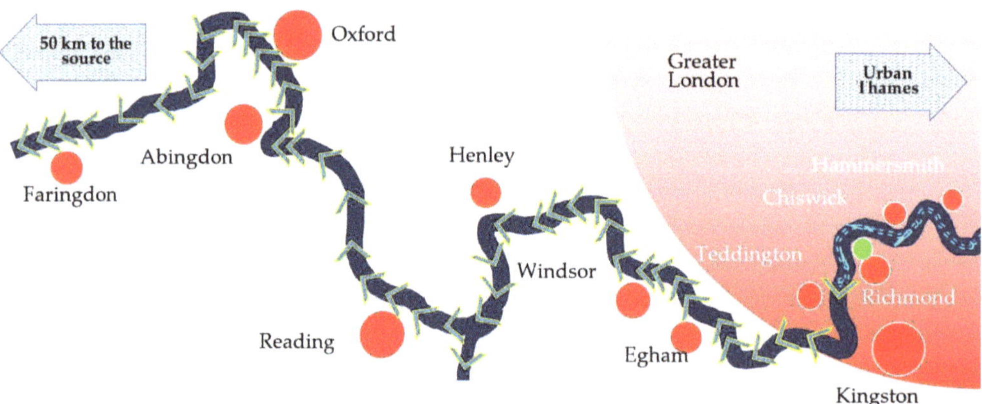

Figure 2. Natural and engineered components of the Thames catchment system. Non-tidal reaches of the Thames with 45 locks (arrowheads). The Teddington lock and the weir mark the tidal river limits. The double-dashed line shows tidal areas (brackish water). The Royal Botanic Gardens (Kew) are indicated in green.

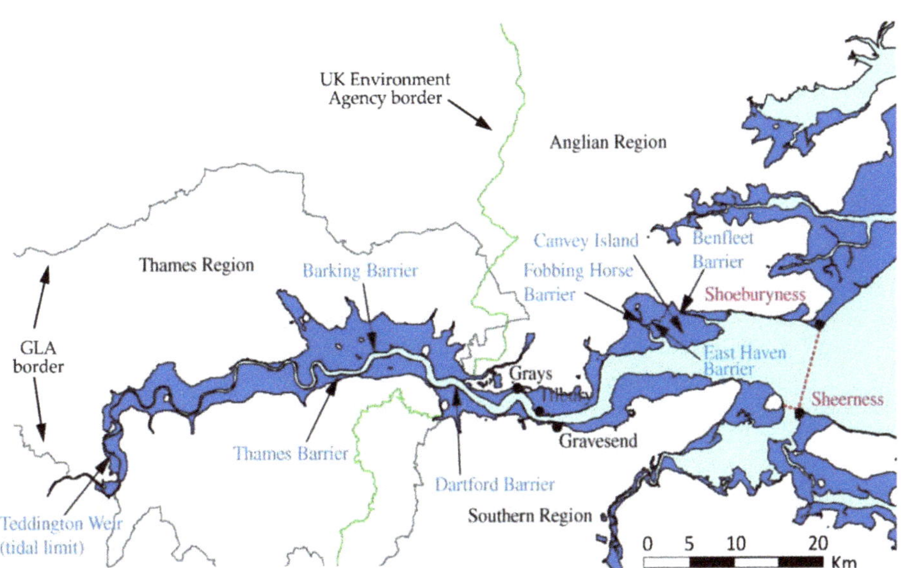

Figure 3. The tidal Thames flood-plain with defences labelled in light blue. The Greater London Area (GLA) is indicated with a grey border. The red dotted line shows the lower limit of the flood protection system. The green boundary delineates the operational area of the UK Environment Agency at the time (2005). Modified from [1], reproduced with permission.

2. Ecosystem Services

2.1. Recreation, Transport and Water

Historically, water depth and navigability were maintained for industrial barge transport. More recently, the Canal and Rivers Trust (CRT) estimated that on average some 246,000 boating enthusiasts visit the canal and river network each week [3]. Wash from boats, even small craft, can greatly affect aquatic life, mobilise sediment and damage riverbanks. However, boating groups, anglers' associations and other groups participate in maintaining the river and positively influence development and restoration work. The upper reaches are monitored for changes to bathymetry and depth for possible dredging needs, and dredging is also conducted in the lower reaches and in the estuary. Large silt banks have built up in downstream stretches. They change morphology, reduce flow capacity, and accumulate xenobiotics such as metals that adhere (adsorb) to the particles and increase the larval survival in invasive species [4]. The sediment is largely anoxic with occasional black, sulphide deposits [5] inhibiting oxygenation by worms or colonization by plants. However, the lack of oxygen in combination with the increased organic matter aids in the reduction of nitrate and preserves natural revetments (e.g., wood or spiling) that would otherwise rapidly degrade. Deposits of nutrients, particularly phosphorus, accumulate in the mud and can be remobilized through disturbance caused by construction work, boat traffic or dredging in the river, but also promote carbon sequestration and contaminant uptake through increased vegetation [6,7]. Damage caused by construction work is sometimes traded-off against a contribution (funds) towards restoration, but nutrient release increases algal growth and aids phosphorous-loving (and excreting) aquatic species belonging to the Ponto–Caspian group such as *Dreissena polymorpha* (Pallas, 1771) [8,9].

Large abstractions are taken upstream of Teddington for agriculture and drinking water and major inputs of wastewater occur as sewage treatment effluent. Increasing domestic and industrial water use means that effluent from sewage treatment plants can comprise a major component of stream flow downstream of treatment facilities. At times, wastewater can even comprise the majority of the river flow. Water abstraction from above the tidal area and changes in rainfall with increased intensity of storms have exaggerated the extremes in flow regimes with periods of reduced flow, many more ephemeral streams and increased flooding. There are several remaining Thames tributaries, which provide refuge for fish during times of low oxygen in the main channel, but many have been closed off. Large stretches of the banks and islands have been armoured to protect against erosion and the river has generally narrowed. Sediment inputs from agricultural tilling [10] have created mud banks which decrease the volume of the river basin through related algal and silt deposits. Modifications including increased reservoir capacity such as the opening of the Wraysbury reservoir in 1970 provide additional stored capacity to maintain improved flows during periods of drought [11]. Licensing has been brought in to manage abstractions, which frequently increase during periods of drought, reducing flows to the point where they cause stress to the ecosystems [12].

2.2. Fisheries and Ecosystem Decline

Prior to 1800, the river was clean enough to support large populations of many species [13]. Diadromous species which use the ocean and freshwater at different stages of their lifecycle including smelt, salmon, eels, sprat and flounder, were caught for food along the river for centuries. Fishing and fisheries have been an integral part of the Thames community for several hundred years. Fishing communities lived for generations at several locations along the river including Kew and Chiswick [14]. The fishery included coarse fish species which are not typically consumed today such as pike, however, they are still currently important to anglers. In the past, large quantities of shellfish, starfish and shrimp were landed from the estuary [15]. Starfish were caught for use as fertiliser on crops. The river was wider in many places, with extensive gravel banks that once attracted large spawning aggregations of smelt. In recent times, fishing has been reduced to mainly recreational angling for coarse species including carp and pike in the river, and some sea angling in the estuary. Small but important commercial fisheries continue in the estuary for whelks, cockles and oysters. The modern Thames above Teddington is freshwater and contains a typical UK species assemblage for a degraded river (Figure 4) including non-native mussels, fish, invertebrates and invasive aquatic plants [16].

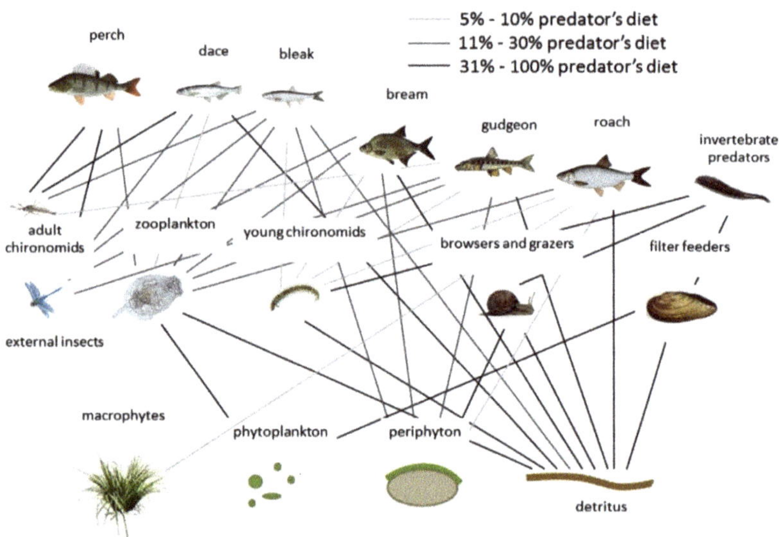

Figure 4. Simplified freshwater food web from the non-tidal part of the Thames. Note that no non-native species are shown. Modified from [16], reproduced with permission.

Water is well oxygenated after passing over the weirs, which causes spikes in saturation in the reaches immediately below and maintaining localised communities of freshwater species, especially in the area above Richmond, including native unionid mussels and, more recently, non-native bivalves. However, these primarily freshwater faunae are sensitive to low flows caused by periods of drought and abstractions, which are taken further upstream. In the past, this has given rise to cyclical declines in many taxa, including caddisflies, mayflies, unionid mussels, isopods and leeches, as occurred, for example, during pronounced drought in 1989 and 1990. Some species of fish including eels and snakehead are able to leave the water to navigate weirs and obstacles. However, barriers still impede progress both upriver and downriver during migration. Potentially invasive bivalves including the zebra mussel, which were predicted to increase substantially, have now declined from this part of the river.

The Teddington weir marks the transition between the purely freshwater and tidal parts of the river. The lock gates are lowered at high slack water and fish passage throughout the river has recently been provided at locks and weirs. Downriver from Teddington, the water has low salinity most of the time, however, this varies with the seasons and flow. Many freshwater fish including barbel, *Barbus barbus* (Linnaeus, 1758), and carp, *Cyprinus carpio* (Linnaeus, 1758), are tolerant of brackish water and can thrive in even the lower reaches. A total of approximately 120 species of fish live in the Thames [3,17,18] but many are rare, non-native (exotic) or not representative of resident populations (Table 1) [3,18,19]. Less than ~20 pollution tolerant species comprise a majority abundance (Table 2) [20].

Table 1. Non-native fish species recorded from the Thames (complete catchment) [1].

Common Name	Species	Source	Suspected Origin [2]
Wel's catfish	*Silurus glanis* (Linnaeus, 1758)	[17]	Stocking
Siberian sturgeon	*Acipenser baerii* (Brandt, 1869)	[3]	Pet trade
Sterlet	*Acipenser ruthenus* (Linnaeus, 1758)	[3]	Pet trade
Short-snouted seahorse	*Hippocampus hippocampus* (Linnaeus, 1758)	[17]	Unknown
Bitterling	*Rhodeus sericeus* (Pallas, 1776)	[17]	Ornamental
Koi carp	*Cyprinus carpio* (Linnaeus, 1758)	[3]	Ornamental
Grass carp	*Ctenopharyngodon idella* (Valenciennes, 1844)	[3]	Stocking
Pumpkinseed	*Lepomis gibbosus* (Linnaeus, 1758)	[3]	Stocking
Sunbleak	*Leucaspius delineatus* (Heckel, 1843)	[17]	Pet trade
Topmouth gudgeon	*Pseudorasbora parva* (Temminck and Schlegel, 1846)	[3]	Pet trade
Zander	*Sander lucioperca* (Linnaeus, 1758)	[3]	Stocking
Goldfish	*Carassius auratus* (Linnaeus, 1758)	[17]	Aquaculture
Goldfish × carp hybrid		[18]	
Orfe	*Leuciscus idus* (Linnaeus, 1758)	[17]	Ornamental
Bream × Orfe hybrid		[18]	
Rainbow trout	*Oncorhynchus mykiss* (Walbaum, 1792)	[17]	Stocking
Brook trout	*Salvelinus fontinalis* (Mitchill, 1814)	[17]	Stocking
Fathead minnow	*Pimephales promelas* (Rafinesque, 1820)	[18]	Ornamental
Guppy	*Poecilia reticulata* (Peters, 1859)	[17]	Ornamental
European catfish	*Silurus glanis* (Linnaeus, 1758)	[17]	Stocking

[1] Unusual, exotic species are mentioned from time to time in the media as curiosities when caught by anglers. Many non-native species originate from imports for food, bait for fishing, and the pet and ornamental fish trade. They are frequently released into surface waters when they become too big for an aquarium or pond. [2] The explicit means of arrival is not usually known. Additional, cryptic species probably exist in the river, however, most are unlikely to sustain reproducing populations.

Table 2. Principal fish species in the tidal Thames, from [20].

Common Name	Species
European smelt	*Osmerus eperlanus* (Linnaeus, 1758)
European eel	*Anguilla anguilla* (Linnaeus, 1758)
Common dace	*Leuciscus leuciscus* (Linnaeus, 1758)
Common goby	*Pomatoschistus microps* (Krøyer, 1838)
Dover sole	*Solea solea* (Linnaeus, 1758)
European seabass	*Dicentrarchus labrax* (Linnaeus, 1758)
European sprat	*Sprattus sprattus* (Linnaeus, 1758)
Flounder	*Platichthys flesus* (Linnaeus, 1758)
Herring	*Clupea harengus* (Linnaeus, 1758)
Stickleback	*Gasterosteus aculeatus* (Linnaeus, 1758)
Pouting	*Trisopterus luscus* (Linnaeus, 1758)
Roach	*Rutilus rutilus* (Linnaeus, 1758)
Sand goby	*Pomatoschistus minutus* (Pallas, 1770)
Whiting	*Merlangus merlangus* (Linnaeus, 1758)

The tidal Thames supports a variety of migratory and non-native birds (swans, ducks, geese and other wildfowl) and a variety of fish (flounder, bass and mullet to name a few) as well as invertebrates and molluscs, including freshwater snails and bivalves. Aquatic mammals such as seals and dolphins mainly use the estuary but will on occasion come upriver as far as Teddington. In addition, there are a variety of other mammals including foxes, water voles, cats and rats. Recent increases in mammal populations in the Thames estuary do not necessarily translate to improved health of the river but may be due to other causes including behavioural changes as marine mammal colonies are known to relocate for reasons including seeking refuge from disturbance. Certainly, bioaccumulation of toxins in fish can be expected to have profound effects on animals higher up the food web, especially as it has been shown that increases in particulate matter in water can increase uptake of toxic chemicals [21]. Sea grass occurs off the Essex coast but is under increasing pressure, especially from sediment [22]. Native and non-native molluscs occur in the estuary, and the most abundant is the invasive slipper limpet, *Crepidula fornicata* (Linnaeus, 1758). Despite recent declines, there are surviving fisheries for the native oyster, *Ostrea edulis* (Linnaeus, 1758), the imported American oyster, *Crassostrea gigas* (Thunberg, 1793), as well as the endemic cockle, *Cerastoderma edule* (Linnaeus, 1758), and whelks.

3. Pollution

The combined sewer network disposes of domestic and industrial waste as well as urban drainage and can release raw sewage, with added hydrogen peroxide, to the river, particularly during periods of heavy rain or flooding due to limited capacity [21]. Historical levels of 150 million tonnes a year were discharged in the 1850s. Around 40 million tonnes were still being released to the river in 2011 but recent improvements have reduced this to 18 million tonnes, the majority entering the river higher in the tidal reaches at the Hammersmith, Lots Road and Western Pumping stations [17,23,24]. In the London area, many tributaries have been lost and built over, with some assimilated into the sewer system [25]. Nutrient inputs from farming and livestock stimulate algal growth and related bacterial and viral communities [26]. These amplify greenhouse gases released by the river and promote growth and pathogenicity in microbial communities. Microbial assemblages are affected by antibiotics released in effluent from sewage treatment works [27,28]. Antibiotics may also trigger toxin release in Harmful Algal Blooms (HABs) by initiating lysis, causing a cascading reaction throughout the bloom in response to the presence of cyanotoxin in the water [29–31]. Oxygen depletion occurs in the lower reaches of rivers due to bacterial respiration associated with the consumption of organic material and the subsequent senescence and decomposition. Tidal areas of the river Thames are affected by oxygen depletion, partially because of the periodic sewage release (Figure 5, Supplementary Figure S1) [32] and algal growth stimulated by an excess of nutrients, especially at the mouth of the estuary, which is a common and increasing feature of many rivers [33,34]. Other pollutants enter the water from historic (unlined) landfills (Figure 6, Supplementary Figure S1) [35,36], which become grossly exposed when subjected to increasing erosion [36].

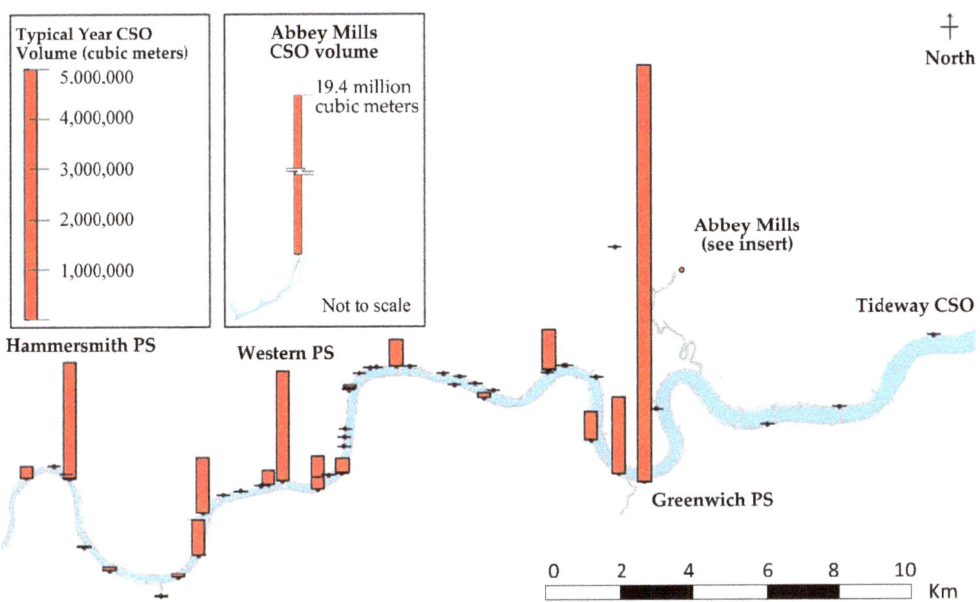

Figure 5. Typical annual sewage releases (prior to 2013) from combined sewer system outfalls and pumping stations (PS) in the upper tidal Thames. CSO stands for combined sewage overflow. Red bars indicate release volume in cubic metres; all bars are shown to the same scale, except for the Abby Mills CSO. Pumping stations raise sewage to higher elevation levels to facilitate gravitational flow in the system of pipes towards the estuary. Modified from [32], reproduced with permission.

Rainfall can more than double flows in the sewer network, which is connected to the river at multiple locations through outfalls. In some Thames tributaries, such as the river Lee, flows are greatly reduced due to water abstractions for agriculture and drinking water. Downstream of these locations, especially at times of low flow, the river can be over 40% effluent. Sewage treatment does not remove all pollutants from water. Some pharmaceuticals, which are specifically designed to be effective at a low dose, including psychotropic drugs, many hormones, antibiotics, pain killers and antidepressants remain in the effluent [31,37,38]. Many are known to have detrimental effects on aquatic animals [39,40] and can degrade to potentially more harmful transformation products through biotic and abiotic processes [41,42]. After treatment, wastewater and its associated sludge is frequently reused for crop irrigation and fertilisation, which allows contaminants to circulate and further accumulate in the environment. In a recent study, sediment samples collected from multiple locations in the tidal Thames revealed contamination from persistent organic pollutants such as PCBs and heavy metals [43]. Soil samples taken from an island in the river showed increasing contaminant concentrations from a 20 cm to 40 cm depth, which is suggestive of accumulation in riverbanks and other areas subjected to flooding (Supplementary Figures S2 and S3, Supplementary Table S1). Similar findings have been reported elsewhere [44].

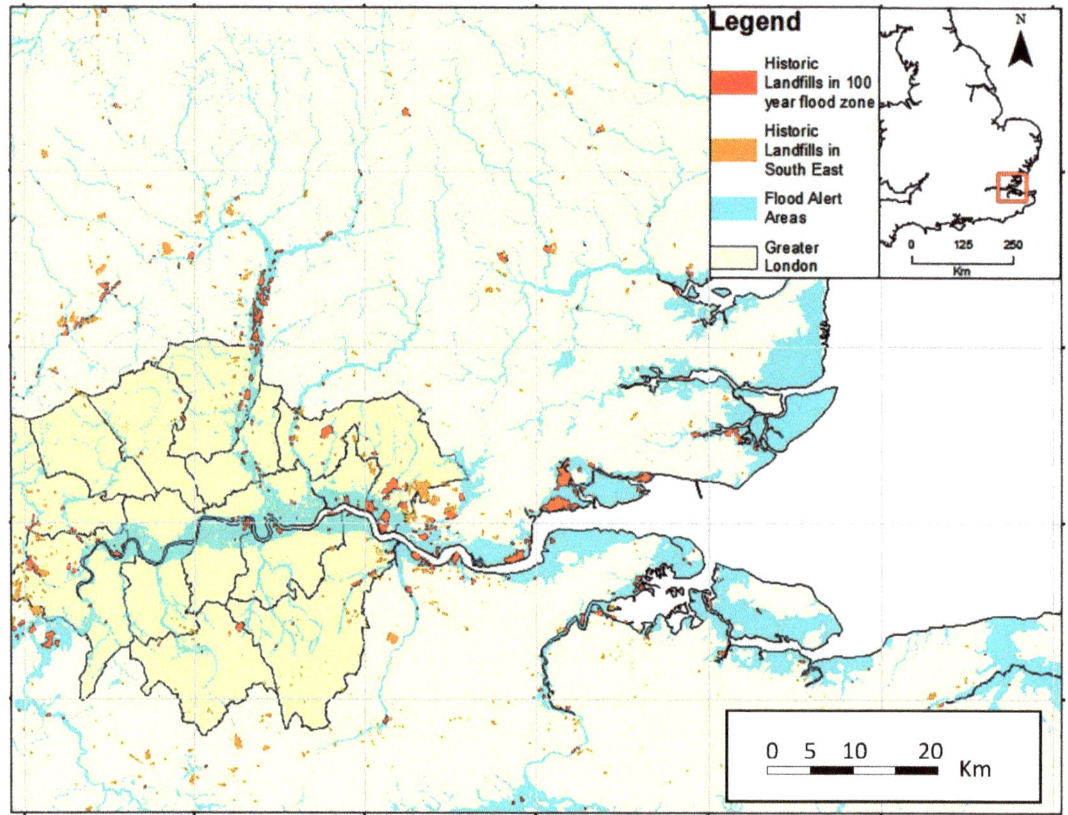

Figure 6. Historic landfills in the Thames Estuary. Blue areas are designated at risk of flooding by the Environment Agency, the red and yellow areas are historic landfills that are leaching contaminants. Landfills are also vulnerable to subsidence thereby amplifying the effects of sea-level rise and flooding. Reproduced from [36] under a Creative Commons Attribution Generic License.

The Thames, as well as other rivers, now contain large quantities of nano-plastics, micro-fibres and other micro-plastics which cannot feasibly be removed but aggregate with biogenic particles and eventually settle into the sediment [45] or are flushed into the ocean. However, these are continually replenished by others through aerial deposition as well as run-off from the land and roads [46,47]. Materials released into the river, including rubbish and sewage, especially in the upper regions of the tidal Thames, do not completely flush but move back and forth with the tides and may take several weeks or even months to leave the estuary during times of low flow [48]. Most of the organic material decomposes within a few days, but fats take much longer. 'Fatbergs', solid masses of combined fats and other waste including synthetic (plastic) 'wet-wipes', can form hard accretions in sewers several hundred metres long and harbour toxic chemicals and antibiotic-resistant pathogens, which can be released directly into the river through outfalls [28,49].

In the early 19th century, the Thames was used as an open sewer. The fast growth of the London population led to major increases in the amount of sewage discharged to the Thames and tributaries, severely reducing the oxygen content of the water. A campaign to improve public health led to the opening of the London sewer system in 1870, which improved the situation. However, the system became overwhelmed again in the first half of the 20th century and stretches of the tidal Thames turned anoxic with widespread hypoxia, resulting in a major depletion of aquatic life. Prior to upgrades of the sewer network in the 20th century, the river was declared 'biologically dead'. Between the early 1960's and late 1970's, the river began to recover from severe pollution, and much was made of the return of many North Sea coastal fish species to the Thames, especially salmon [50,51]. This reflected contemporary ideas regarding the reversible nature of environmental degradation and the belief that the river could be returned to pristine conditions one species at a time with technical improvements of the system. The return of fish to the river was described as a 'recovery', but changes following previous improvements to the London sewage system were not measured and comprehensive before and after data were not collected [5].

4. Biodiversity Loss

Many native species of all taxa are in decline and local extinction may be inevitable for populations of native mussels, salmonids (smelt, trout and salmon), several plants, gastropods and some types of river flies. Continuing occurrence does not necessarily mean that there is potential to recover species. Population decline beyond a certain minimum level of genetic diversity might take decades to reverse. Deteriorating environmental conditions and extremes of weather further impede recovery and could make it unattainable. For example, fluctuation in eel populations is a typical indicator of population decline in species with high reproductive capacity [52]. This, as well as the precipitous declines in river flies [53] and the continuing decline in native mussels [54,55], is to be expected and provide typical examples of the urban river syndrome. Biodiversity loss can occur through the extinction of an individual species but also through the chronic decline in the abundance of subpopulations. Species are not generally highly localized except perhaps for the smelt spawning grounds at Kew, and unionid mussels at Richmond; other species such as cockles and flounder occur in large numbers. Whilst being isolated in different local habitats at different times of year and varying stages of their lifecycle, many species manage to survive the occasional serious pollution events and escape mass die-offs.

5. Non-Native Species

The Thames is home to an increasing number of non-native species [56] including a Ponto–Caspian assemblage and elements of American and Asian freshwater assemblages. Around 100 have been recorded from the Thames [57] and there will certainly be more cryptic species that have gone undetected. Potentially, many more non-native plants and animals occur on land and in gardens within the catchment area as the ecosystem adapts to environmental change. The Thames has been described as one of the most highly invaded rivers in the world [57].

Range shifts due to climate change have been underway for at least several decades [58]. Species such as the spotted bass, native to North Africa, are now established in the estuary. Understanding how non-native species can make communities more resilient (in addition to the problems they may cause) is increasingly important. Deliberate introductions, frequently through oversight, but mainly on purpose, compose a significant part of ecosystems worldwide. There are many examples in the literature regarding deliberate introductions for conservation (assisted migration) or pest control purposes [59–62].

Food webs and other interactions are changing but need to be investigated as well. The populations of European-wide invasive species are typically seen to decline after varying periods of time (some in the short-term, others require a longer period of perhaps several years) to reduced levels following the initial population eruptions [63]. This may be due to integration with local food webs and acclimatisation to predators, disease and other factors constraining population size. However, short-term responses may be unhelpful in the long-term management of non-native species since it is a form of disturbance and can therefore provide additional opportunities for future colonisation, e.g., through the provision of additional habitat or ecological niche.

Pollution and reductions in oxygen saturation may have aided pollution-tolerant species from a competitive perspective, especially if they are air breathing or able to temporarily leave the water to avoid oxygen depleted areas of the river. These include the Chinese mitten crab, *Eriocheir sinensis* (H.Milne Edwards, 1853), eels *Anguilla anguilla* (Linnaeus, 1758), potential future colonisers including the walking catfish, *Clarias batrachus* (Linnaeus, 1758), already recorded in the Thames, and perhaps the Chinese river mussel, *Sinanodonta woodiana* (I.Lea, 1834), which is more pollution tolerant than the native unionid species. Some crabs including *Eriocheir sinensis* (H. Milne Edwards, 1853) are also known to have a high tolerance for ammonia possibly providing an advantage in water polluted by sewage [64].

Following disturbance due to climate or land use change, non-native species are sometimes better adapted, whereas native species may take long periods to recover if they do at all. The simplification of communities through declines in abundance and diversity means that non-native species sometimes increase biodiversity, and this has created conflicts between ecologists who wish to restore indigenous communities and those more interested in increasing biodiversity generally [65,66]. It is typical for an invasive species to acclimatise to local conditions through integration with food webs and adaptation to local climate, pathogens, predators and competition. This process may in turn include the alteration of the environment by the organism in terms of increased nutrient availability. The quagga mussel, *Dreissena bugensis* (Andrusov, 1897), for example, occurs in dense beds which increase phosphorous content of the water, which, when combined with changes to habitat caused by the establishment of large colonies, is believed to assist in the settlement of additional members of the Ponto–Caspian assemblage [67]. Shells of the American slipper Limpet, *Crepidula fornicata* (Linnaeus, 1758), form banks in the Thames Estuary. They produce copious amounts of pseudo-faeces which smother native oysters, transform the substrate, and enrich the nitrogen content of the water column [68]. Their reproductive cycle, as with the Chinese mitten crab, involves long residence times for larvae in the estuary, meaning that population dynamics are similarly regulated by phenology, nutrient levels, particulates (which increase larval survival), temperature and flows.

Reporting on invasive and non-native species populations is highly variable. For example, considerable attention has been paid to species that are invasive elsewhere such as the Chinese mitten crab and Dreissenid mussels [67], but relatively little consideration has been given to the vast slipper limpet population in the estuary. With hindsight, perhaps a disproportionate emphasis has been placed on charismatic species such as salmon and smelt and to the conservation of the slow-growing native unionid mussels, with monitoring of their limited return to the Thames following engineering work to improve the sewer system beginning in the 1950's [69]. These anomalies probably related to factors other than a comprehensive prioritisation strategy.

6. Past Surveys, Current Management, Sampling and Assessments

A programme of sampling species collected from cooling water intakes at London power stations (e.g., Lots Road, Brunswick Wharf, Blackwall Point) was initiated in the second half of the 20th century in cooperation with the Central Electricity Generating Board [15,70]. This followed major improvements to environmental conditions following the second period of restoration of the London sewer system in the 1960's. However, monitoring in the tidal Thames has been somewhat sporadic. Some work has been carried out by the Environment Agency, various ecological contractors, the Zoological Society London (ZSL) and others who have conducted monitoring of juvenile fish as well as certain historically important fish species including smelt, and used traps located in the river to assess eels. On occasion, seine netting is conducted in the river but there is no complete, long-term dataset and continuing work remains patchy. Since 1974, starting initially at West Thurrock Power Station, the water intake and, on occasion, outfall resting tank from cooling water systems at power stations in the Thames have been used for semi-quantitative assessment of aquatic species populations [5,71]. Water from a sub-tidal collection point in the river is passed through a screen filter before entering the station and items are collected in a pit. Since the rate of water intake is known, sampling for a specified period provides an indication of the abundance of animals in the immediate vicinity of the intake in terms of animals per litre. This system is convenient but has been problematic because operators were not trained to correctly identify to the species level. The location of the intakes is subject to bias attributable to attraction to warm water effluent from the power station (particularly when using condensers in the past), as well as migratory patterns which can cause animals to form spawning aggregations or accumulate as they converge from tributaries and progress downstream. In addition, larger animals are able to escape being sucked in at the intake and this was confirmed using trawls that illustrated that the number of larger fish were being underestimated [5].

The ZSL surveys benthic species from time to time at a location approximately 11 km upriver from Chiswick [51,52] and does have conservation programmes in place for certain species in the Thames including eels, smelt and seals. Notwithstanding this, the inherent value of the river within the overall context of London as well as the associated biota should merit a progressive and systematic monitoring programme conducted to the highest standards as one component of a comprehensive adaptation plan.

7. Remediation, Restoration and Monitoring

The Tideway project includes the expansion of the Beckton sewage treatment plant (downstream of the Thames Barrier), currently the largest sewage treatment plant in Europe. Increases in effluent may create problems in the estuary since not all harmful chemicals (including pharmaceuticals) or plastics are removed in the processing of sewage, and there may be challenges in the disposal of increased volumes of sludge (biosolids) produced by the treatment process. At the current time, biosolids from Beckton are recycled and used as fertilizer, although some waste is incinerated and disposed of as ash. Sludge contains nano and micro-plastics as well as pharmaceuticals [44,72,73], drug residues and other chemicals which can be blown or washed into the river as run-off from fields and landfills. The new Tideway Tunnel system will reduce sewage and effluent releases into the upper tidal Thames with most effluent entering the river closer to the estuary at Beckon. However, complete flushing will still be dependent on the tides, and some releases in the upper reaches will continue, especially during storms. These major changes to nutrient and run-off pollution, with a general shift from London down river towards the estuary will affect diverse elements of the ecosystem in both positive and negative ways providing an opportunity for a Before-After-Control-Impact (BACI) survey. Some species such as the non-native Asiatic clam will benefit from cleaner water but might also be affected by reduced nutrients and associated algal content. This will be the third major technological improvement to the Thames and the London sewer system after the 1850–1900 and the 1950–1980 sensitisation improvements [15]. However, it is now well understood that purely engineered solutions, while they may be robust, are not resilient but rather brittle and prone to failure. The Tideway project should improve dissolved oxygen levels within the upper tidal reaches. That will aide juvenile fish and fish species with higher oxygen requirements such as salmonids (salmon, smelt and shad), eels and to a lesser extent coarse species including pike, dace and zander (an introduced species), which are predominantly freshwater but do also occur in the upper reaches of the tidal part of the river. However, as with previous improvements, such changes could increase the receptiveness of the Thames ecosystem and open the door to further non-native species. The system can also respond in unexpected ways and it may be that established non-native or endemic species could become invasive and their numbers increase greatly. The pressures that will result, for example, dissolved oxygen increase and organic matter reductions, work selectively by boosting or inhibiting individual species through a complex interaction of effects. The Chinese mitten crab, for example, might be negatively affected by the reduction of nutrients and as a consequence, of the algae on which the crabs primarily feed.

There have been large population fluctuations within the tidal area and significant change is underway as estuarine species shift. However, a comprehensive Thames monitoring plan is missing. This would appear out of sync with regards to the Tideway project, for which a BACI survey would no doubt be very useful in assessing its effects on the ecology of the river. The absence of consistent monitoring practices is a surprising feature of the recent history of the Thames given the cost of improvement projects, long-term conservation work and the associated conservation areas. In fact, the general condition of the river and foreshore fall well below that of the surrounding areas, indicating that restoration, monitoring and stewardship within the overall context of the immediate neighbourhood have been disproportionately low. Sampling was initiated as a means of evaluating improvements resulting from modifications to the sewer system in the middle of the 20th century [5]. However, it was not part of a formal monitoring protocol and important information on the recovery of the river was lost, including data regarding the return of parasites [74].

8. Conclusions

Two major cycles of drastic deterioration of the Thames ecosystem and depletion of aquatic life, followed by technology-driven restorations over the last two centuries brought about strong ecosystem disturbances. Under the continuous pressure of anthropogenic impact, replacement of the Thames ecosystem rather than recovery becomes the most likely outcome. The Thames is home to an increasing number of non-native species, including pollution-tolerant species, whilst many native species undergo a continuing process of decline and extinction. Before engaging in the wholesale destruction of non-native species (at significant cost) it is increasingly important to understand the possible future state of the ecosystem, especially given that once a species is lost it cannot be retrieved. Therefore, it is possible to damage future ecosystems through the arbitrary destruction of well adapted, non-native species as they respond to climate change and migrate to more suitable environmental conditions. Species persistence must also be viewed from within the context of ecosystem restoration. The ecology has changed and cannot be expected to revert to its previous condition as most of the changes are irreversible. For example, even brief periods of reduced oxygen levels in the ocean and coastal waters can have lasting effects on behaviour and reproduction through epigenetic effects on the current and future generations [75]. However, even in relatively undisturbed environments, restoration has not resulted in a return to prior pristine condition [76]. In recognition of these issues, some restoration practitioners in the field of woodlands have been working to develop a future-proof ecosystem using species that are not endemic, i.e., using transplants, while others have considered more heat-tolerant engineered organisms. Such self-assembly is already underway with many non-native species successfully integrating to existing ecosystems, particularly in the oceans [77,78].

A more holistic view of restoration should be adopted with the emphasis on improving resilience and aiding adaptation of the complete socio-techno-ecological system, including anthropogenic impacts. The adaptive capacities of these ecosystems must be maintained by preserving a balance among heterogeneity, modularity and redundancy, tightening feedback loops to provide incentives for sound stewardship. Similar issues relate to flood and coastal defences, which must be developed as part of a complete socio-technical solution [79]. It has therefore become increasingly important to collect data on species abundance within comprehensive monitoring programmes. The use of 'citizen science' and crowdsourcing has been largely insufficient and a larger, higher quality effort is required. It should be centred around organisations which embody high levels of expertise such as the National Biodiversity Network (NBN) and universities. In the case of the Thames, this means regular surveys and sampling with integration between other monitoring programmes around the country. Intelligent monitoring of species that may cause population eruptions, such as the Chinese mitten crab, especially following years of low flow, can provide an early warning of potential pest outbreaks in accordance with the governments' strategy around 'sleeper species' [80].

Supplementary Materials: The following are available online at https://www.mdpi.com/article/10.3390/su13116045/s1, Figure S1: Relative positions of each of the maps shown in Figures 2, 3, 5 and 6 of the manuscript; Figure S2: Supplemental information on contaminants (Persistent Organic Pollutant (PCB)); Figure S3: Supplemental information on contaminants (single cores); Table S1: Supplemental information on contaminants (Mean pollutant concentrations).

Funding: This research received no external funding.

Conflicts of Interest: The funders had no role in the design of the study; in the collection, analyses or interpretation of data; in the writing of the manuscript; or in the decision to publish the results.

References

1. Lavery, S.; Donovan, B. Flood Risk Management in the Thames Estuary Looking Ahead 100 Years. *Philos. Trans. R. Soc. A Math. Phys. Eng. Sci.* **2005**, *363*, 1455–1474. [CrossRef] [PubMed]
2. Dawson, R.J.; Hall, J.W.; Bates, P.D.; Nicholls, R.J. Quantified Analysis of the Probability of Flooding in the Thames Estuary under Imaginable Worst-Case Sea Level Rise Scenarios. *Int. J. Water Resour. Dev.* **2005**, *21*, 577–591. [CrossRef]
3. Canal & Rivers Trust: Fish Species. Available online: https://canalrivertrust.org.uk/enjoy-the-waterways/fishing/fish-species (accessed on 15 May 2021).
4. Deaker, D.J.; Agüera, A.; Lin, H.A.; Lawson, C.; Budden, C.; Dworjanyn, S.A.; Mos, B.; Byrne, M. The Hidden Army: Corallivorous Crown-of-Thorns Seastars Can Spend Years as Herbivorous Juveniles. *Biol. Lett.* **2020**, *16*, 20190849. [CrossRef] [PubMed]
5. Attrill, M.J. *A Rehabilitated Estuarine Ecosystem: The Environment and Ecology of the Thames Estuary*; Kluwer Academic Publishers: London, UK, 1998; pp. 116–121.
6. Zhang, H.; Han, J.; Chen, F.; Yuan, Q. Review on Plants Selection and Application Effects in Ecological Floating Beds Based on Water Purification. *IOP Conf. Ser. Earth Environ. Sci.* **2021**, *647*, 012186. [CrossRef]
7. Wei, Z.; Van Le, Q.; Peng, W.; Yang, Y.; Yang, H.; Gu, H.; Lam, S.S.; Sonne, C. A Review on Phytoremediation of Contaminants in Air, Water and Soil. *J. Hazard. Mater.* **2021**, *403*, 123658. [CrossRef]
8. Li, J.; Ianaiev, V.; Huff, A.; Zalusky, J.; Ozersky, T.; Katsev, S. Benthic Invaders Control the Phosphorus Cycle in the World's Largest Freshwater Ecosystem. *Proc. Natl. Acad. Sci. USA* **2021**, *118*, e2008223118. [CrossRef]
9. Vanderploeg, H.A.; Nalepa, T.F.; Jude, D.J.; Mills, E.L.; Holeck, K.T.; Liebig, J.R.; Grigorovich, I.A.; Ojaveer, H. Dispersal and Emerging Ecological Impacts of Ponto-Caspian Species in the Laurentian Great Lakes. *Can. J. Fish. Aquat. Sci.* **2002**, *59*, 1209–1228. [CrossRef]
10. Montgomery, D.R. Soil Erosion and Agricultural Sustainability. *Proc. Natl. Acad. Sci. USA* **2007**, *104*, 13268–13272. [CrossRef] [PubMed]
11. Matrosov, E.S.; Harou, J.J.; Loucks, D.P. A Computationally Efficient Open-Source Water Resource System Simulator-Application to London and the Thames Basin. *Environ. Model. Softw.* **2011**, *26*, 1599–1610. [CrossRef]
12. Department for Environment Food & Rural Affairs (DEFRA). Water Abstraction Plan. Available online: https://www.gov.uk/government/publications/water-abstraction-plan-2017/water-abstraction-plan (accessed on 15 May 2021).
13. Griffiths, R. An Essay to Prove That the Jurisdiction and Conservacy of the River of Thames. Printed for T. Longman in Paternoster-Row. 1758. Available online: https://thames.me.uk/PDF/1758RiverThamesGriffiths.pdf (accessed on 18 March 2020).
14. Bolton, D.K.; Croot, P.E.; Hicks, M.A. A History of the County of Middlesex: Volume 7, Acton, Chiswick, Ealing and Brentford, West Twyford, Willesden. Available online: https://www.british-history.ac.uk/vch/middx/vol7 (accessed on 15 May 2021).
15. Wood, L.B. *The Restoration of the Tidal Thames.*; Hilger: Bristol, UK, 1982; pp. 32–52.
16. Lombardo, A.; Franco, A.; Pivato, A.; Barausse, A. Food Web Modeling of a River Ecosystem for Risk Assessment of Down-the-Drain Chemicals: A Case Study with AQUATOX. *Sci. Total Environ.* **2015**, *508*, 214–227. [CrossRef]
17. National Biodiversity Network (NBN) Atlas. Available online: https://nbnatlas.org/ (accessed on 10 April 2021).
18. Jackson, M.C.; Grey, J. Electronic Supplementary Material Accelerating Rates of Freshwater Invasions in the Catchment of the River Thames List of Freshwater Non-Indigenous Species Recorded as Established in the Thames Catchment. Available online: https://www.academia.edu/2768923/Electronic_Supplementary_Material_for_Jackson_and_Grey_2013_Accelerating_rates_of_freshwater_invasions_in_the_catchment_of_the_River_Thames (accessed on 5 April 2021).
19. ZSL (Zoological Society London) Guidance Document Conservation of Tidal Thames Fish through the Planning Process. 2016. Available online: http://www.lbhf.gov.uk/sites/default/files/section_attachments/guidance_document_conservation_of_tidal_thames_fish_through_the_planning_process_october_2016.pdf (accessed on 15 May 2021).
20. The Zoological Society London. Available online: https://www.zsl.org/ (accessed on 15 May 2021).
21. Taylor, K.G.; Owens, P.N. Sediments in Urban River Basins: A Review of Sediment-Contaminant Dynamics in an Environmental System Conditioned by Human Activities. *J. Soils Sediments* **2009**, *9*, 281–303. [CrossRef]
22. Green, A.E.; Unsworth, R.K.F.; Chadwick, M.A.; Jones, P.J.S. Historical Analysis Exposes Catastrophic Seagrass Loss for the United Kingdom. *Front. Plant Sci.* **2021**, *12*, 629962. [CrossRef] [PubMed]
23. Tideway Tideway-Reconnecting London with the Tidal Thames. *Thames Tideway*, 2019. Available online: https://www.tideway.london/ (accessed on 18 March 2020)
24. Department for Environment Food & Rural Affairs (DEFRA). Creating a River Thames Fit for Our Future: A Strategic and Economic Case for the Thames Tunnel. Available online: https://assets.publishing.service.gov.uk/government/uploads/system/uploads/attachment_data/file/471847/thames-tideway-tunnel-strategic-economic-case.pdf (accessed on 15 May 2021).
25. Talling, P. *London's Lost Rivers*; Random House Books: London, UK, 2011; pp. 10–50.

26. Haas, A.F.; Fairoz, M.F.M.; Kelly, L.W.; Nelson, C.E.; Dinsdale, E.A.; Edwards, R.A.; Giles, S.; Hatay, M.; Hisakawa, N.; Knowles, B.; et al. Global Microbialization of Coral Reefs. *Nature Microbiol.* **2016**, *1*, 16042. [CrossRef] [PubMed]
27. Garcia-Armisen, T.; Vercammen, K.; Passerat, J.; Triest, D.; Servais, P.; Cornelis, P. Antimicrobial Resistance of Heterotrophic Bacteria in Sewage-Contaminated Rivers. *Water Res.* **2011**, *45*, 788–796. [CrossRef]
28. Rosi, E.J.; Bechtold, H.A.; Snow, D.; Rojas, M.; Reisinger, A.J.; Kelly, J.J. Urban Stream Microbial Communities Show Resistance to Pharmaceutical Exposure. *Ecosphere* **2018**, *9*, e02041. [CrossRef]
29. Plaas, H.E.; Paerl, H.W. Toxic Cyanobacteria: A Growing Threat to Water and Air Quality. *Environ. Sci. Technol.* **2021**, *55*, 44–64. [CrossRef]
30. Aguilera, A.; Klemenčič, M.; Sueldo, D.J.; Rzymski, P.; Giannuzzi, L.; Martin, M.V. Cell Death in Cyanobacteria: Current Understanding and Recommendations for a Consensus on Its Nomenclature. *Front. Microbiol.* **2021**, *12*, 631654. [CrossRef] [PubMed]
31. Pan, M.; Lyu, T.; Zhan, L.; Matamoros, V.; Angelidaki, I.; Cooper, M.; Pan, G. Mitigating Antibiotic Pollution Using Cyanobacteria: Removal Efficiency, Pathways and Metabolism. *Water Res.* **2021**, *190*, 116735. [CrossRef]
32. *Thames Tideway Tunnel Application for Development Consent: Resilience to Change*; Thames Tideway: London, UK, 2013; p. 6.
33. Wurtsbaugh, W.A.; Paerl, H.W.; Dodds, W.K. Nutrients, Eutrophication and Harmful Algal Blooms along the Freshwater to Marine Continuum. Wiley Interdiscip. *Rev. Water* **2019**, *6*, e1373.
34. Diaz, R.J.; Rosenberg, R. Spreading Dead Zones and Consequences for Marine Ecosystems. *Science* **2008**, *321*, 926–929. [CrossRef]
35. Brand, J.H.; Spencer, K.L.; O'shea, F.T.; Lindsay, J.E. Potential Pollution Risks of Historic Landfills on Low-Lying Coasts and Estuaries. *Wiley Interdiscip. Rev. Water* **2018**, *5*, e1264. [CrossRef]
36. O'Shea, F.T.; Cundy, A.B.; Spencer, K.L. The Contaminant Legacy from Historic Coastal Landfills and Their Potential as Sources of Diffuse Pollution. *Mar. Pollut. Bull.* **2018**, *128*, 446–455. [CrossRef] [PubMed]
37. Thomas, K.V.; Hurst, M.R.; Matthiessen, P.; McHugh, M.; Smith, A.; Waldock, M.J. An Assessment of in Vitro Androgenic Activity and the Identification of Environmental Androgens in United Kingdom Estuaries. *Environ. Toxicol. Chem.* **2002**, *21*, 1456–1461. [CrossRef] [PubMed]
38. Letsinger, S.; Kay, P.; Rodríguez-Mozaz, S.; Villagrassa, M.; Barceló, D.; Rotchell, J.M. Spatial and Temporal Occurrence of Pharmaceuticals in UK Estuaries. *Sci. Total Environ.* **2019**, *678*, 74–84. [CrossRef]
39. Nielsen, S.V.; Frausing, M.; Henriksen, P.G.; Beedholm, K.; Baatrup, E. The Psychoactive Drug Escitalopram Affects Foraging Behavior in Zebrafish (Danio Rerio). *Environ. Toxicol. Chem.* **2019**, *38*, 1902–1910. [CrossRef] [PubMed]
40. Polverino, G.; Martin, J.M.; Bertram, M.G.; Soman, V.R.; Tan, H.; Brand, J.A.; Mason, R.T.; Wong, B.B.M. Psychoactive Pollution Suppresses Individual Differences in Fish Behaviour. *Proc. R. Soc. B Biol. Sci.* **2021**, *288*, 20202294. [CrossRef]
41. Yin, L.; Wang, B.; Yuan, H.; Deng, S.; Huang, J.; Wang, Y.; Yu, G. Pay Special Attention to the Transformation Products of PPCPs in Environment. *Emerg. Contam.* **2017**, *3*, 69–75. [CrossRef]
42. Fatta-Kassinos, D.; Vasquez, M.I.I.; Kümmerer, K. Transformation Products of Pharmaceuticals in Surface Waters and Wastewater Formed during Photolysis and Advanced Oxidation Processes-Degradation, Elucidation of Byproducts and Assessment of Their Biological Potency. *Chemosphere* **2011**, *85*, 693–709. [CrossRef]
43. Vane, C.H.; Turner, G.H.; Chenery, S.R.; Richardson, M.; Cave, M.C.; Terrington, R.; Gowing, C.J.B.B.; Moss-Hayes, V. Trends in Heavy Metals, Polychlorinated Biphenyls and Toxicity from Sediment Cores of the Inner River Thames Estuary, London, UK. *Environ. Sci. Process. Impacts* **2020**, *22*, 364–380. [CrossRef]
44. Lu, Q.; Jürgens, M.D.; Johnson, A.C.; Graf, C.; Sweetman, A.; Crosse, J.; Whitehead, P. Persistent Organic Pollutants in Sediment and Fish in the River Thames Catchment (UK). *Sci. Total Environ.* **2017**, *576*, 78–84. [CrossRef]
45. Michels, J.; Stippkugel, A.; Lenz, M.; Wirtz, K.; Engel, A. Rapid Aggregation of Biofilm-Covered Microplastics with Marine Biogenic Particles. *Proc. R. Soc. B Biol. Sci.* **2018**, *285*, 20181203. [CrossRef]
46. Kole, P.J.; Löhr, A.J.; Van Belleghem, F.G.A.J.; Ragas, A.M.J. Wear and Tear of Tyres: A Stealthy Source of Microplastics in the Environment. *Int. J. Environ. Res. Public Health* **2017**, *14*, 1265. [CrossRef] [PubMed]
47. Brahney, J.; Mahowald, N.; Prank, M.; Cornwell, G.; Klimont, Z.; Matsui, H.; Prather, K.A. Constraining the Atmospheric Limb of the Plastic Cycle. *Proc. Natl. Acad. Sci. USA* **2021**, *118*, e2020719118. [CrossRef] [PubMed]
48. Wallingford, H.R. Thames 2D Base Model 2012. Available online: https://www.pla.co.uk/assets/NA012_0609_MAL_Thames_2D_base_model.pdf (accessed on 15 May 2021).
49. Nogales, B.; Lanfranconi, M.P.; Piña-Villalonga, J.M.; Bosch, R. Anthropogenic Perturbations in Marine Microbial Communities. *FEMS Microbiol. Rev.* **2010**, *35*, 275–298. [CrossRef]
50. Andrews, M.J.; Rickard, D.G. Rehabilitation of the Inner Thames Estuary. *Mar. Pollut. Bull.* **1980**, *11*, 327–332. [CrossRef]
51. Andrews, M.J. Thames estuary: Pollution and recovery. In *Effects of Pollutants at the Ecosystem Level*; Sheehan, P.J., Miller, D.R., Butler, G.C., Bourdeau, P.H., Eds.; John Wiley & Sons: New York, NY, USA, 1984; pp. 195–227.
52. Worden, L.; Botsford, L.W.; Hastings, A.; Holland, M.D. Frequency Responses of Age-Structured Populations: Pacific Salmon as an Example. *Theor. Popul. Biol.* **2010**, *78*, 239–249. [CrossRef]
53. Measham, N. Riverfly Census 2015. Salmon Trout Conserv. UK. 2015. Available online: https://salmon-trout.org/wp-content/uploads/2017/08/2015-Report_Riverfly-Census.pdf (accessed on 15 May 2021).

54. Ainscough, J.; Barker, J.; Pecorelli, J. Freshwater Bivalve Survey in the Upper Tidal Thames. UK Eur. Conserv. Program. 2015. Available online: https://www.zsl.org/sites/default/files/media/2015-03/ReportonZSLsFreshwaterBivalveSurveyintheUpperTidalThames2015.pdf (accessed on 15 May 2021).
55. Pecorelli, J. Freshwater Mussel Survey in the Upper Tidal Thames. 2018. Available online: https://www.zsl.org/sites/default/files/media/2018-06/INNSSurveyReport_ZSL2017_Final_07.02.18_0.pdf (accessed on 15 May 2021).
56. Cohen, A.N.; Carlton, J.T. Accelerating Invasion Rate in a Highly Invaded Estuary. *Science* **1998**, *279*, 555–558. [CrossRef] [PubMed]
57. Jackson, M.C.; Grey, J. Accelerating Rates of Freshwater Invasions in the Catchment of the River Thames. *Biol. Invasions* **2013**, *15*, 945–951. [CrossRef]
58. Pecl, G.T.; Araújo, M.B.; Bell, J.D.; Blanchard, J.; Bonebrake, T.C.; Chen, I.-C.; Clark, T.D.; Colwell, R.K.; Danielsen, F.; Evengård, B.; et al. Biodiversity Redistribution under Climate Change: Impacts on Ecosystems and Human Well-Being. *Science* **2017**, *355*, eaai9214. [CrossRef] [PubMed]
59. Etterson, J.R.; Cornett, M.W.; White, M.A.; Kavajecz, L.C. Assisted Migration across Fixed Seed Zones Detects Adaptation Lags in Two Major North American Tree Species. *Ecol. Appl.* **2020**, *30*, e02092. [CrossRef]
60. Sáenz-Romero, C.; Mendoza-Maya, E.; Gómez-Pineda, E.; Blanco-García, A.; Endara-Agramont, A.R.; Lindig-Cisneros, R.; López-Upton, J.; Trejo-Ramírez, O.; Wehenkel, C.; Cibrián-Tovar, D.; et al. Recent Evidence of Mexican Temperate Forest Decline and the Need for Ex Situ Conservation, Assisted Mi-gration, and Translocation of Species Ensembles as Adaptive Management to Face Projected Climatic Change Impacts in a Megadiverse Country. *Can. J. For. Res.* **2020**, *50*, 843–854. [CrossRef]
61. Sáenz-Romero, C.; O'neill, G.; Aitken, S.N.; Lindig-Cisneros, R. Assisted Migration Field Tests in Canada and Mexico: Lessons, Limitations, and Challenges. *Forests* **2021**, *12*, 9. [CrossRef]
62. Bladon, A.; Donald, P.; Collar, N.; Deng, J.; Dadacha, G.; Wondafrash, M.; Green, R. Climatic change and extinction risk of two globally threatened Ethiopian endemic bird species. *PLoS ONE* **2021**, *16*, e0249633. [CrossRef]
63. Simberloff, D.; Gibbons, L. Now You See Them, Now You Don't!-Population Crashes of Established Introduced Species. *Biol. Invasions* **2004**, *6*, 161–172. [CrossRef]
64. Weihrauch, D.; Morris, S.; Towle, D.W. Ammonia Excretion in Aquatic and Terrestrial Crabs. *J. Exp. Biol.* **2004**, *207*, 4491–4504. [CrossRef] [PubMed]
65. Briggs, J.C. Rise of Invasive Species Denialism? A Response to Russell and Blackburn. *Trends Ecol. Evol.* **2017**, *32*, 231–232. [CrossRef]
66. Crowley, S.L.; Hinchliffe, S.; Redpath, S.M.; McDonald, R.A. Disagreement About Invasive Species Does Not Equate to Denialism: A Response to Russell and Blackburn. *Trends Ecol. Evol.* **2017**, *32*, 228–229. [CrossRef] [PubMed]
67. Gallardo, B.; Aldridge, D.C. Is Great Britain Heading for a Ponto–Caspian Invasional Meltdown? *J. Appl. Ecol.* **2015**, *52*, 41–49. [CrossRef]
68. Fitzgerald, A. Slipper Limpet Utilisation and Management 2007. Available online: http://www.shellfish.org.uk/files/Literature/Projects-Reports/0701-Slipper_Limpet_Report_Final_Small.pdf (accessed on 15 May 2021).
69. Griffiths, A.M.; Ellis, J.S.; Clifton-Dey, D.; Machado-Schiaffino, G.; Bright, D.; Garcia-Vazquez, E.; Stevens, J.R. Restoration versus Recolonisation: The Origin of Atlantic Salmon (*Salmo Salar* L.) Currently in the River Thames. *Biol. Conserv.* **2011**, *144*, 2733–2738. [CrossRef]
70. Wheeler, A. *The Tidal Thames: The History of a River and Its Fishes*; Routledge & Kegan Paul: London, UK, 1980; pp. 89–108.
71. Clark, P.F.; Rainbow, P.S.; Robbins, R.S.; Smith, B.; Yeomanst, W.E.; Thomast, M.; Dobson, G. The Alien Chinese Mitten Crab, Eriocheir Sinensis (Crustacea: Decapoda: Brachyura), in the Thames Catchment. *J. Mar. Biol. Ass. UK* **1998**, *78*, 1215–1221. [CrossRef]
72. Smith, S.R. Organic Contaminants in Sewage Sludge (Biosolids) and Their Significance for Agricultural Recycling. *Philos. Trans. R. Soc. A Math. Phys. Eng. Sci.* **2009**, *367*, 4005–4041. [CrossRef]
73. Richardson, S.D.; Ternes, T.A. Water Analysis: Emerging Contaminants and Current Issues. *Anal. Chem.* **2018**, *90*, 398–428. [CrossRef]
74. Munro, M.; Whitfield, P.; Lee, S. Host-parasite interactions: Case studies of parasitic infections in migratory fish. In *A Rehabilitated Estuarine Ecosystem: The Environment and Ecology of the Thames Estuary*; Attrill, M.J., Ed.; Springer US: Boston, MA, USA, 1998; pp. 141–167.
75. Breitburg, D.; Levin, L.A.; Oschlies, A.; Grégoire, M.; Chavez, F.P.; Conley, D.J.; Garçon, V.; Gilbert, D.; Gutiérrez, D.; Isensee, K.; et al. Declining Oxygen in the Global Ocean and Coastal Waters. *Science* **2018**, *359*, eaam7240. [CrossRef] [PubMed]
76. Levin, S.A.; Lubchenco, J. Resilience, Robustness, and Marine Ecosystem-Based Management. *Bioscience* **2008**, *58*, 27–32. [CrossRef]
77. Paniw, M.; James, T.D.; Archer, C.R.; Römer, G.; Levin, S.; Compagnoni, A.; Che, J.; Bennett, C.J.M.; Mooney, A.; Childs, D.Z.; et al. The Myriad of Complex Demographic Responses of Terrestrial Mammals to Climate Change and Gaps of Knowledge: A Global Analysis. *J. Anim. Ecol.* 2021, Online ahead of print. [CrossRef]
78. Chaudhary, C.; Richardson, A.J.; Schoeman, D.S.; Costello, M.J. Global Warming Is Causing a More Pronounced Dip in Marine Species Richness around the Equator. *Proc. Natl. Acad. Sci. USA* **2021**, *118*, e2015094118. [CrossRef] [PubMed]

79. Mastrángelo, M.E.; Pérez-harguindeguy, N.; Enrico, L.; Bennett, E.; Lavorel, S.; Cumming, G.S.; Abeygunawardane, D.; Amarilla, L.D.; Burkhard, B.; Egoh, B.N.; et al. Key Knowledge Gaps to Achieve Global Sustainability Goals. *Nat. Sustain.* **2019**, *2*, 1115–1121. [CrossRef]
80. Department for Environment Food & Rural Affairs (DEFRA). The National Adaptation Programme and the Third Strategy for Climate Adaptation Reporting. 2018. Available online: https://assets.publishing.service.gov.uk/government/uploads/system/uploads/attachment_data/file/727252/national-adaptation-programme-2018.pdf (accessed on 15 May 2021).

 *sustainability*

MDPI

Article

Hydro-Economic Water Allocation Model for Water Supply Risk Analysis: A Case Study of Namhan River Basin, South Korea

Gimoon Jeong and Doosun Kang *

Department of Civil Engineering, Kyung Hee University, 1732, Deogyeong-daero, Giheung-gu, Yongin-si, Seoul 17104, Korea; gimoon1118@gmail.com
* Correspondence: doosunkang@khu.ac.kr

Abstract: Rational water resource management is used to ensure a stable supply of water by predicting the supply of and demand for future water resources. However, rational water allocation will become more difficult in the future owing to the effects of climate change, causing water shortages and disputes. In this study, an advanced hydro-economic water allocation and management model (WAMM) was introduced by improving the optimization scheme employed in conventional models and incorporating the economic value of water. By relying upon economic valuation, the WAMM can support water allocation efforts that focus not only on the stability but also on the economic benefits of water supply. The water supply risk was evaluated following the different objective functions and optimization methods provided by the WAMM using a case study of the Namhan River basin in South Korea under a climate change scenario over the next 30 years. The water shortages and associated economic damage were compared, and the superior ability of WAMM to mitigate future water shortages using economic valuation and full-step linear programming (FSLP) optimization was demonstrated. It is expected that the WAMM can be applied to help resolve water shortages and disputes among river basin units under severe drought conditions.

Keywords: climate change; hydro-economic water allocation; optimal water allocation; risk analysis; water supply

Citation: Jeong, G.; Kang, D. Hydro-Economic Water Allocation Model for Water Supply Risk Analysis: A Case Study of Namhan River Basin, South Korea. *Sustainability* **2021**, *13*, 6005. https://doi.org/10.3390/su13116005

Academic Editors: Andrzej Wałęga and Alban Kuriqi

Received: 31 March 2021
Accepted: 23 May 2021
Published: 26 May 2021

Publisher's Note: MDPI stays neutral with regard to jurisdictional claims in published maps and institutional affiliations.

1. Introduction

Water resources are essential for human life and should be stably secured and rationally allocated. According to the World Bank [1], water resource management (WRM) refers to a series of processes, including planning, developing, and managing water resources, related to both water quantity and quality, and across all water uses by consumers. UNESCO [2] reported that global water consumption is expected to increase gradually because of population growth and lifestyle changes, while conflicts over water are expected to intensify owing to climate change (e.g., during times of frequent drought and limited water supply in some areas). According to Adetoro et al. [3], many studies have been conducted worldwide to mediate water disputes from social and engineering perspectives. Such studies include various types of water allocation models presented as a part of engineering solutions to these disputes. The primary objective of a water allocation model is to present the most efficient supply plan for limited water resources according to the priorities or water rights of various water demand sectors. These demand sectors account for municipal, agricultural, and industrial water usage. The allocation results may differ in terms of fairness, economic feasibility, or social importance. The utilization of water allocation models has evolved with their complexity, from early models allocating water resources based on quantitative efficiency to hydro-economic models seeking to maximize socio-economic benefits.

Major studies related to the development and utilization of water allocation models include the introduction of one of the earliest models, MITSIM (Massachusetts Institute of Technology River Basin Simulation Model), by Strzepek and Lenton [4]. Subsequently,

Strzepek [5] improved the water allocation simulation function and used it to evaluate the stability of agricultural water supply in the Swedish South-Western Skane Basin. Shafer and Labadie [6] developed the MODSIM (Modified SIMYLD) model, another typical early water allocation model. Later, optimal water allocation models equipped with optimization modules based on linear programming or quadratic programming were developed to maximize water allocation efficiency. Such models include the AQUATOOL, developed by Andreu et al. [7], and the water evaluation and planning system (WEAP), initiated by Raskin et al. [8] and further developed by Yates et al. [9], to provide enhanced user-friendliness. In terms of quantitative water management, the AQUARIUS model developed by Diaz et al. [10] is considered a representative optimal water allocation model for the maximization of water supply reliability and minimization of water shortages. Notably, the WEAP model was incorporated into the Korea-water evaluation and planning system (K-WEAP) model to strengthen its applicability in Korea and has been utilized in various nationwide water management plans [11]. In addition, the Riverware model developed by Zagona et al. [12] further enhanced the optimal water allocation model using an optimization system based on a prioritized set of objectives implemented by linear pre-emptive goal programming. The California value integrated network (CALVIN) model developed by Howitt et al. [13] is a critically important, globally well-known, hydro-economically integrated model that considers not only the water requirements in its optimal solution but also the economic value of water resources based on data collected over 50 years in California, USA. The CALVIN water allocation model has been mainly used to inform water supply plans that minimize water supply costs and maximize the associated economic benefits. Newlin et al. [14] and Pulido-Velazquez et al. [15] presented various possibilities for utilizing the CALVIN model to establish state-wide water resources plans in southern California.

The environmental functions of water resources were divided into regulation, habitat, production, and information by Costanza et al. [16] and de Groot et al. [17]. As reported by UNSD (United Nations Statistics Division) [18], the economic valuation of water resource services is challenging; therefore, numerous researchers have attempted to define their economic value based on measurable variables such as changes in quantity (e.g., supply and regulation), quality (e.g., treatment and generation), ecological services (e.g., refugia and pollination), and cultural features (e.g., recreation). However, since the economic value of water resources mainly depends on national and regional policies and environments, their value may be quantified based on the conditions in the area of interest and period. For instance, many different water resource valuation studies have been conducted in South Korea. Hwang et al. [19] calculated the potential value of groundwater for drought preparation using the contingent valuation method (CVM). Park et al. [20] evaluated the economic value of rainfall in the spring season in terms of dam reservoir water resource utilization during droughts. Moreover, a study by K-water [21] employed various methods to quantitively evaluate the overall value of South Korea's water resources. The K-water study calculated the municipal water supply benefit (based on the willingness to pay, and investigated using the CVM), the industrial water supply benefit (based on the value of the marginal product (VMP) and estimated using the production function of major industrial sectors), and the agricultural water supply benefit (based on a farm crop budget analysis). Lim and Lee [22] presented the agricultural water supply benefit based on the estimated VMP of rice crops in South Korea. Since 2011, the Korea Environment Institute (KEI) has organized, classified, and presented various valuation cases for water resources using the environmental valuation information system (EVIS) [23].

Climate change represents an environmental factor that has a direct impact on the security of water resources. Numerous climate change scenarios have been created globally using various general circulation models (GCMs) to predict rainfall, temperature, and humidity. Using data from these models, multiple studies have predicted the damage arising from potential water shortages due to climate change. Krol et al. [24] developed and applied a model to explain the relationship between climate change and water use, focusing on agriculture in northeast Brazil. Ray et al. [25] and Ashofteh et al. [26] analyzed the water supply risk by applying long-term climate change scenarios to water supply networks. Liu et al. [27] and Basupi and Kapelan [28] compared water shortages under different climate change conditions and suggested a plan for utilizing decision-making tools and a framework to establish an optimal water supply strategy.

Here, a new hydro-economic water allocation model was examined, the water allocation and management model (WAMM), developed by Jeong et al. [29]. Equipped with an advanced optimization module, the WAMM provides enhanced functionality compared with conventional models and incorporates the economic value of water to optimize water allocation. In this case study, the WAMM was applied to the Namhan River basin in South Korea for long-term (30-year) water allocation planning using a future climate change scenario; the water allocation results obtained according to various water supply objectives and optimization methods were compared.

The remainder of this paper is structured as follows. Section 2 introduces and provides details of the WAMM, including the improved optimization algorithm and calculation of the economic value of water resources. It also introduces the process used to establish the climate change scenarios. Section 3 introduces the case study river basin and summarizes the economic value of the water resources in the study area, as well as the relevant climate change scenarios. Section 4 quantitatively compares and analyzes the simulation results of the WAMM according to the type of optimization method and objective function applied. Finally, Section 5 presents the conclusions and plausible future studies.

2. Materials and Methods

A water allocation model is an engineering support tool that can facilitate the rational establishment of a water supply and demand plan within a given river basin. Figure 1 illustrates a simplified river basin in which municipal, industrial, agricultural water demands, and the maintenance flow rate are collected from the mainstream. The water flows in the river and the water storage in reservoirs are controlled by dams. The flow rate of the river and the water available in reservoirs can be estimated based on the water resources available within the relevant basin. Then, the actual available supply that can be supplied to demand sites is determined based on river maintenance flow and restricted water storage capacity in reservoirs. A water allocation plan becomes important when the available water supply does not meet the demand and needs to be determined by considering the importance, water rights, and economic feasibility of each demand site.

The WAMM was developed based on components and input data similar to those of the K-WEAP model that is currently utilized to generate water allocation plans in South Korea. It is equipped with a graphical user interface (GUI) for user convenience, and the water supply system components and input data are provided through a dedicated input module in the form of an Excel spreadsheet. Figure 2a shows the datasheet of the input module and user-interface screens in the WAMM. The WAMM manages water allocation simulations as an individual project that includes all the specifications of system components, operational information, and simulation results needed for establishing a water supply plan. Figure 2b illustrates the control panel for the water allocation strategies and results analysis screens. Notably, the WAMM also provides an advanced optimization module that enables users to select optimization methods and objective functions to apply. The structure of the improved linear programming-based optimization module contained within the WAMM, as well as the development of a water allocation strategy that considers both the reliability and economic benefits of the water supply are described in detail in the next section.

2.1. Full-Step Linear Programming (FSLP) in the WAMM

It was relatively simple to establish a water supply plan in the past. However, as the number and complexity of the factors that must be considered have increased, the importance of water allocation plans and the role of related models have gradually increased as well. As discussed in Section 1, the majority of conventional water allocation models rely on optimal water allocation based on linear programming (LP), such as the AQUATOOL [7] and WEAP [8] models, or quadratic programming (QP), such as the AQUARIUS [10] model. As a mathematical optimization method providing a global optimal solution, LP is still widely used, despite only being applicable to problems with objectives and constraints that can be formulated as linear functions.

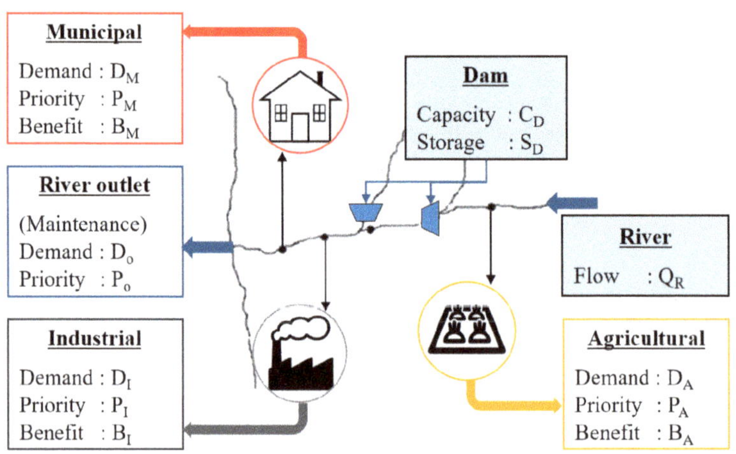

Figure 1. Scheme of water allocation in a river basin.

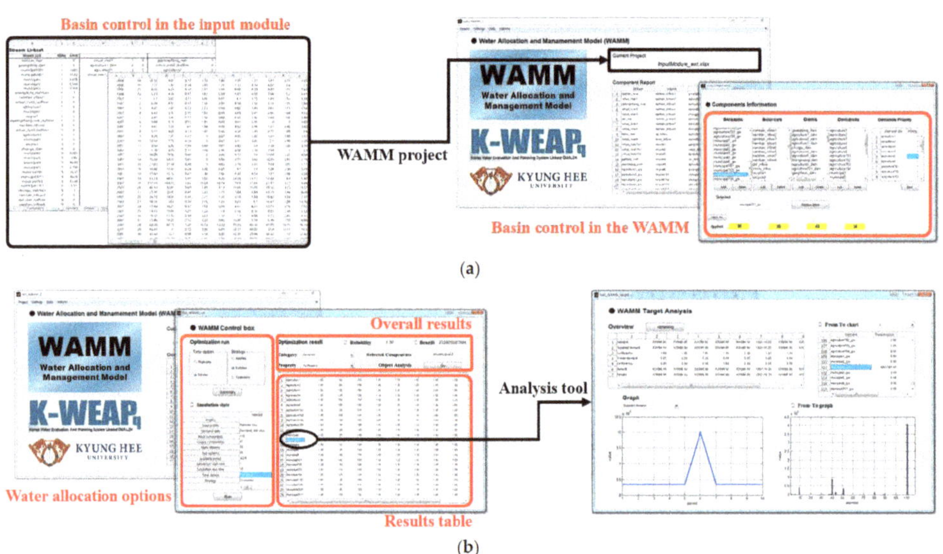

Figure 2. Graphical user interface (GUI) of the water allocation and management model (WAMM). (**a**) Project management window; (**b**) allocation management and results window.

In optimization problems such as water resource planning and allocation, the minimization of costs and maximization of benefits, or the maximization of target performance, are used as objective functions. In the previously described K-WEAP model, the maximization of water reliability (defined as the ratio of water supply to water demand) is used as an objective function. In the K-WEAP model and the WAMM, a function setting the water supply priority for each demand site is provided to reflect the importance of water use. This can be simply expressed as an objective function that maximizes the reliability of water supply based on a set of weighted values as follows:

$$Maximize: Reliability = \sum_{i=1}^{N_d} W_i \frac{S_i}{D_i}, \qquad subject\ to: S_i \leq D_i \tag{1}$$

where N_d is the number of demand sites in the basin, W_i is the weight according to the priority of demand site i, S_i is the water supplied to demand site i (m^3), and D_i is the water demand of demand site i (m^3).

For the hydro-economic water allocation model, which allocates water by taking into account the economic value of supplied water, optimization to maximize benefits can be performed by setting the economic value as an objective function. In this study, the economic objective function can be formulated to maximize the benefits of the water supply by weighting the water resources in terms of the economic value of the unit water supply as follows:

$$Maximize: Benefit = \sum_{i=1}^{N_d} V_i S_i, \qquad subject\ to: S_i \leq D_i \tag{2}$$

where V_i is the economic benefit in Korean Won (KRW) of the supplied water to the demand site i (KRW/m^3). Note that 1000 KRW can be approximated as 1 US dollar.

To establish a rational water supply plan, various socio-economic impacts must be considered using different types of objective functions. Here, two objective functions were constructed and implemented to facilitate optimal water allocation in the WAMM: (1) the reliability (stability) of the water supply (Equations (1)) and (2) the economic benefits of the water supply (Equation (2)).

Traditional water allocation models typically utilize single-step linear programming (SSLP) to optimize the water allocation for each simulation time step. As indicated in Figure 3a, a separate optimization is sequentially performed along the simulation time period, in which the allocation results (e.g., individual supply from source to demand and reservoir storage) of the previous time step are provided as initial values in the subsequent simulation time step. However, this sequential approach has a drawback as the decisions made in the earlier time step cannot be modified, resulting in a potential water shortage in subsequent time steps. Therefore, the WAMM uses full-step linear programming (FSLP) to optimize the entire simulation period at once, mitigating such limitations. As shown in Figure 3b, water allocations over the entire simulation periods were determined simultaneously through a single optimization run. This made it possible to allocate water such that the shortage occurring over the entire simulation period was minimized, realizing water allocations that actively and flexibly responded to changing environments.

For this flexible response, the reliability and benefits of the water supply were calculated for the entire simulation period and not the unit simulation period. Hence, the modified forms of the objective functions for FSLP are as follows:

$$Maximize: Reliability = \sum_{t=1}^{T} \sum_{i=1}^{N_d} W_i \frac{S_{i,t}}{D_{i,t}}, \qquad subject\ to: S_{i,t} \leq D_{i,t} \tag{3}$$

$$Maximize: Benefit = \sum_{t=1}^{T} \sum_{i=1}^{N_d} V_i S_{i,t}, \qquad subject\ to: S_{i,t} \leq D_{i,t} \tag{4}$$

where T is the total number of simulation periods in which water is allocated, $S_{i,t}$ is the water supplied to demand site i at simulation time t (m^3), and $D_{i,t}$ is the water demand of demand site i at simulation time t (m^3).

2.2. Economic Valuation of Water Resources in the WAMM

The value of a resource can be defined as the monetary value of the benefit or loss arising from its presence or absence, or a change in its state. In particular, water resources have the characteristics of public goods and behave differently from goods in the general market economy, making it difficult to devise an effective valuation method. Accordingly, the WAMM employed the results of previous studies that evaluated and applied the economic value of water resources for economic analysis as follows.

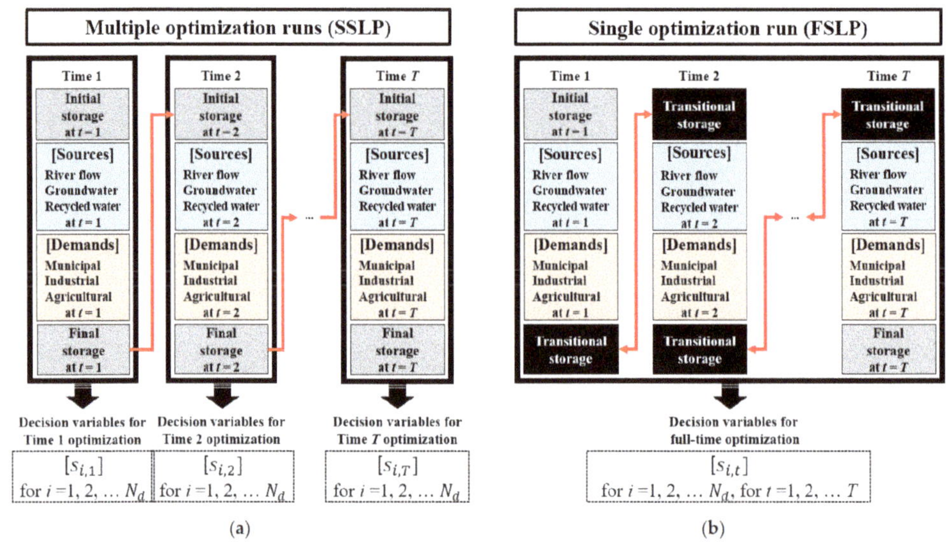

Decision variables for Time 1 optimization: $[s_{i,1}]$ for $i = 1, 2, \ldots N_d$

Decision variables for Time 2 optimization: $[s_{i,2}]$ for $i = 1, 2, \ldots N_d$

Decision variables for Time T optimization: $[s_{i,T}]$ for $i = 1, 2, \ldots N_d$

Decision variables for full-time optimization: $[s_{i,t}]$ for $i = 1, 2, \ldots N_d$, for $t = 1, 2, \ldots T$

(a)   (b)

Figure 3. Comparison of long-term water allocation optimization. (**a**) Single-step linear programming (SSLP) optimization framework and (**b**) full-step linear programming (FSLP) optimization framework.

To estimate the economic value of municipal water, K-water [21] calculates the benefit of municipal water supplied to the user by estimating the user's willingness to pay (WTP) for unit demand. The WTP indicates how much value the consumer places on a particular product. In this study, the water supply benefit of municipal water was calculated using the measured WTP and the estimated demand function from the K-water study [21]. Here, the demand function defines the relationship between price and demand to represent the change in the quantity of a product the consumer intends to purchase according to the price of the product. In general, when other supply conditions are the same, the curve of the demand function gradually decreases with increasing price.

To estimate the economic value of municipal water in Korea, the reduction in water consumption corresponding to an increase in water price was first investigated based on the maximum WTP for municipal water demand using a consumer survey. Then, the final supply benefit function was derived by calculating the market demand function reflecting the individual responses in the form of the WTP. In this study, the dependent variable of the supply benefit function is the WTP for municipal water (i.e., the value of unit municipal water). The average monthly water demand per person, average monthly income per person, and the number of persons for each household were selected as explanatory variables. Hence, the linear municipal water supply benefit function can be expressed as

$$P_M = \alpha_0 + \alpha_1 Q + \alpha_2 I + \alpha_3 F + \epsilon \tag{5}$$

where P_M is the value of unit municipal water per person, measured as the WTP (KRW/m^3/person); Q is the average monthly demand for municipal water per person (m^3/month

/person); I is the average monthly income per person (KRW/month/person); F is the number of persons per household; α_0, α_1, α_2, and α_3 are the coefficients of the regression equation; and ϵ is the error variable of the regression equation.

After excluding the income variable from Equation (5) through a t-test, K-water [21] derived the supply benefit function for municipal water by performing a regression analysis based on the historical data and yielded:

$$P_M = 516.336 - 2.716Q + 65.137F \tag{6}$$

To estimate the economic value of industrial water in Korea, K-water [21] calculated the marginal product output according to the industrial water supply for a given production function. The production function is a mathematical representation of the relationship between input sources and output values of industrial products. K-water [21] expressed this function using the Cobb–Douglas production function estimation method with the production value as a dependent variable, while the input labor, manufacturing cost, amount of tangible fixed assets at year-end, and water supply were taken as explanatory variables, as follows:

$$\ln P_I = \ln \beta_0 + \beta_1 \ln L + \beta_2 \ln W_I + \beta_3 \ln M_I + \beta_4 \ln K \tag{7}$$

where P_I is the production value of industrial products (KRW); L is the number of workers (person); W_I is the industrial water supply quantity (m^3); M_I is the unit manufacturing cost (KRW/unit product); K is the balance of tangible assets at the end of the year (KRW/year); and β_0, β_1, β_2, and β_3 are the coefficients of the regression equation.

A regression analysis of the average technical statistics for all industrial sectors surveyed by the National Water Resources Plan 2011–2020 of South Korea [30] resulted in the following industrial water production function:

$$\ln P_I = 2.02 + 0.73 \ln L + 0.05 \ln W_I + 0.27 \ln M_I + 0.05 \ln K \tag{8}$$

Thus, the value of unit industrial water (KRW/m^3) can be calculated by dividing the calculated production value of industrial products determined using Equation (8) with the quantity of industrial water supplied.

To estimate the economic value of agricultural water in Korea, Lim and Lee [22] calculated its economic benefits based on a production function correlating the agricultural water input with the annual rice production. This process was the same as that used to analyze the VMP of industrial water. Lim and Lee [22] expressed the economic value of agricultural water using the Cobb–Douglas production function estimation method based on the production value of rice as a dependent variable and the total agricultural water supply, agricultural management cost, irrigated farmland area, and drought year correction coefficient as explanatory variables as follows:

$$\ln P_A = \ln \gamma_0 + \gamma_1 \ln A + \gamma_2 \ln W_A + \gamma_3 \ln M_A + \gamma_4 D \tag{9}$$

where P_A is the production value of rice (KRW); A is the area of irrigated farmland (1000 ha); W_A is the agricultural water supply quantity (million m^3); M_A is the agricultural management cost (KRW); D is the drought year correction coefficient; and γ_0, γ_1, γ_2, and γ_3 are the coefficients of the regression equation.

A regression analysis of the total amount of rice production and agricultural infrastructure status in South Korea over 35 years was then applied to obtain the agricultural water production function:

$$\ln P_A = 2.97 + 0.10 \ln A + 0.28 \ln W_A + 0.36 \ln M_A + 0.18 D \tag{10}$$

Thus, the value of unit agricultural water (KRW/m^3) can be calculated by dividing the calculated production value of rice determined using Equation (10) with the quantity of agricultural water supplied.

The economic value of various water uses, determined according to the methods described above, can be used to quantify the direct benefit generated from the water supply for each sector. Therefore, in the event of a water supply shortage, the potential economic damage in each demand sector can be calculated and water can be allocated accordingly to minimize overall economic loss (or to maximize overall economic benefit).

2.3. Climate Change Scenarios

2.3.1. Changes in Water Supply

Cho et al. [31] analyzed 52 GCMs (including the Representative Concentration Pathway (RCP) 4.5 and 8.5 scenarios) and evaluated their fitness for application in South Korea through the climate change adaptation for water resources (CCAW) research project of South Korea. Seventeen GCMs were evaluated as having better availability based on their standardized skill scores and were divided into wet, average, and dry scenarios. In the present research, the RCP 8.5 climate change scenario was selected. This scenario is generated from the Hadley Center Global Environment Model version 2 Atmosphere Ocean (HadGEM2-AO) [32] and is the most suitable scenario for representing typical climate change in the study area [31].

Meteorological information, such as daily precipitation and temperature for the next 30 years, were provided from the model, and the river basin flow was estimated using rainfall-runoff models, such as the Hydrological Simulation Program-FORTRAN (HSPF) [33], Soil and Water Assessment Tool (SWAT) [34], and Precipitation-Runoff Modeling System (PRMS) [35], based on meteorological information in South Korea [36]. Lee et al. [36] applied the three rainfall-runoff models to the four biggest river basins in South Korea and reported that the PRMS model is the most suitable for the Han River basin. Therefore, using the pentad (5-day) averaged flow rate resulting from the PRMS model, the river basin flow according to climate change in the case study basin was estimated and applied as input to the WAMM. Note that the river basin flow represents the available water resources in the study basin.

2.3.2. Changes in Water Demand

Agricultural demand, municipal/industrial (combined) demand, and river maintenance flow are considered water demand sectors in the WAMM. Based on the water demand in 2015, in this study, the changes in population and farmland over the next 30 years were applied, as suggested by the Korea Statistical Yearbook [37], as well as the population projections by province [38], to the water demand change scenario. Based on farmland area statistics in the study area in 2015–2020, farmland area was extrapolated from 2015 to 2045, with a gradual decrease of 15%. Conversely, according to national population projections by province [38], the population in the study area is expected to increase by 12.8% from 2015 to 2036, then decrease by 2.2% from 2036 to 2045. Furthermore, three long-term water demand projections (low demand, average demand, and high demand) were suggested by the National Water Resources Plan 2011–2020 of South Korea (MLTM, Ministry of Land, Transport and Maritime Affairs) [30] to account for the uncertainty of water supply conditions. According to the high demand projection [30], municipal, industrial, and agricultural water demands are expected to increase by 0.66%, 11.07%, and 4.85%, respectively. In the present study, among the three demand projections, the high demand condition was adopted with farmland area and population trends for the estimation of municipal, industrial, and agricultural water demands, for use as WAMM inputs.

3. Case Study

3.1. Application Area (Namhan River Basin, South Korea)

Figure 4 shows the case study area, the Namhan River basin in South Korea, in which approximately 1.5 million people live (investigated in the year 2018), including seven rural sub-basins, to which WAMM performance was applied and analyzed under the climate change scenarios. The total length and area of the river basin are 375 km and 12,407 km^2 (2447, 1773, 2483, 1614, 524, 1491, and 2072 km^2 for the 1001 to 1007 sub-basins). In this case study, water was allocated to meet the requirements in June when water shortage has been frequently observed in the study area. The water allocation period was limited to June (frequent drought month) and evaluated using a pentad (5-day) simulation time step over the next 30 years (with 2015 as the base year); that is, June was divided into six time intervals for every simulation year. Thus, each simulation was conducted independently for each year.

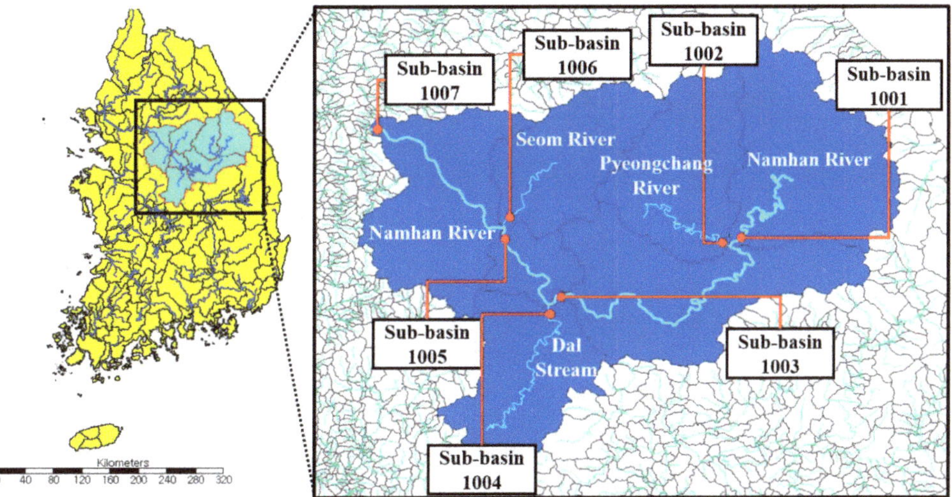

Figure 4. Case study area with seven sub-basins of the Namhan River basin, South Korea.

River flow, groundwater, and dam and agricultural reservoirs were considered water supply sources in the case study basin, while agricultural, municipal/industrial (combined), and river maintenance flow were considered demand sites. The amount of reclaimed water, an additional water supply source, was dynamically determined based on the allocated water at each of the municipal/industrial demand sites with specific reclamation rates. In South Korea, according to the MLTM [30], 65% of the demand supplied to municipal and industrial sites is generally discharged into the river downstream of the demand sites. In the case of agricultural demand, 35% of the demand contributes to water resources gradually; therefore, the net demand (i.e., 65% of the total demand) was applied as the actual agricultural demand without the water reclamation process.

A schematic of the water supply system in the Namhan River basin is shown in Figure 5. The base demands of the municipal/industrial and agricultural demand sites in June 2015 are summarized in Table 1. In total, 29 demand sites, including 10 agricultural sites, 14 municipal/industrial sites, and 5 river maintenance flow gauges, with identical priorities for reliability-based water allocation, use water in the case study basin. The Namhan River is the main stream draining the basin and it has three branches: the Pyeongchang River, Dal Stream, and Seom River. The water flow is controlled by the Gwangdong Dam on the Pyeongchang River before it joins the Namhan River and then by the Chungju Dam on the Namhan River. Next, the Dal stream joins the Namhan River, followed by the

Seom River. The Hoengseong Dam is located upstream on the Seom River and controls its flow. After the joining of the Seom River, three large weirs—the Gangcheon Weir, Yeoju Weir, and Ipo Weir—control the flow of the mainstream of the Namhan River downstream. In addition, agricultural water is supplied by several agricultural reservoirs distributed throughout the basin. Consequently, it is possible to store and supply water in the case study basin through a total of six main water storage facilities (i.e., three dam reservoirs and three weir pools) and 10 agricultural reservoirs in addition to the river flow.

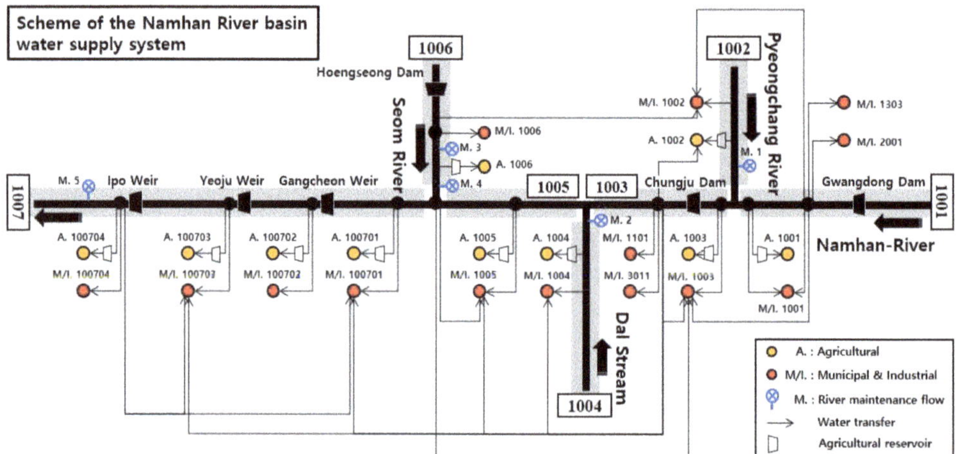

Figure 5. Schematic of the Namhan River basin water supply network.

Table 1. Average water demand at individual sites in the Namhan River basin (in June 2015).

Sub-Basin	Demand Sites	Water Demand (1000 m^3)	Sub-Basin	Demand Sites	Water Demand (1000 m^3)
External transmission	M/I. 1303	245	1005	A. 1005	11,280
	M/I. 2001	780		M/I. 1005	615
	M/I. 1101	225	1006	A. 1006	18,559
	M/I. 3011	3170		M/I. 1006	6095
1001	A. 1001	3900	1007	A. 100701	7649
	M/I. 1001	3090		M/I. 100701	2820
1002	A. 1002	7361		A. 100702	3318
	M/I. 1002	955		M/I. 100702	1960
1003	A. 1003	11,227		A. 100703	49,057
	M/I. 1003	4175		M/I. 100703	6880
1004	A. 1004	28,585		A. 100704	4584
	M/I. 1004	4815		M/I. 100704	2875

Note: A: agricultural water demand; M/I: municipal and industrial water demand.

Every municipal/industrial and agricultural demand site has access to separate groundwater sources nearby. In the WAMM, the recharge of groundwater is not simulated, but only initial groundwater storage is set and supplied to demand sites. However, in the target season (June), little groundwater use was recorded, and so the effect of groundwater supply was not specified in the present case study. Furthermore, Figure 5 shows five flow gauges (M.1 to M.5) for river maintenance flow over the entire river basin, which have flow requirements of 7.73, 5.03, 23.1, 7.57, and 32.5 m^3/s, based on maintenance demand in MLTM [30]. The river maintenance flow requires the minimum flow rates for eco-environmental preservation and induces the discharge of reservoirs and release downstream without actual water consumption.

3.2. Economic Value of Water Resources in the Namhan River Basin

The economic values of municipal, industrial, and agricultural water supply in Namhan River basin were estimated and provided to the WAMM. The economic value of municipal water was estimated to be 1283 KRW/m^3 by substituting the municipal water use status reported in the National Water Resources Plan 2011–2020 of South Korea (MLTM) [30] in the municipal water supply benefit function (Equation (6)). The economic values of industrial and agricultural water were estimated to be 5583 KRW/m^3 and 384 KRW/m^3, respectively, by applying the data used to derive the respective production functions in previous studies (Equations (8) and (10) suggested in [21,22]). Here, the economic value of river maintenance flow was not considered. As the WAMM considers the demand sites of both municipal and industrial water together, their economic value (Table 2) was applied by considering the proportions of municipal and industrial water use at each demand site.

Table 2. Economic valuation of water supply to municipal/industrial water demand sites.

Demand Site	Municipal Water Demand Ratio (%)	Industrial Water Demand Ratio (%)	Economic Value of Supply (KRW/m^3)
M/I. 1001	38	62	3949
M/I. 1002	92	8	1627
M/I. 1003	69	31	2616
M/I. 1004	80	20	2143
M/I. 1005	63	37	2874
M/I. 1006	92	8	1627
M/I. 100701	61	39	2960
M/I. 100702	61	39	2960
M/I. 100703	61	39	2960
M/I. 100704	61	39	2960
M/I. 1101	100	0	1283
M/I. 1303	100	0	1283
M/I. 2001	100	0	1283
M/I. 3011	100	0	1283

3.3. Application of Climate Change Scenario in the Namhan River Basin

3.3.1. River Basin Flow Projection

The predicted precipitation and maximum temperatures in the seven sub-basins in June for the next 30 years are illustrated in Figure 6. Using the PRMS rainfall-runoff simulation, the river basin flow in June for each sub-basin was estimated for the next 30 years (independent year-by-year simulation; Figure 7). It should be noted that the initial storage in the reservoirs at the beginning of June for the next 30 years is fixed as the base value of the standard year of 2015. Sub-basin flows generally change following consistent trends across the years but deviate considerably between wet and drought years. The total river basin flow of the entire Namhan River basin in June exhibits an average of 405 million m^3/month, a maximum of 1455 million m^3/month (2018), and a minimum of 29 million m^3/month (2036) over the next 30 years. Based on comparisons of predicted and gauged rainfall in 2015–2018, the climate change scenarios represent relatively wet conditions in the 1001–1004 sub-basins, and dry conditions in the 1005–1007 sub-basins.

3.3.2. Water Demand Projection

The water demand in June was predicted for each demand site over the next 30 years (Figure 8) based on the water demand change scenario of the Namhan River basin. The agricultural water demand in June was predicted to gradually decrease over the next 30 years from a maximum of 170 million m^3/month in 2015 to a minimum of 144 million m^3/month in 2045 owing to a decrease in the farmland area. On the other hand, the demand for municipal and industrial water in June was predicted to increase from a minimum of 52 million m^3/month in 2015 to a maximum of 57 million m^3/month in 2045, and the annual deviation was predicted to be relatively small.

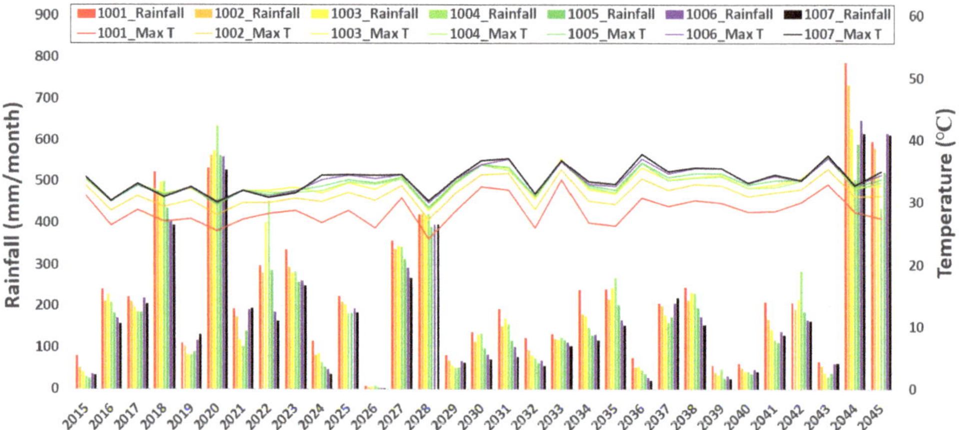

Figure 6. Rainfall and temperature predictions in the Namhan River sub-basins (in June).

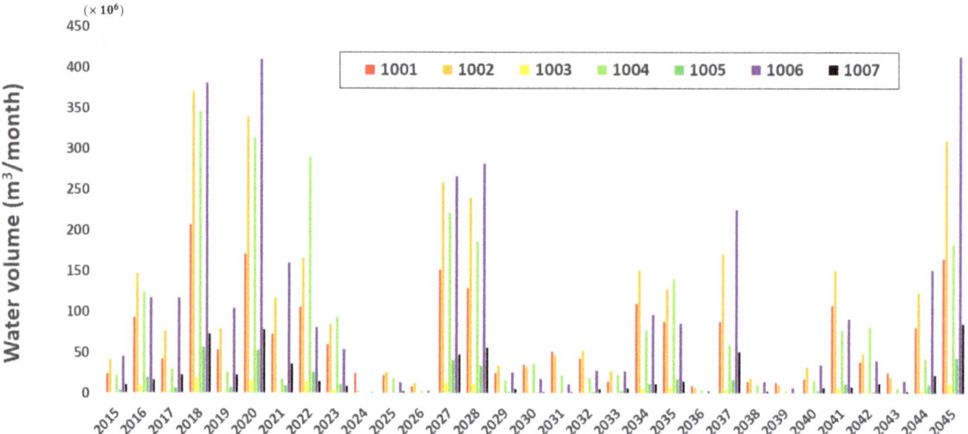

Figure 7. Runoff predictions in the Namhan River sub-basins (in June).

The estimated river basin flow and water demand for the next 30 years were preliminarily compared for the entire Namhan River basin (Figure 9). The total river basin flow shows significant annual variation compared with the water demand. As the river basin flow in June was sometimes remarkably lower than the demand during the 30-year simulation period, the predictions show that water shortage events will occur throughout the basin. However, these results do not include water resources that can be utilized from reservoirs (e.g., dams, weirs, and agricultural reservoirs) in the basin. Thus, actual water shortages are likely to primarily occur in the most notable drought years, including 2024–2026, 2036, and 2038–2039, which are predicted to have insufficient river flow.

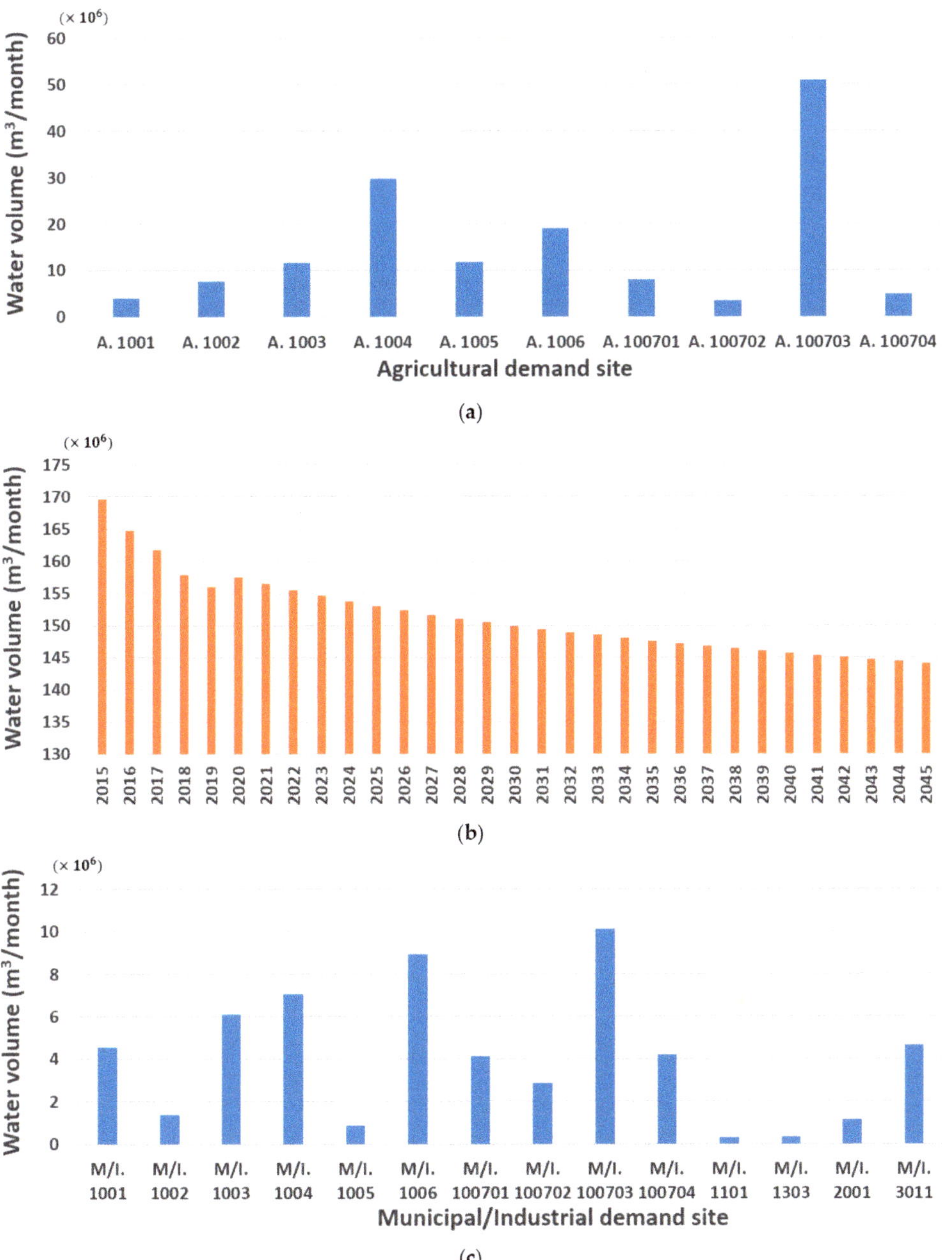

Figure 8. *Cont.*

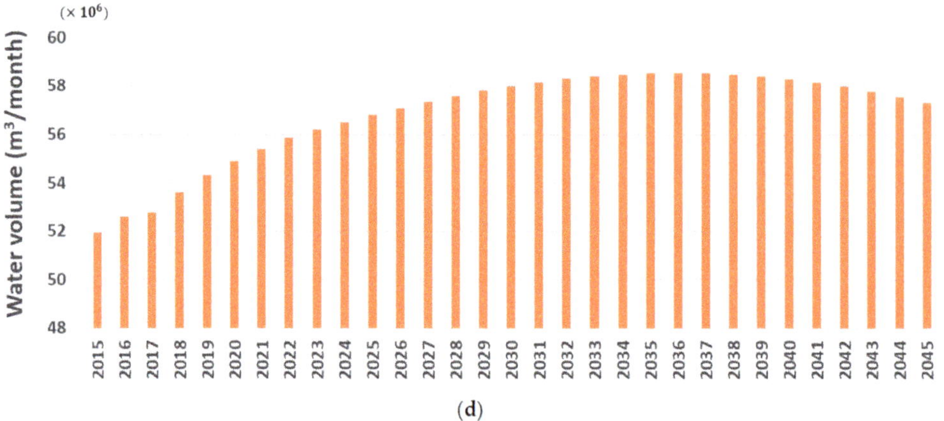

Figure 8. Water demand prediction in the Namhan River basin (in June). (**a**) Agricultural water demand of each demand site (30-year average); (**b**) total agricultural water demand changes over 30 years; (**c**) municipal/industrial water demand of each demand site (30-year average); and (**d**) total municipal/industrial water demand changes over 30 years.

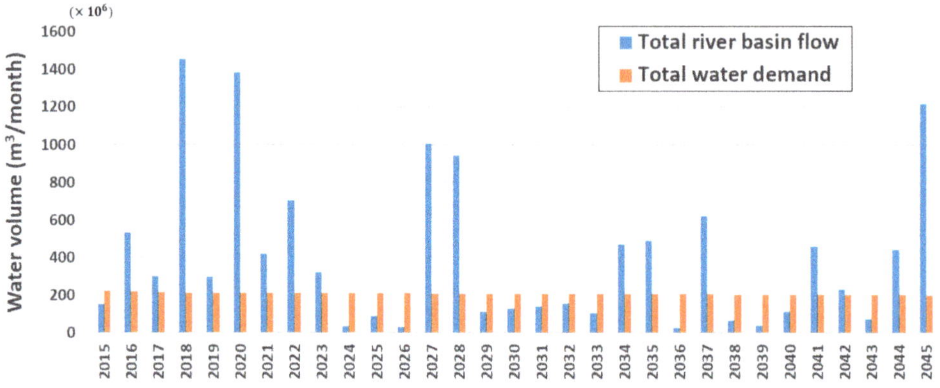

Figure 9. Comparison of predicted river flow and water demand in the Namhan River basin.

4. Application Results

Optimal water allocation under the applied climate change scenario was simulated using the WAMM for the Namhan River basin over the next 30 years. By applying two optimization methods (SSLP and FSLP) and two objective functions (supply reliability and economic benefit), the allocation results were compared to demonstrate the relative effectiveness of the allocation methods embedded in the WAMM.

4.1. Water Allocation Using SSLP

4.1.1. Reliability-Based Water Allocation Using SSLP

First, optimal water allocation was conducted to maximize the water supply reliability as the objective function using the SSLP method (Figure 10). According to the WAMM, water shortages with an average, maximum, and minimum values of 9.7 million m^3/month, 32.3 million m^3/month (2024), and 0.7 million m^3/month (2016), respectively, will occur in the case study area in June over the next 30 years. These water shortages correspond to average, maximum, and minimum economic losses of 5.0 billion KRW/month, 16.9 billion KRW/month (2024), and 0.9 billion KRW/month (2016), respectively.

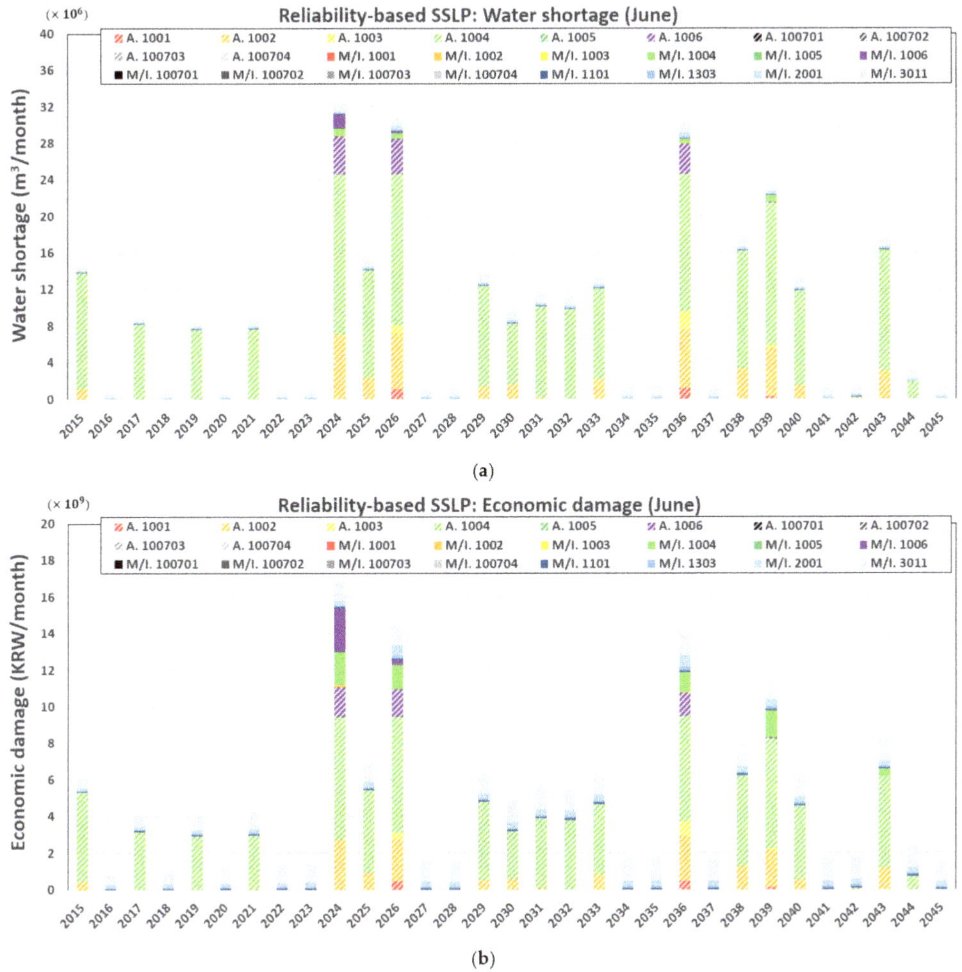

Figure 10. Water allocation simulation results obtained using reliability-based single-step linear programming (SSLP). (**a**) Water shortage and (**b**) economic damage.

Within the simulation period (30 years), water shortages occur every year at demand site M/I. 3011, where the increasing demand cannot be met owing to the restriction of the transmission capacity. In addition, agricultural sites such as A. 1002, A. 1004, and A. 1006 also incur severe water shortages during drought years. These water shortages are particularly significant in 2024, 2026, and 2036 when the shortages exceed approximately 30 million m^3/month. As shown in Figure 10, agricultural water shortages are dominant (accounting for approximately 86% of the shortages occurring in the basin over 30 years) compared with municipal and industrial water shortages (14%). However, this larger proportion of shortages account for only approximately 63% of the economic damage, while the economic damage in the municipal and industrial sectors is approximately 37%. This result is related to the different economic values of water according to its use.

4.1.2. Economy-Based Water Allocation Using SSLP

The results of the reliability-based water allocation analysis confirm that the trends of water shortage and economic loss in the basin do not necessarily match because the economic value of water is different according to its use. Here, the allocation of water to maximize economic benefit while using the SSLP method is considered.

The results of the water allocation simulation yielded the distribution of water shortages and economic damages in June (Figure 11). The overall water shortage and economic damage are lower than those based on the reliability-based SSLP water allocation. There are two reasons for the lower water shortage. First, the increase in supply to the municipal and industrial water demand sites results in increased production and use of reclaimed water. Second, the river maintenance flow is utilized for supplying demand sectors because the economic value of the maintenance flow is not considered in the economy-based SSLP under drought conditions.

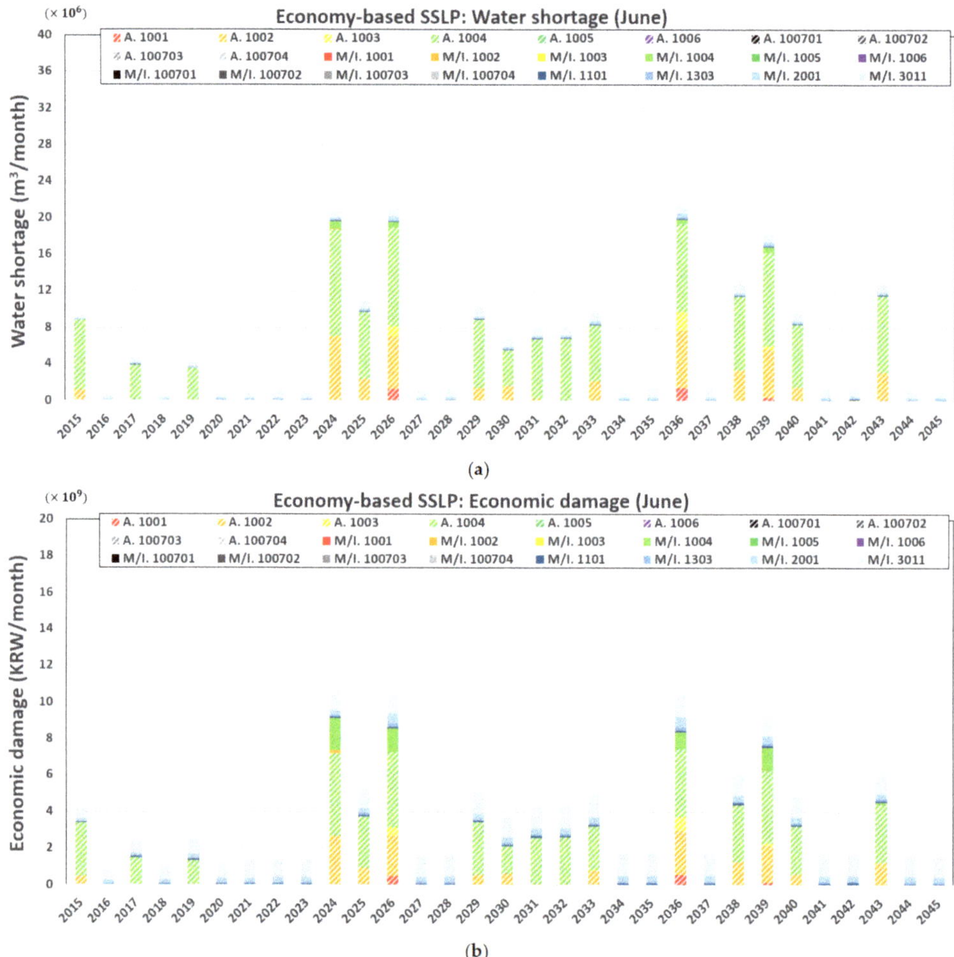

Figure 11. Water allocation simulation results obtained using economy-based single-step linear programming (SSLP). (**a**) Water shortage and (**b**) economic damage.

The water shortage pattern at site M/I. 3011 for the economy-based water allocation observed in Figure 11 is the same as that observed for the reliability-based water allocation in Figure 10. However, the municipal and industrial water shortages in 2024 and 2026 were resolved to a great extent by using the economy-based water allocation; that is, most of the municipal and industrial water shortage events were resolved, except for cases in which the capacity of the supply conduit pipe was limited. In summary, when using economy-based water allocation, the average, maximum, and minimum water shortages in June will be 6.7 million m^3/month, 21.5 million m^3/month (2036), and 0.7 million m^3/month (2016), respectively, which is equivalent to average, maximum, and minimum economic damages of 3.8 billion KRW/month, 10.6 billion KRW/month (2024), and 0.9 billion KRW/month (2016), respectively. Compared with the reliability-based SSLP water allocation results (Section 4.1.1), the economy-based SSLP water allocation method reduces maximum the water shortage and economic damage by approximately 10.8 million m^3/month and 6.2 billion KRW/month, respectively, and the average water shortage and economic damage over 30 years by 3.0 million m^3/month and 1.2 billion KRW/month, respectively.

Notably, the severity of economic damage due to water shortage does not seem to be directly related to the severity of water shortages. The most severe water shortage will occur in 2036, but the most severe economic loss will occur in 2024. In other words, water shortage-induced damage and economic damage exhibit different patterns.

4.2. Water Allocation Using FSLP

4.2.1. Reliability-Based Water Allocation Using FSLP

Next, FSLP-based water allocation was performed (i.e., allocating water over the entire simulation period) and the results were compared with those of the conventional SSLP.

The reliability-based FSLP water allocation simulation provided water shortage and economic loss results for June (Figure 12). As expected, the results of the FSLP water allocation demonstrated a significant reduction in the overall water shortage compared to the results of the SSLP. This reduction is the effect of utilizing storage facilities when simultaneously considering the entire simulation period (all six time steps of June). In particular, in SSLP, regardless of the water shortage occurring in other time steps, the remaining water in the current time step is discharged as is or stored mainly in downstream storage facilities. In FSLP, when water shortage is expected in future time steps, the remaining water in the current time step is secured in an appropriate storage facility where possible, improving allocation results through more flexible preparation for expected water shortage events.

Analyses of the simulation results in Figure 12 reveals that, for reliability-based FSLP water allocation, the average, maximum, and minimum water shortage in June is 2.5 million m^3/month, 8.3 million m^3/month (2024), and 0.7 million m^3/month (2016), respectively; the average, maximum, and minimum economic damage is 2.03 billion KRW/month, 4.38 billion KRW/month (2024), and 0.9 billion KRW/month (2016), respectively. Compared to the reliability-based SSLP water allocation (Section 4.1.1), the FSLP-based allocation considerably increases water allocation efficiency by reducing the maximum water shortage and economic damage by approximately 24.0 million m^3/month and 12.5 billion KRW/month, respectively, and the average water shortage and economic damage over 30 years by 7.2 million m^3/month and 3.0 billion KRW/month, respectively.

4.2.2. Economy-Based Water Allocation Using FSLP

Lastly, economy-based FSLP water allocation was performed and the results of water shortage and economic damage are shown in Figure 13. Compared to the economy-based SSLP (Section 4.1.2), which maximizes economic benefits with the conventional SSLP method, the FSLP reduced the maximum water shortage and economic damage by approximately 13.2 million m^3/month and 6.3 billion KRW/month, respectively, and the average water shortage and economic damage over 30 years by 4.2 million m^3/month and 1.8 billion KRW/month, respectively.

The results of a comparison between the reliability-based FSLP (Section 4.2.1) and economy-based FSLP are as follows. As the reliability-based allocation already maximizes the quantitative efficiency of water supply, the total water shortage was the same throughout the simulation period for both approaches. However, using the economy-based FSLP, the economic damage was slightly reduced by supplying water preferentially to demand sites M/I. 1001 and M/I. 100701–M/I.100704, which have larger proportions of industrial, water rather than supplying demand sites M/I. 1002 and M/I. 1006, in which the proportion of municipal water is higher than industrial water. In summary, the water shortages resulting from the economy-based FSLP water allocation were the same as those resulting from the reliability-based FSLP, but with slightly lower average, maximum, and minimum economic damages of 2.02 billion KRW/month, 4.36 billion KRW/month (2024), and 0.9 billion KRW/month (2016), respectively.

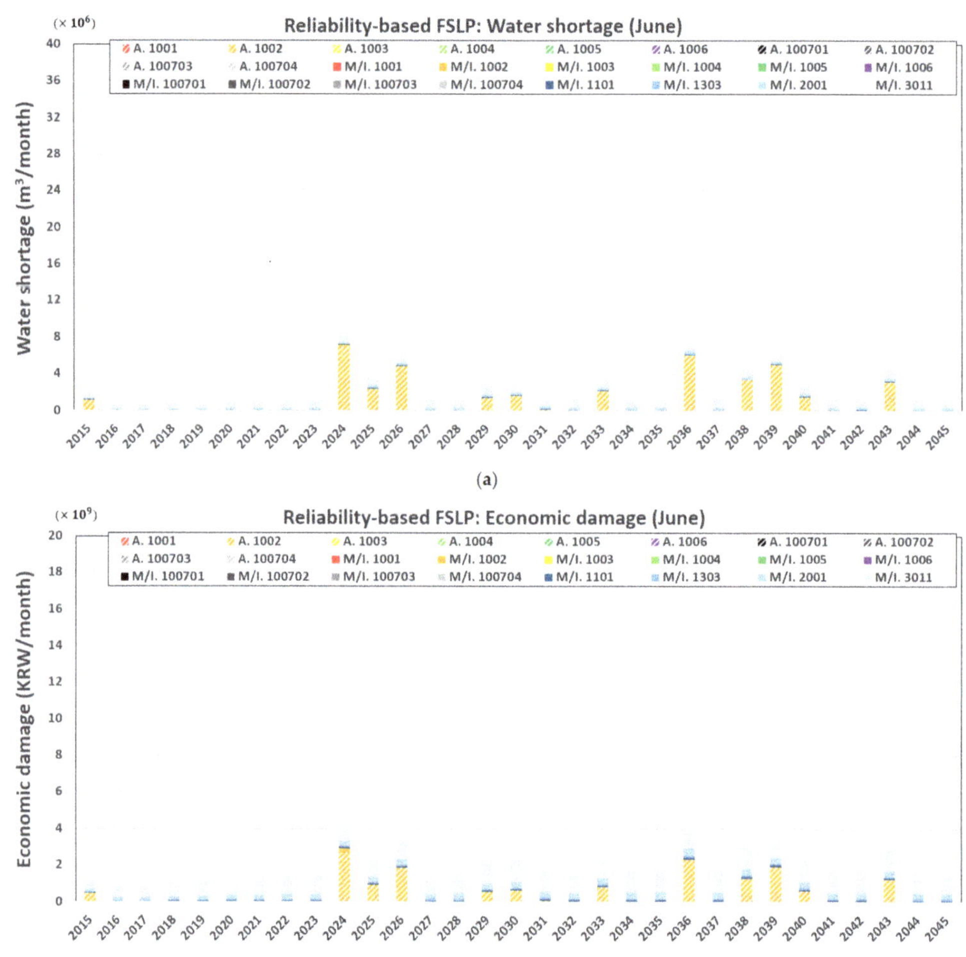

Figure 12. Water allocation simulation results obtained using reliability-based full-step linear programming (FSLP). (**a**) Water shortage and (**b**) economic damage.

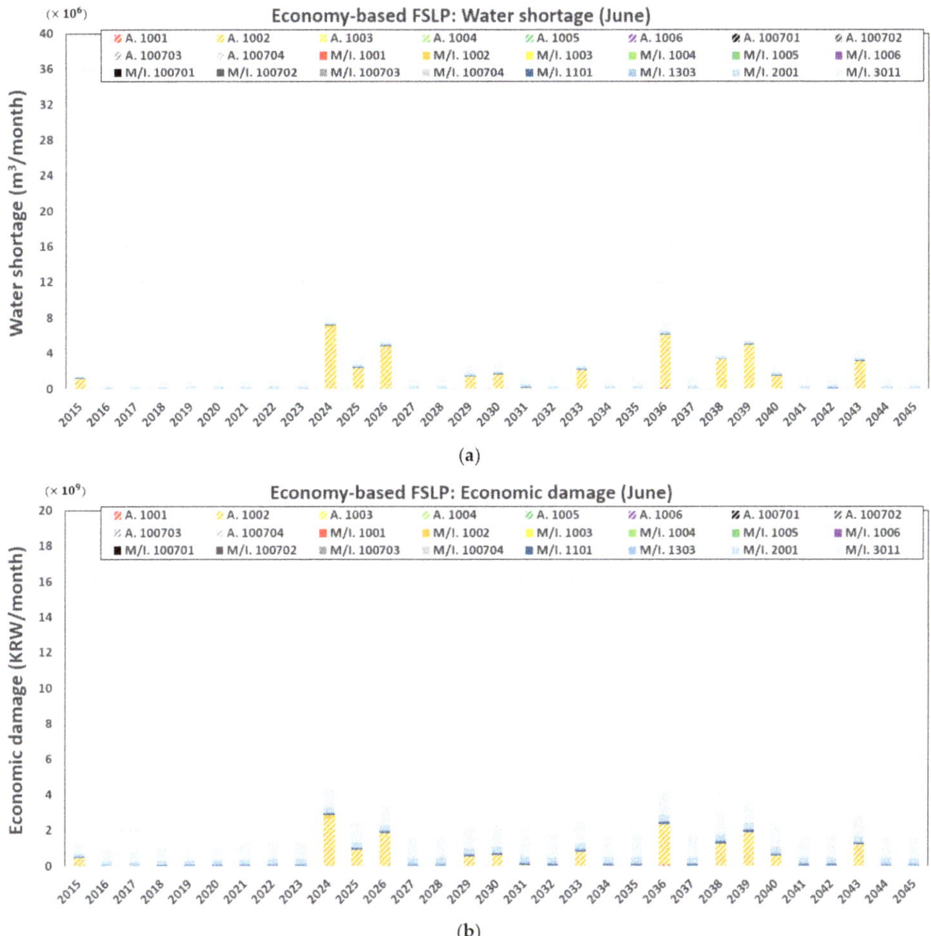

Figure 13. Water allocation simulation results obtained using economy-based full-step linear programming (FSLP). (**a**) Water shortage and (**b**) economic damage.

5. Discussion

Conventional water allocation models such as MODSIM [6] and WEAP [8] were developed decades ago but many studies are still widely applying them to water resource management by combining them with optimization schemes and various objective functions. Chou et al. [39] conducted a representative case study in optimizing water resources operation by combining existing models and advanced optimization algorithms. In this study, water allocation results using the reliability-based SSLP of the WAMM represent the conventional approach of WEAP [8]; however, the results using FSLP or economy-based allocation offer an alternative decision for better water resources management. The results, which are based on assumptions such as climate change and water demand scenarios, reveal both water shortage risks and associated economic considerations. Moreover, water supply efficiency according to the objectives and optimization methods of water allocation can be comparatively analyzed.

Table 3 provides a comparative analysis of the four water allocation simulation results evaluated in this study. In terms of the objective function, the economy-based water allocation approach reduced both water shortage and economic damage compared with the reliability-based allocation approach. In terms of the optimization method, the FSLP significantly reduced water shortages compared with the SSLP. Therefore, a water allocation method that considers the economic value of water resources using an FSLP optimization method capable of simultaneously considering the entire simulation period is the most effective approach for the case study basin. Finally, the 30-year average water supply reliability of the reliability-based SSLP water allocation (i.e., the conventional method) was 95%, which improved to 97% when using the economy-based SSLP water allocation and to 99% when using FSLP water allocation.

Table 3. Comparison of full-step linear programming (FSLP) and single-step linear programming (SSLP) water allocation results.

Water Allocation Method	Total Reliability	Average		Maximum		Minimum	
		Water Shortage (10^6 m^3/Month)	Economic Damage (10^9 KRW/Month)	Water Shortage (10^6 m^3/Month)	Economic Damage (10^9 KRW/Month)	Water Shortage (10^6 m^3/Month)	Economic Damage (10^9 KRW/Month)
Reliability-based SSLP	0.95	9.7	5.0	32.3	16.9	0.7	0.9
Economy-based SSLP	0.97	6.7	3.8	21.5	10.6	0.7	0.9
Reliability-based FSLP	0.99	2.5	2.03	8.3	4.38	0.7	0.9
Economy-based FSLP	0.99	2.5	2.02	8.3	4.36	0.7	0.9

6. Conclusions

In this study, a new hydro-economic water allocation and management model, WAMM, was tested by improving the optimization algorithm of conventional water allocation models and considering the economic value of water resources. The WAMM was applied for water allocation simulation considering the impact of climate change and water demand scenarios on the Namhan River basin in South Korea over the next 30 years. A climate change-influenced water supply scenario was constructed based on the precipitation and maximum temperature predictions of the RCP 8.5 climate change scenario derived from HadGEM2-AO GCM. Also, a demand scenario was built based on a demand forecast obtained according to the changes in population and farmland and accounted for the uncertainty of conditions. To evaluate the two objective functions (reliability and economic benefit) and the two optimization methods (SSLP and FSLP) within the WAMM, water supply from the basin over the next 30 years was allocated to four different cases and the allocation results were comparatively analyzed. The results can be summarized as follows.

1. The results of the reliability-based SSLP allocation show that, according to the considered climate change scenario, the water shortage in June over the next 30 years will be an average and maximum of 9.7 million m^3/month and 32.3 million m^3/month (2024), respectively, corresponding to average and maximum economic damages of 5.0 billion KRW/month and 16.9 billion KRW/month (2024), respectively.

2. The results of the economy-based SSLP allocation show a lower total water shortage because water allocation focused on municipal and industrial sectors and the use of reclaimed water increased; moreover, the priority of river maintenance flow demand was lowered. Over the next 30 years, average and maximum water shortages in June of 6.7 million m^3/month and 21.5 million m^3/month (2036), respectively, were obtained, corresponding to average and maximum economic damages of 3.8 billion KRW/month and 10.6 billion KRW/month (2024), respectively. Therefore, compared to the reliability-based SSLP method, in which the objective function of water allocation is simply based on the supply quantity, water allocation considering the economic value of the water reduces both the water shortage and economic loss to a certain degree.

3. The results of the reliability-based FSLP allocation confirm that the total water shortage was considerably reduced by flexibly utilizing storage facilities to respond to water shortages that are predicted to develop throughout the entire simulation period. The average and maximum water shortages in June were found to be 2.5 million m^3/month and 8.3 million m^3/month (2024), respectively, corresponding to aver-

age and maximum economic damages of 2.03 billion KRW/month and 4.38 billion KRW/month (2024), respectively.

4. Economy-based FSLP allocation indicates the same quantity of water shortage as the reliability-based FSLP allocation but preferentially supplies water to municipal/industrial demand sites that comprise a high proportion of industrial water demand. This difference resulted in average and maximum economic damages of 2.02 billion KRW/month and 4.36 billion KRW/month (2024), respectively, over the 30-year period, which are slightly less than the corresponding values obtained using the reliability-based FSLP.

The WAMM has a few important limitations to take note of when applying climate change scenarios to each sub-basin; primarily, the scenarios and economic value of water are somewhat roughly aggregated. However, the water shortage-induced economic damage due to climate change was still successfully minimized and quantified using the model. In future work, the WAMM will be modified to more comprehensively consider water sources such as the conjunctive use of surface water, groundwater, and reclaimed water. Furthermore, more precise climate change scenarios and water valuations would also contribute to specified water resources management. Once effectively derived and demonstrated, the WAMM is expected to offer a robust decision-making tool to resolve water shortages and disputes among river basin units.

Author Contributions: Conceptualization, G.J. and D.K.; methodology, G.J. and D.K.; software, G.J.; validation, D.K.; formal analysis, G.J. and D.K.; writing—original draft preparation, G.J.; writing—review and editing, D.K.; supervision, D.K.; funding acquisition, D.K. All authors have read and agreed to the published version of the manuscript.

Funding: This work was financially supported by the Korea Ministry of Environment (MOE) as the Graduate School specializing in Climate Change.

Data Availability Statement: The data presented in this study are available on request from the corresponding author. The data are not publicly available due to privacy.

Conflicts of Interest: The authors declare no conflict of interest.

References

1. The World Bank. "Water Resources Management." World Bank 20 September 2017. Available online: www.worldbank.org/en/topic/waterresourcesmanagement#2 (accessed on 7 May 2021).
2. UNESCO, UN-Water. United Nations World Water Development Report 2020: Water and Climate Change, Paris, UNESCO. 2020. Available online: http://www.unwater.org/publications/world-water-development-report-2020/ (accessed on 7 May 2021).
3. Adetoro, A.A.; Ngidi, M.; Nyam, Y.S.; Orimoloye, I.R. Temporal evaluation of global trends in water footprint, water sustainability and water productivity research. *Sci. Africa* **2021**, *12*, e00732.
4. Strzepek, K.M.; Lenton, R.L. Analysis of Multipurpose River Basin Systems: Guidelines for Simulation Modelling. In *Technical Report No. 236 Ralph M. Parsons Laboratory for Water Resources and Hydrodynamics, Department of Civil Engineering*; Massachusetts Institute of Technology: Cambridge, MA, USA, 1978.
5. Strzepek, K.M.; MITSIM-2 a Simulation Model for Planning and Operational Analysis of River Basin Systems. 1981. WP-81-124. Available online: http://pure.iiasa.ac.at/id/eprint/1637/ (accessed on 20 August 2020).
6. Shafer, J.; Labadie, J. Synthesis and Calibration of a River Basin Water Management Model. Completion Report No. 89, Colorado Water Resources Research Institute, Colorado State University, Ft. Collins. 1978. Available online: http://former.iemss.org/sites/iemss2006//papers/w5/MODSIMpaper.pdf (accessed on 20 August 2020).
7. Andreu, J.; Capilla, J.; Sanchis, E. AQUATOOL: A computer-assisted support system for water resources research management including conjunctive use. In *Decision Support Systems. NATO ASI Series (Series G: Ecological Sciences)*; Loucks, D.P., da Costa, J.R., Eds.; Springer: Berlin/Heidelberg, Germany, 1991; Volume 26, pp. 333–355.
8. Raskin, P.; Hansen, E.; Zhu, Z.; Stavisky, D. Simulation of water supply and demand in the Aral Sea Basin. *Water Int.* **1992**, *17*, 55–67. [CrossRef]
9. Yates, D.; Sieber, J.; Purkey, D.; Huber-Lee, A. WEAP21—A demand-, priority-, and preference-driven water planning model: Part 1: Model characteristics. *Water Int.* **2005**, *30*, 487–500. [CrossRef]
10. Diaz, G.E.; Brown, T.C.; Sveinsson, O. AQUARIUS: A modeling system for river basin water allocation. General Technical Report RM-GTR-299. In *U.S. Department of Agriculture, Forest Service*; Rocky Mountain Forest and Range Experiment Station: Fort Collins, CO, USA, 1997.

11. Choi, S.; Lee, D.; Moon, J.; Kang, S. Application of integrated water resources evaluation and planning system (K-WEAP). *J. Korea Water Resour. Assoc.* **2010**, *43*, 625–633. [CrossRef]
12. Zagona, E.A.; Fulp, T.J.; Shane, R.; Magee, T.; Goranflo, H.M. Riverware: A generalized tool for complex reservoir system modeling. *J. Am. Water Resour. Assoc.* **2001**, *37*, 913–929. [CrossRef]
13. Howitt, R.E.; Lund, J.R.; Kirby, K.W.; Jenkins, M.W.; Draper, A.J.; Grimes, P.M.; Ward, K.B.; Davis, M.D.; Newlin, B.D.; van Lienden, B.J.; et al. Integrated Economic-Engineering Analysis of California's Future Water Supply. Project Completion Report, Department of Civil and Environmental Engineering, University of California, Davis. 1999. Available online: https://www.researchgate.net/profile/Marion_Jenkins/publication/238672816_Integrated_Economic-Engineering_Analysis_of_California\T1\textquoterights_Future_Water_Supply/links/54871eba0cf2ef34478eb92e/Integrated-Economic-Engineering-Analysis-of-Californias-Future-Water-Supply.pdf (accessed on 20 August 2020).
14. Newlin, B.D.; Jenkins, M.W.; Lund, J.R.; Howitt, R.E. Southern California water markets: Potential and limitations. *J. Water Resour. Plan. Manag.* **2002**, *128*, 21–32. [CrossRef]
15. Pulido-Velazquez, M.; Jenkins, M.W.; Lund, J.R. Economic values for conjunctive use and water banking in southern California. *Water Resour. Res.* **2004**, *40*. [CrossRef]
16. Costanza, R.; d'Arge, R.; de Groot, R.; Farber, S.; Grasso, M.; Hannon, B.; Raskin, R.G. The value of the world's ecosystem services and natural capital. *Nature* **1997**, *387*, 253–260. [CrossRef]
17. De Groot, R.S.; Wilson, M.A.; Boumans, R.M. A typology for the classification, description and valuation of ecosystem functions, goods and services. *Ecol. Econ.* **2002**, *41*, 393–408. [CrossRef]
18. Portela, R.; Bezerra, M.O.; Alam, M.; Shaad, K.; Banerjee, O.; Honzák, M. Discussion Paper 8: Water Supply Services: Biophysical Modeling and Economic Valuation in Ecosystem Accounting. Proceedings of Expert Meeting on Advancing the Measurement of Ecosystem Services for Ecosystem Accounting, New York, NY, USA, 22–24 January 2019; Subsequently Revised. Version of 15 March 2019. Available online: https://seea.un.org/events/expert-meeting-advancing-measurement-ecosystem-services-ecosystem-accounting (accessed on 7 May 2021).
19. Hwang, S.; Um, M.; Kim, T. The valuation of the reliability of municipal water supply using contingent valuation method in Korea. *Environ. Resour. Econ. Rev.* **1999**, *8*, 109–126.
20. Park, S.; Ryoo, K.; Kim, J.; Kim, B. The case study of economic value assessment of spring rainfall in the aspect of water resources. *J. Environ. Sci. Int.* **2014**, *23*, 193–205. [CrossRef]
21. K-water. Improving Measures of Feasibility Study for Water Resources Projects. Research Report, K-Water. 2008. Available online: https://scienceon.kisti.re.kr/srch/selectPORSrchReport.do?cn=TRKO201800028815 (accessed on 20 August 2020).
22. Lim, J.; Lee, M. Marginal benefit-cost analysis of irrigation water in rice production. *J. Agric. Sci. Chungnam Nat'l Univ.* **2001**, *28*, 132–146.
23. KEI (Korea Environment Institute). EVIS (Environment Valuation Information System). Available online: http://evis.kei.re.kr/research (accessed on 20 August 2020).
24. Krol, M.; Jaeger, A.; Bronstert, A.; Güntner, A. Integrated modelling of climate, water, soil, agricultural and socioeconomic processes: A general introduction of the methodology and some exemplary results from the semi-arid north-east of Brazil. *J. Hydrol.* **2006**, *328*, 417–431. [CrossRef]
25. Ray, P.A.; Kirshen, P.H.; Watkins, D.W., Jr. Staged climate change adaptation planning for water supply in Amman, Jordan. *J. Water Resour. Plan. Manag.* **2012**, *138*, 403–411. [CrossRef]
26. Ashofteh, P.S.; Haddad, O.B.; Marino, M.A. Risk analysis of water demand for agricultural crops under climate change. *J. Hydrol. Eng.* **2015**, *20*, 04014060. [CrossRef]
27. Liu, Y.; Guo, H.; Zhang, Z.; Wang, L.; Dai, Y.; Fan, Y. An optimization method based on scenario analysis for watershed management under uncertainty. *Environ. Manag.* **2007**, *39*, 678–690. [CrossRef] [PubMed]
28. Basupi, I.; Kapelan, Z. Evaluating flexibility in water distribution system design under future demand uncertainty. *J. Infrastruct. Syst.* **2015**, *21*, 04014034. [CrossRef]
29. Jeong, G.; Choi, S.; Kang, D. Development and application of hydro-economic optimal water allocation and management model. *J. Korea Water Resour. Assoc.* **2019**, *52*, 707–718.
30. MLTM (Ministry of Land, Transport and Maritime Affairs). National Water Resources Plan (2011–2020), Republic of Korea. 2011. Available online: https://www.kwater.or.kr/gov3/sub03/annoView.do?seq=1889&s_mid=54 (accessed on 20 August 2020).
31. Cho, J.; Jung, I.; Park, J.; Lee, J. Development of AR5 Representative GCMs Selection Technique and Generation of Downscaled Climate Change Scenarios, 2020, CCAW-TR-01, Climate Change Adaptation for Water Resources, Republic of Korea. Available online: https://drive.google.com/file/d/13RY1wOfh8tTtOaRXP75LU3OzkPYrt7qi/view (accessed on 20 August 2020).
32. Collins, W.J.; Bellouin, N.; Doutriaux-Boucher, M.; Gedney, N.; Halloran, P.; Hinton, T.; Martin, G. Development and evaluation of an Earth-System model–HadGEM2. *Geosci. Model Dev. Discuss* **2011**, *4*, 997–1062.
33. Duda, P.; Kittle, J., Jr.; Gray, M.; Hummel, P.; Dusenbury, R.; Decatur, G.; Kinerson, R. An Interactive Windows Interface to HSPF (WinHSPF). In User's Manual, AQUA TERRA Consultants, 2001. Available online: https://textarchive.ru/c-1882560-pall.html (accessed on 20 August 2020).
34. Neitsch, S.L.; Arnold, J.G.; Kiniry, J.R.; Williams, J.R.; King, K.W. Soil Water Assessment Tool Theoretical Document, Version 2005. Grassland, Soil and Water Research Laboratory, Agricultural Research Service, 808, 2005. Available online: https://swat.tamu.edu/media/1292/SWAT2005theory.pdf (accessed on 20 August 2020).

35. Markstrom, S.L.; Regan, R.S.; Hay, L.E.; Viger, R.J.; Webb, R.M.; Payn, R.A.; LaFontaine, J.H. PRMS-IV. The Precipitation-Runoff Modeling System, Version 4. US Geological Survey Techniques and Methods, (6-B7). 2015. Available online: https://pubs.er.usgs.gov/publication/tm6B7 (accessed on 20 August 2020).
36. Lee, D.; Choi, S.; Kang, S. Results of the National Climate Change Impact Assessment Based on AR5, 2020, CCAW-TR-17, Climate Change Adaptation for Water Resources, Republic of Korea. Available online: https://drive.google.com/file/d/1mBH0L0kJny8FquUiOkzBXJasCjKv8PlY/view (accessed on 20 August 2020).
37. Korea Statistics. Korea Statistics Yearbook, Republic of Korea. 2017. Available online: https://www.bok.or.kr/eng/bbs/B0000289/view.do?nttId=10046080&menuNo=400366&pageIndex=1 (accessed on 20 August 2020).
38. Korea Statistics. Population Projections by Province: 2017–2067, Republic of Korea. 2019. Available online: http://kostat.go.kr/portal/eng/pressReleases/8/8/index.board (accessed on 20 August 2020).
39. Chou, F.N.F.; Linh, N.T.T.; Wu, C.W. Optimizing the management strategies of a multi-purpose multi-reservoir system in Vietnam. *Water* **2020**, *12*, 938. [CrossRef]

sustainability

MDPI

Review

Sustainable Use of Geosynthetics in Dykes

Pietro Rimoldi [1], Jonathan Shamrock [2], Jacek Kawalec [3] and Nathalie Touze [4,*]

[1] Geosynthetic Consultant, 20121 Milano, Italy; pietro.rimoldi@gmail.com
[2] Tonkin & Taylor Ltd., Auckland 1142, New Zealand; jshamrock@tonkintaylor.co.nz
[3] Department of Geotechnics & Roads, Faculty of Civil Engineering, Silesian University of Technology, 44-100 Gliwice, Poland; jacek.kawalec@polsl.pl
[4] SDAR, Université Paris-Saclay INRAE, 78352 Jouy-en-Josas, France
* Correspondence: nathalie.touze@inrae.fr

Abstract: Dykes, or levees, are structures designed and constructed to keep the water in a river within certain bounds in the event of a flood. In relation with climate change, more frequent floods, of higher intensity, can be expected due to anthropogenic emissions of greenhouse gases into the atmosphere. The objective of this review paper is to address the many ways in which geosynthetics contribute to sustainable construction of dykes and thus to water systems management. This review paper, prepared by the four Technical Committees and the Sustainability Committee of the International Geosynthetics Society, briefly describes geosynthetics and their function, dykes and dyke failure modes, before presenting the main focus of the use of geosynthetics for the design and construction of durable dykes to ensure the protection of life and infrastructure. The optimization of dyke construction with geosynthetics to increase their resilience not only results in performance advantages, but also in economic advantages. The way geosynthetics can contribute to mitigating greenhouse gas emissions for a sustainable river management is discussed. This is done not only by allowing more economic construction methods to be implemented, but also solutions with increased resilience to face the extreme stresses related to climate change, while at the same time bringing about a positive contribution to the reduction of greenhouse gas emissions during the construction process itself. Finally, it is shown that by following state of the art standards and design practice any possible risk associated with the use of geosynthetics in dykes can be mitigated.

Keywords: geosynthetics; dykes; levees; protection; reinforcement; stabilization; drainage; erosion control; barriers; emission reductions

Citation: Rimoldi, P.; Shamrock, J.; Kawalec, J.; Touze, N. Sustainable Use of Geosynthetics in Dykes. *Sustainability* **2021**, *13*, 4445. https://doi.org/10.3390/su13084445

Academic Editors: Andrzej Wałęga and Alban Kuriqi

Received: 4 March 2021
Accepted: 9 April 2021
Published: 15 April 2021

Publisher's Note: MDPI stays neutral with regard to jurisdictional claims in published maps and institutional affiliations.

1. Introduction

To protect the land and to allow for beneficial uses such as irrigation or navigation, special measures are often necessary to keep the water in a river within certain bounds [1]. Longitudinal dykes constitute one of the most often used active structural methods to control the course of a river [1,2]. A dyke is an embankment constructed to prevent flooding, keep out the sea or confine a river to a particular course, usually only temporarily charged by floods. Dykes are commonly made of different natural materials such as soil and rock, often supplemented by other materials, such as geosynthetics. Dykes are generally categorized into river dykes (or levees) and sea dykes [3]. River dykes are the focus of this review paper.

Since the Intergovernmental Panel on Climate Change (IPCC) [4,5] was established in 1988, vast and constant research has been performed on the effects on natural systems, such as rivers, of the increasing content of anthropogenic greenhouse gases in the atmosphere. Changes in weather patterns will be one of the principal effects of climate change, and one of the expected changes is more extreme weather conditions. This is expected to be of considerable consequence as it impacts on the vulnerability of communities living in low lying areas exposes them to environmental risks and effects of flooding and causes changes to land use on a large scale [6].

One particular objective of adaptation to climate change is minimizing any negative impacts of climate change on social and ecological systems. This involves enhancing the resistance of the social and ecological systems—also called resilience, e.g., resistance to extreme weather events of: (1) infrastructures supplying both the public and the economy, and (2) structures, for the purposes of, e.g., work, health, energy, transport as well as living [7].

In recent decades several major flood events have shown the vulnerability of flood protection structures all around the world. Frequently, the overtopping of flood protection dykes has caused total failure of the structure. In the aftermath of past disastrous flood events in Germany and other European countries, it became evident that dykes are part of society's infrastructure—the improvement of old dykes and the construction of new ones has become essential since the 1990s [8]. Structures such as dykes need a maximum of safety, the flood disasters of the past years were far more destructive than they should have been [1]. The river dykes affected were often too low, or in too poor condition to resist water height, water pressure, or the duration of the flood period [9].

The improvement of dykes cross-sections by using different geosynthetics has developed to be state-of-the-art as geosynthetic solutions, used with natural materials, have proven to provide strength and flexibility, imperviousness and drainage, durability and robustness or to control degradation [1,3,8,10–12]. These technologies bring not just structural defense but more time for evaluating risk and providing emergency response to populated areas that are threatened by rising water levels [8,11]. They can be designed to control the interaction of water and soil according to the individual and local requirements to allow for an excellent execution of waterways and flood protection structures [1].

In dykes, drains affect the pore water flow to avoid internal and surface erosion. Impervious elements prevent an interaction of soil and water. If the water flow cannot be modified, structures and soil have to be strengthened to be able to withstand unfavorable actions of the water [1].

As literature on the proper use of geosynthetics in dykes is scarce and as no synthesis has been presented to date, according to the authors' knowledge, the four Technical Committees of the International Geosynthetics Society (IGS) working respectively on hydraulics, reinforcement, stabilization and barrier systems have decided, in a joint effort with the Sustainability Committee of the IGS, to prepare the review presented in this paper.

The appropriate use of geosynthetics in dykes in relation to filtration and separation, to the management of the drainage of water within the structure over time, strengthening or steepening the structure with reinforcement and stabilization, minimizing impacts of erosion on the structure and enhancing barriers to water flow will be emphasized.

Before presenting an overview of the main applications of geosynthetics in preventing the failure of dykes, thus ensuring their longevity in Section 4, Section 2 of this paper will give a brief overview of geosynthetics and their functions and Section 3 will introduce the structure of earthen dykes and the failure mechanisms that can be encountered.

In compiling the references for this review, it was found that some reference documents on dykes give a brief insight into possible uses of geosynthetics, their functions and their applications [3,12]. As stated above, a state-of-the-art review, as the one presented in this review paper, does not exist according to the authors' knowledge, specifically one that places an emphasis on the sustainable use of geosynthetics in dykes.

Adding resilience to flood protection structures is critical to future risk mitigation as building higher and stronger structures to prevent overtopping waves, storm surge, and flood waters is more costly [13]. Section 5 will emphasize that geosynthetics solutions are also cost effective.

Finally, a focus will be given to the reduction of greenhouse gas emissions resulting from the use of geosynthetic solutions compared to traditional solutions, evidencing how, in addition to improving the resilience of structures, the use of geosynthetics also positively affects the environment by contributing to a reduction in greenhouse gases emissions of constructing and maintaining these structures, as demonstrated in Section 6 of this paper.

2. Geosynthetics and Their Applications

After more than 60 years of successful experience, geosynthetics are very well established for many applications in hydraulic engineering including dykes, and the possible uses are growing continuously [14].

ISO 10318 [15] defines the various types of geosynthetics and their functions.

A geosynthetic [15] is defined as a product, at least one of whose components is made from a synthetic or natural polymer, in the form of a sheet, a strip, or a three-dimensional structure, used in contact with soil and/or other materials in geotechnical and civil engineering applications. Geosynthetics have pervaded geotechnical engineering to the point where it is no longer possible to practice geotechnical engineering without geosynthetics. Various families of geosynthetics can be defined depending on the functions they fulfil: barrier, drainage, filtration, protection, reinforcement, stabilization, separation, and surface erosion control [15].

The barrier function [15] consists of preventing or limiting the migration of fluids. Geosynthetic barriers (GBRs) are geosynthetic materials that fulfill this function. A geosynthetic barrier is defined as a low-permeability geosynthetic material used in geotechnical and civil engineering applications with the purpose of reducing or preventing the flow of fluid through the construction [15]. GBRs fall into three categories according to the material that fulfils the barrier function: (i) clay geosynthetic barriers (GBR-C) whereby the barrier function is implemented by clays (also called geosynthetic clay liners (GCL)), (ii) bituminous geosynthetic barriers (GBR-B) whereby the barrier function is implemented by bitumen, and (iii) polymeric geosynthetic barriers (GBR-P) whereby the barrier function is implemented by a polymer. GBR-B and GBR-P are also called geomembranes.

The principal other functions that other families of geosynthetics can fulfil are drainage, filtration, protection, reinforcement, separation, stabilization and surface erosion control. The various functions are defined as follows [15]:

- Drainage is the collection and transportation of precipitation, groundwater and/or other fluids in the plane of a geosynthetic material,
- Filtration is the restraining of uncontrolled passage of soil or other particles subjected to hydro-dynamic forces, while allowing the passage of fluids into or across a geosynthetic material,
- Protection is the prevention or limitation of local damage to a given element or material by the use of a geosynthetic material,
- Reinforcement is the use of the stress-strain behavior of a geosynthetic material to improve the mechanical properties of soil or other construction materials,
- Separation is the prevention from intermixing of adjacent dissimilar soils and/or fill materials by the use of a geosynthetic material,
- Stabilization is improvement of the mechanical behavior of an unbound granular material by including one or more geosynthetic layers such that deformation under applied load is reduced by minimizing movements of the unbound granular material,
- Surface erosion control is the use of a geosynthetic material to prevent or limit soil or other particle movements at the surface of, for example, a slope.

Various materials within the family of geotextiles and related products can fulfil the seven functions described above [1]. A geotextile is defined [15] as a planar, permeable, polymeric (synthetic or natural) textile material, which may be nonwoven, knitted, or woven and that is used in contact with the soil and/or other materials in geotechnical and civil engineering applications.

Geotextile-related products [15] are planar, permeable, polymeric (synthetic or natural) material used in contact with the soil and or other materials in geotechnical and civil engineering applications, and that do not comply with the definition of a geotextile.

Of the various families of related products, those that are especially emphasized in this paper are geocomposites, geogrids, geostrips, geocells, geoblankets, geonets, geospacers and geomats.

A geocomposite [15] is a manufactured and assembled material, at least one of whose components is a geosynthetic product.

A geogrid [15] is a planar, polymeric structure consisting of a regular open network of integrally connected, tensile elements that may be linked by extrusion, bonding, or interlooping or interlacing and whose openings are larger than the constituents.

A geostrip [15] is a polymeric material in the form of a strip of width not more than 200 mm, used in contact with soil and/or other materials in geotechnical and civil engineering applications.

A geocell [15] is a three-dimensional, permeable, polymeric (synthetic or natural) honeycomb, or similar cellular structure, made of linked strips of geosynthetics.

A geoblanket [15] is a permeable structure of loose, natural or synthetic fibres and geosynthetic elements bonded together to form a continuous sheet.

A geonet [15] is a geosynthetic consisting of parallel sets of ribs overlying and integrally connected with similar sets at various angles;

A geospacer [15] is a three-dimensional polymeric structure with an interconnected air space in between.

A geomat [15] is a three-dimensional, permeable structure, made of polymeric filaments, and/or other elements (synthetics or natural) mechanically, and/or thermally and/or otherwise bonded. When geomats are placed on the ground surface, and filled with topsoil and seeds, provide better vegetative entanglement, improved shear resistance of the root system, and result in a more durable surface, they are referred to as turf reinforcement mats (TRM).

For all these materials and the many applications of geosynthetics, experience has been gained, tests have been developed and standardized, at the international level in ISO TC 221, CEN TC 189 and ASTM D35, and regulations and recommendations have been written.

3. Dykes Structure and Failure Mechanisms

3.1. Insight in Dykes Structure

Dykes (also called levees) are the last artificial defense against floods. For dykes to be effective, they need to be continuously maintained and some fundamental rules are generally followed in their construction.

A dyke consists of several parts as shown in Figure 1. Often the inner core of the embankment is made up of low permeability soil and granular structural fill, in various arrangements, while the outer part is covered with a growth medium or topsoil layer that is covered with perennial vegetation to limit erosion.

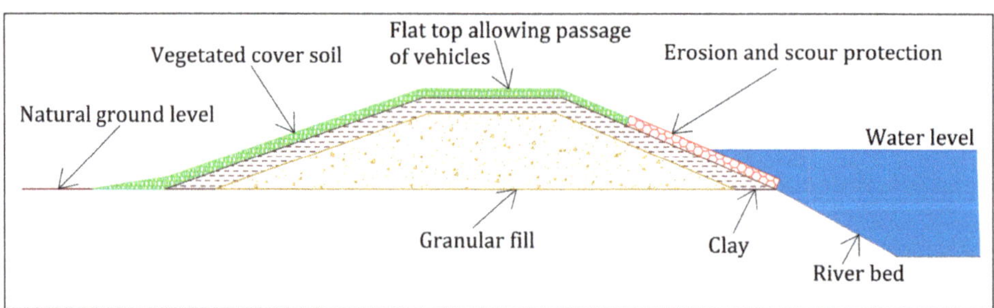

Figure 1. Typical cross-section of a river dyke.

The granular structural fill supports the low permeability soil and is often needed to prevent burrowing animal weakening the structure by digging burrows that could open passages through the embankment resulting in serious damage in the event of a flood. The surface vegetation gives the embankment protection against erosion by rainfall and flood waters. To prevent animals from finding a suitable environment for excavating their burrows, it is essential that the embankment is covered only with grass and that no bushes or trees are allowed to grow on them. The top of the dyke is often a flat area that allows for access for maintenance inspections. The side exposed to the river, or riverside, often has additional erosion protection measures to resist the high flow and energy of a flood, such as large riprap rocks. The side away from the river, or landside, would typically only be vegetated, and it thus the most vulnerable area of the dyke.

3.2. Failure Mechanisms in Earthen Structures

A flood defense structure can fail due to a number of causes [8,16,17]. These causes are also known as failure mechanisms. The most common potential failure mechanisms in earthen dykes are:

- Overflow flooding caused by a water level in the river that is higher than the crest of the dyke, leading to overtopping. This can cause the structure to fail on the landside due to progressive erosion (Figure 2a).
- Overtopping: Progressive erosion of crest, landside slope and/or toe due to the force of the water in the event of overtopping or overflow (Figure 2b).
- Macro-instability in riverside slope consisting in sliding of riverside slope when the outer water level falls sharply after a high water event (outward macro-instability) (Figure 2c).
- Macro-instability in landside slope, consisting in sliding either due to water pressure exerted against the structure and increased pore pressure in the subsurface, or due to infiltration of the overflowing water when high water levels are combined with overtopping (Figure 2d)
- Micro-instability in the landside (or riverside) slope due to outward seepage through the structure (Figure 2e).
- Erosion of riverside slope due to wave action or currents (Figure 2f).
- Piping as a result of seepage through the subsurface [17], carrying sand particles with it and undermining the levee and siphoning beyond the foot of an embankment (Figure 3).
- Excavation of dykes by rodents (Figure 4).

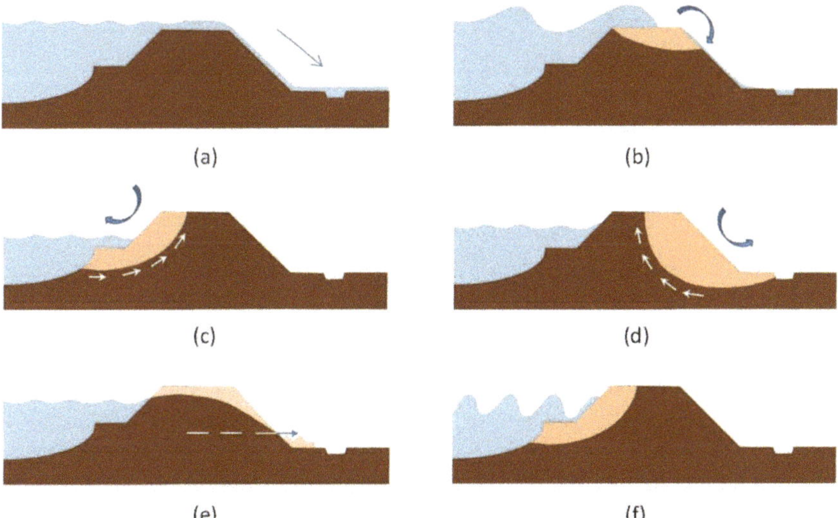

(a)

(b)

(c)

(d)

(e)

(f)

Figure 2. Failure caused by: (**a**) overflow, (**b**) overtopping, (**c**) macro-instability in riverside slope, (**d**) macro-instability in landside slope, (**e**) micro-instability caused by water filtration in the dyke body, and (**f**) erosion of riverside slope (adapted from [16]).

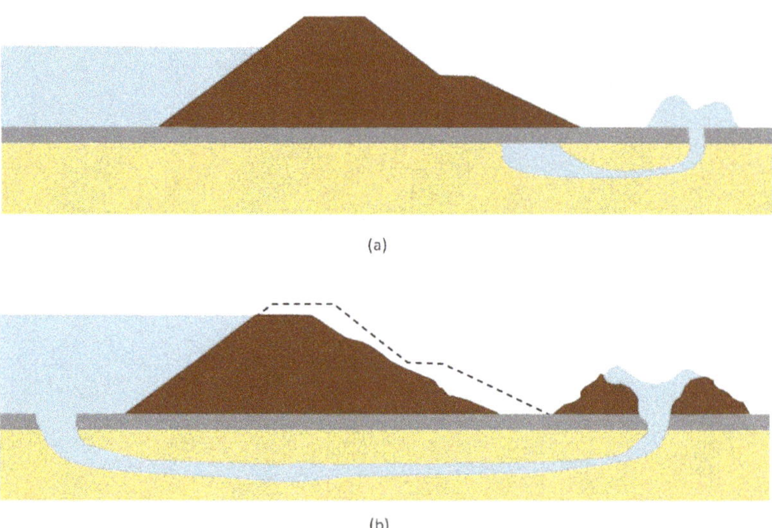

Figure 3. Failure caused by uplift and piping. (**a**) First and (**b**) second phase of the breaking of an embankment by siphoning (adapted from [17]).

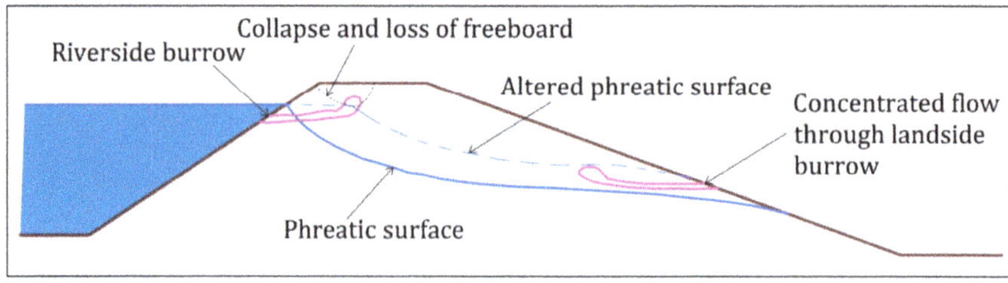

Figure 4. Failure caused by excavation by rodents (adapted from [16]).

Figure 5 show examples of failure of river dykes.

Figure 5. Examples of failure of: (**a**) the dyke of river Po in Mantova province (Northern Italy) during the flood event of October 2000, and (**b**) the dyke of river Arno in Tuscany (Italy) during the flood event of October 1992 (adapted from [17]).

3.3. The Length Effect

The 'length effect' has major implications for the failure probability of a dyke segment: each segment consists of a contiguous series of flood defense structures that are the components of a serial system, similar to the links in a chain [16]. A chain is as only strong as its weakest link: if one link fails, the entire system fails. The longer a dyke, the greater the probability that there will be a weak spot somewhere. This is referred to as the length effect. In practice, the length effect is relatively strong for geotechnical failure mechanisms such as macro-instability and piping. A clear example of the length effect is shown in Figure 5a.

The length effect can be minimized by connecting adjacent segments with longitudinal reinforcement and drainage systems.

3.4. Increasing Strength

The composition of the soil (subsurface), dimensions (including the height and slope) and the revetment (facing on the surface) determine the dyke resistance to failure. Figure 6 shows an example of a dyke profile. The required protective height and width of a flood defense structure is determined by a number of factors. Overflow and overtopping, stability and uplift can all damage the revetment and erode the underlying structure, and can thus potentially cause a breach. These mechanisms also have a negative impact in terms of sliding in the top layer and macro-stability (Figure 6).

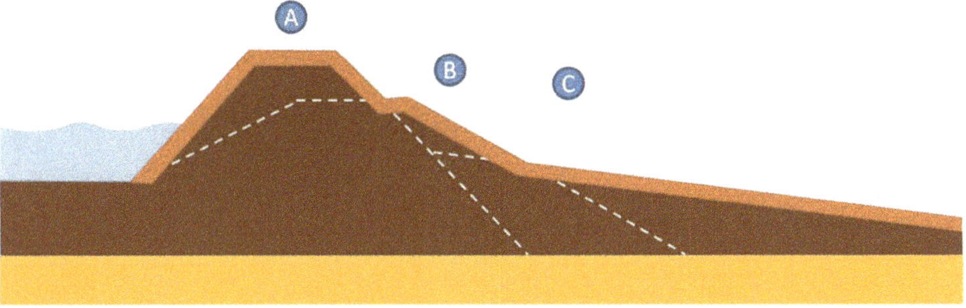

Figure 6. Example of the design of the dyke profile based on three failure mechanisms. The thick orange line envelopes the solutions to the three failure mechanisms, and shows the design profile. The white dash lines show the initial profiles. (**A**) Overtopping (greater height, less steep landside slope, rougher slope). (**B**) Stability (less steep landside slope, widening of levee base). (**C**) Uplift and piping (heavier top layer and longer seepage length) (adapted from [16]).

There is however not always enough room to make an ideal dyke profile entirely of soil, for example when there are buildings, watercourses or other obstacles that are expensive or impossible to move. In such situations the flood defense structure will have to be strengthened using methods that take up very little space, such as reinforced soil structures with geosynthetics. Geosynthetic drainage systems can reduce groundwater pressure in the structure and allow the soil to retain more strength; geosynthetic barriers can replace thick natural low permeability soil with significantly thinner layers, thus reducing the structures footprint and increasing its impermeability (see Figure 7).

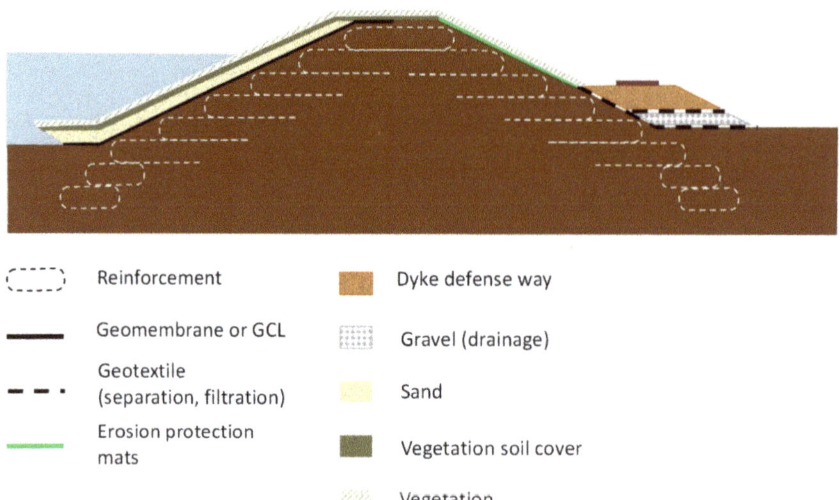

Figure 7. Application of geosynthetics for dykes, schematic examples (adapted from [2,8]).

4. Main Applications of Geosynthetics in River Dykes

4.1. Geotextiles for Filtration

Filtration is one of the most common functions fulfilled by geotextiles in river dykes: it involves intricate interaction mechanisms between soil particles and geotextile fibers, hence it is certainly one of the most complex among all the functions fulfilled by geotextile products. Significant studies and research have been carried out around the world to evaluate the behavior of geotextiles used for soil filtration, both in the laboratory and in the field, with the ultimate goal to improve design criteria [18,19]. Geotextiles have better uniformity than granular filters. Geotextiles are subject to manufacturing tolerances under factory conditions, while granular filters are subject to the natural variability of soils and segregation during placement. Geotextiles rely on their extensibility and strength to remain continuous during placement and subsequent deformation. Nonwoven geotextile filters can be used both around the landside gravel drain and under the riprap protecting the upper portion of the riverside slope. Geotextiles used as filters in river dykes behave very similarly to geotextiles used in earth dams. The separation function is usually combined in those applications with the filtration function in horizontal, vertical and inclined filters and separation elements [8,20].

4.2. Geosynthetics for Drainage

Geosynthetics for drainage can be found in many applications in dyke design, as illustrated in Figure 8, such as in horizontal drains for the downstream slope, chimney and finger drain, vertical or sloping drain to compliment the barrier lining system inside the dyke body and internal drainage in homogeneous fill.

The US Army Corps of Engineers (USACE) [21] presents many situations where granular drainage layers can be incorporated in dykes, some of which are presented in Figure 9. In all these situations geosynthetics drainage systems can be designed in place of the granular ones.

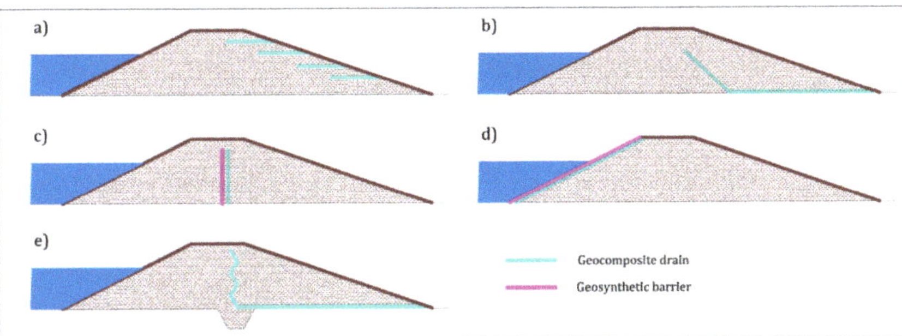

Figure 8. Applications of drainage geocomposites in dykes: (**a**) horizontal drains for the downstream slope, (**b**) chimney and finger drain, (**c**) vertical or (**d**) sloping drain to compliment the barrier lining system inside the dyke body, and (**e**) internal drainage in homogeneous fill (adapted from [21]).

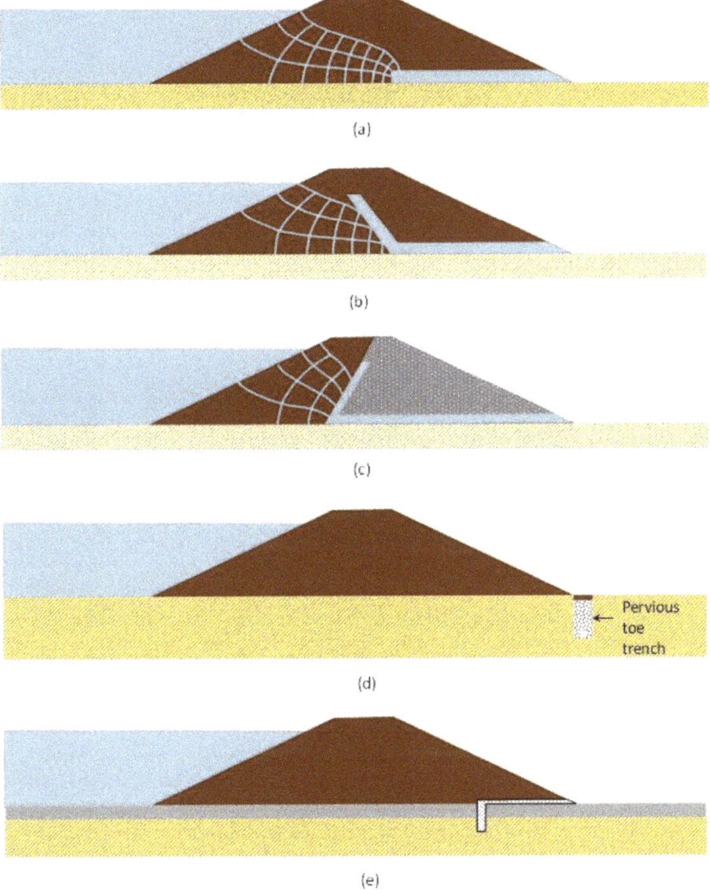

Figure 9. Use of horizontal and inclined drainage layers to control seepage through an embankment (**a**) horizontal drainage layer, (**b**) inclined drainage layer for a homogeneous embankment, (**c**) inclined drainage layer for a zone-embankment, (**d**) typical pervious toe trench with collector pipe, and (**e**) pervious toe trench located beneath landward slope (adapted from [21]).

Geosynthetics for drainage are typically geocomposite drains, i.e., a combination of geosynthetics manufactured into a single product, usually comprising a drainage core with a geotextile filter factory bonded on each face. The drainage core is a 3 dimensional element (geonet, geomat, geospacer) (Figure 10), able to convey water flow along its plane.

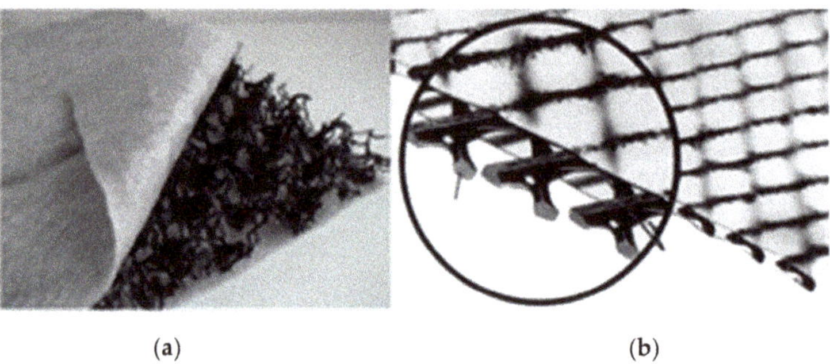

(**a**)   (**b**)

Figure 10. Examples of (**a**) geocomposite drains with geomat and (**b**) geonet core.

4.3. Geosynthetics for Reinforcement

Geosynthetics can be used for increasing the stability of a dyke, or increasing its slope angles and thus reducing its footprint, by building reinforced fill structures. Engineering design procedures and construction methods are well established for these geosynthetic reinforcement applications. Reinforced fill is comprised of three basic components: fill, geosynthetic reinforcement and facing. The fill is usually a relatively clean granular soil material. The reinforcement is usually laid in horizontal layers. The facing is connected to the reinforcement and retains the face of the fill; it is usually made up of precast concrete elements, or by wrapping-around the geosynthetic reinforcement. Geogrids, woven/knitted geotextiles, and geostrips (Figure 11) are typically used for reinforcement in dyke construction.

Figure 11. Examples of (from left to right) woven geotextile, extruded geogrid, woven geogrid, bonded geogrid, geostrip.

The selection of the type of reinforcement and its specifications are undertaken based on design, as the reinforced soil body behaves similar to a structural element, on which the stability of the dyke ultimately depends.

Geosynthetic reinforcement can be incorporated at the base of embankments to aid in construction, reduce potential for foundation failure and excessive deformation, facilitate embankment construction on sloping ground, and to construct steeper embankment slopes.

Dykes frequently must be constructed across soft and compressible soil, with the potential for embankment failures (Figure 12). Basal geosynthetic reinforcement placed at the bottom of the embankment improves stability and may reduce the required width of the embankment (Figure 13), thereby reducing foundation preparation and embankment soil volumes.

Dykes can also be constructed across hill slopes. The natural soil conditions, steepness of the hillside, weight of the embankment, and seepage into the embankment can decrease embankment stability (Figure 14). Geosynthetic reinforcement can be incorporated in the embankment to improve stability (Figure 15). Geosynthetic reinforcement placed along the dyke axis can also be used to minimize the length effect [22].

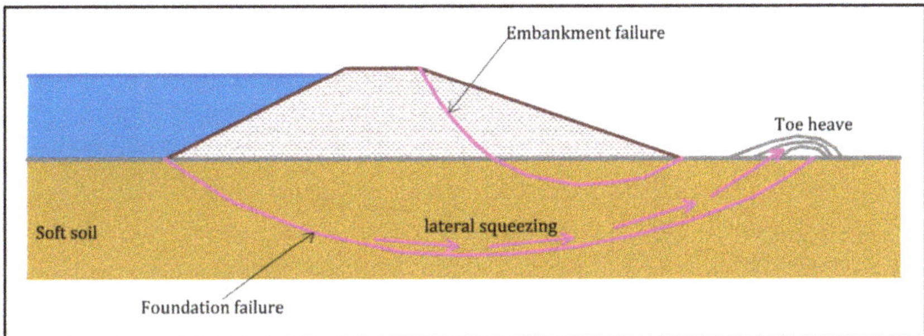

Figure 12. Potential dyke failure when embankment is constructed on soft soil (adapted from [22]).

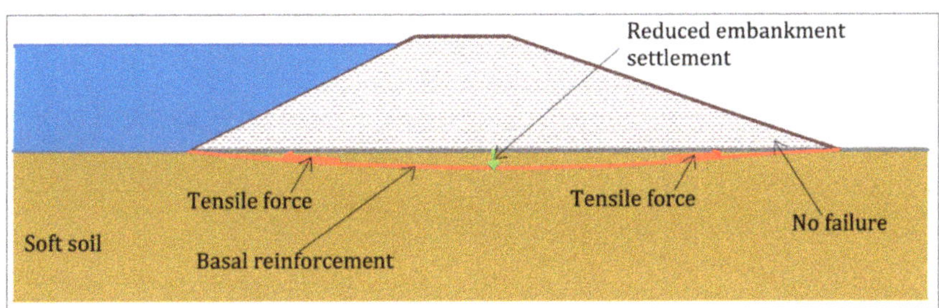

Figure 13. Dyke failure prevented by using geosynthetic basal reinforcement. Tension that develops in the reinforcement (red arrows) provides the required stability and reduce embankment settlements (adapted from [22]).

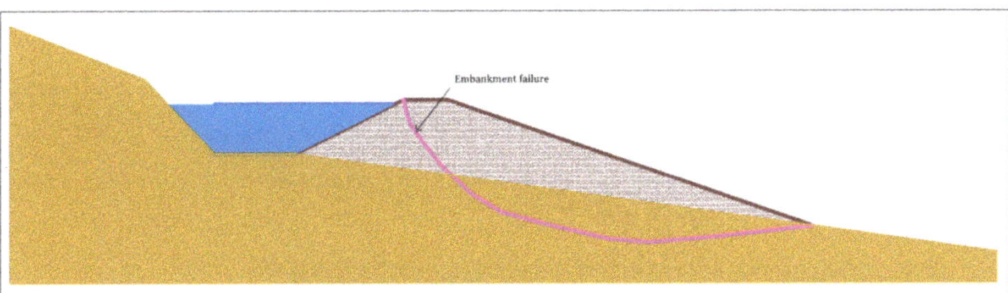

Figure 14. Potential failure of a dyke constructed on a hillside (adapted from [22]).

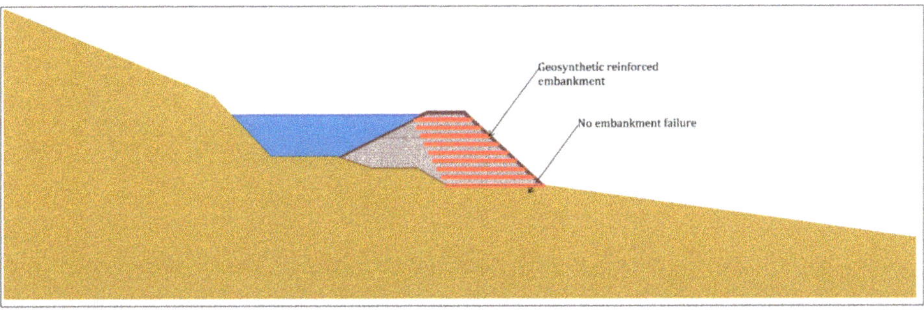

Figure 15. The tensile strength provided by geosynthetic reinforcement can prevent the failure of a dyke constructed on a hillside, while reducing its footprint (adapted from [22]).

In many situations to increase the resistance of a dyke requires the enlargement of the existing embankment (Figure 16) [21]. The required embankment widening can be reinforced with geosynthetics. Depending on site conditions, multiple layers of reinforcement can be used to reinforce the embankment slope, which could permit the slope to be steepened, thereby reducing the foundation area to be prepared and the volume of fill in the embankment. The reduced land area required, reduced fill material costs, and increased embankment reliability provided by inclusion of geosynthetic reinforcement make incorporating geosynthetics a logical and economical choice. Figure 17 shows an example of reinforced soil dyke built in Tuscany (Italy).

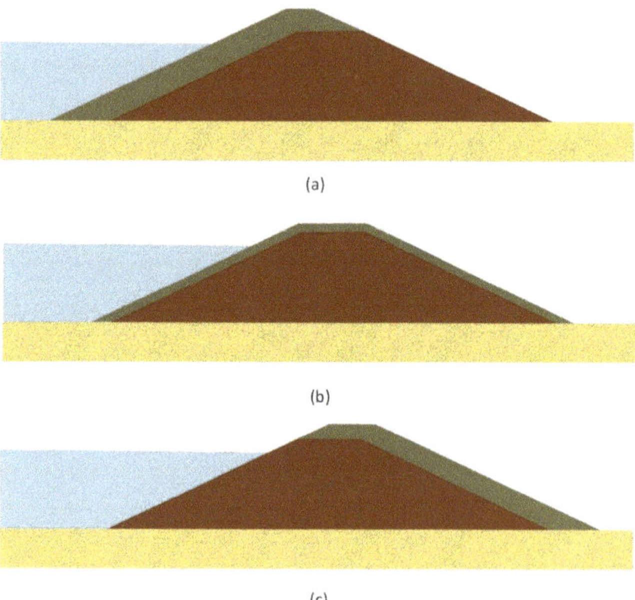

Figure 16. Enlargements of existing dykes (**a**) riverside dyke enlargement, (**b**) straddle dyke enlargement, and (**c**) landside dyke enlargement (adapted from [21]).

Figure 17. Geogrid reinforcement of a dyke with steepened slope in Tuscany (Italy) (**a**) edge of the slope and (**b**) flat area.

While geosynthetic reinforcements placed across the dyke can afford the stability of each cross-section, additional reinforcement placed along the dyke axis can minimize the length effect, as shown in Figure 18. This longitudinal reinforcement can be installed during major refurbishment works on an existing dyke, and can be included in the original design for new dyke construction. The cost of the longitudinal reinforcement is negligible compared to the savings of avoiding potential dyke breaks due to the length effect, and consequent costs.

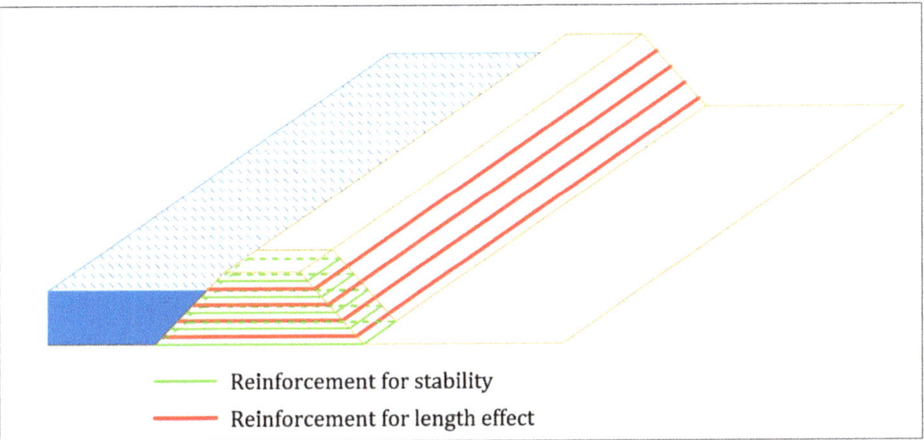

Figure 18. Geosynthetic reinforcement used to afford stability of each cross-section of a dyke and for minimizing the length effect.

4.4. Geosynthetics for Stabilisation

Geosynthetics with stabilization function in dykes are typically used to construct roads and platforms to carry traffic on the top of the structure. As very often dykes are constructed from cohesive soils with a lot of fines, such as clays and silts, they are therefore relatively weak soils so their 'trafficability' is quite low. Any heavy vehicle passing on the top of dyke, especially during wet seasons characterized by increased moisture within top of the structure (spring thaw, heavy rain periods, etc.) will typically result in deep rutting (see Figure 19), making any further vehicle access impossible, especially during periods when the dyke could be experiencing a high-water level on the riverside or when local erosion/shallow slips of dykes occurs that require immediate maintenance to prevent further deterioration. From a management of dykes point of view lack of access for traffic is an unacceptable situation. Emergency vehicle access for urgent repair/maintenance works—which may require access for cranes, excavators, bulldozers or similar type machinery—requires continuous accessibility via a light but stable permanent pavement.

One of the most effective solutions to prevent rutting of the dyke crest is stabilization by geosynthetics of the unbound aggregate layer on the top of dyke, as shown in Figure 20. Geosynthetics for stabilization are in most cases biaxial/multiaxial geogrids or geocells (Figure 21) installed horizontally on the top of dyke structure. The stabilizing effect of geosynthetics is achieved by a reduction of horizontal and vertical movement of aggregate particles which counteracts deformation of that aggregate layer under applied load from vehicle traffic. It results in either prevention of fast rutting and can even provide a full flat profile of aggregate layer over a wet period. Axle load, bearing capacity of soils in the upper part of dyke and type of aggregate are parameters incorporated into design procedure, whilst thickness of aggregate layer and type of geosynthetic are the outcome of it. In individual cases it might be a bespoke solution designed for a specific type of machinery, e.g., heavy crane operating on the top of dyke. Often a geotextile separation layer is also included under the stabilized layer to further increase the redundancy of the system.

Figure 19. Dyke top structure with rutting caused by traffic and with local erosion failures typical for wet seasons.

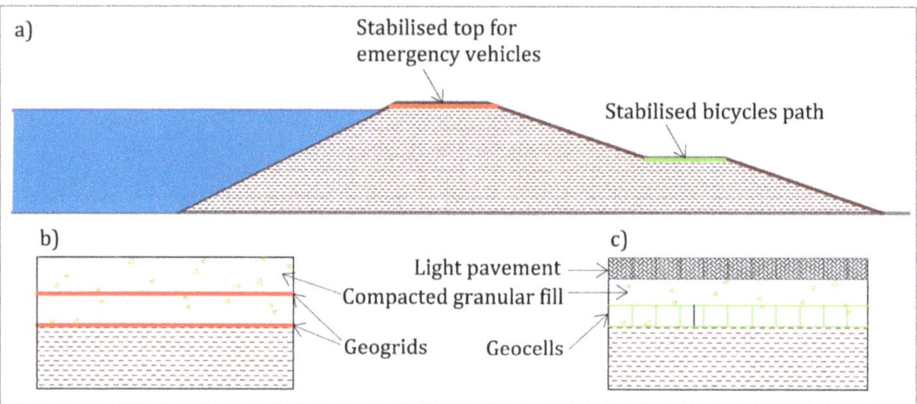

Figure 20. Use of geosynthetics for stabilisation in dykes: (**a**) typical applications, (**b**) typical cross-section with geogrid stabilization, and (**c**) typical cross-section with geocell stabilization.

Figure 21. (a) Typical biaxial geogrid and (b) typical geocell.

Another, very practical application of stabilizing geosynthetics in dykes is the construction of very light pavements for bicycles (Figure 20), which is becoming a very popular solution to attract bikers and to keep populations exercising and healthy.

4.5. Geosynthetics for Surface Erosion Control

Unlike in canals, it is not typical for a dyke to form the bank of a river. In most instances the dyke is placed at a higher level and at a distance from the natural river bed, and the water flow will only reach the river dyke during flood events. Hence the erosion of dykes by the permanent water stream or by the vessels in the river will not be discussed in this paper. Erosion control of riverbeds is discussed in [1,2].

Given these clarifications, in dyke engineering surface erosion control is related to both riverside and landside slopes, and it can be required for two main sources of erosion:

- Erosion caused by rainfall, and
- Erosion caused by overflow and overtopping of the dyke (Figure 2).

Rainfall erosion can occur both on the whole landside slope and riverside slope, and erosion due to overtopping can occur in flooding conditions on the landside of a dyke.

Erosion control against rainfall on the riverside or landside slope of a dyke is exactly the same as on any slope, while erosion control for overtopping is typical of dyke engineering. Many failures of dykes have been triggered by overtopping and consequent failure of the unprotected landside slope. Failure mechanisms due to overtopping have been shown to start at the landside slope, and then progressively involve the dyke body. The need to protect the landside slope is therefore evident.

Overtopping of a dyke can create erosion damage on the crest and on the landside slope, depending on the local boundary conditions. Overtopping often occurs with dyke saturation, which weakens the structure and makes it easier for the water to flow over the crest, subsequently cutting into the landside slope surface or into the toe of the dyke. The primary erosion protection mechanism for dykes against hydraulic loads is grass cover. The resistance of vegetation against erosion depends mainly on the grass, which must be sufficiently dense to prevent soil particles from passing through it and sufficiently robust to prevent bare spots from appearing in the vegetated area. To gain immediate protection and strength, an effective solution can be provided by geosynthetics for erosion protection (geomats, reinforced geomats, geoblankets). In the case where the grass cover does not give full coverage, or where hydraulic impact has locally removed the grass, the erosion control product bridges the bare spots.

Geosynthetics create a physical barrier which can absorb the impact of water and wind on soil, resulting in the prevention of soil loss and enhancing vegetation growth. These products are flexible and environmentally friendly; the installation is easy and fast and they can be applied directly on slopes and along river and canal banks. Examples of geosynthetics for erosion control are given on Figure 22. They promote vegetation growth and increase the long term resistance of the grass cover to the potential hydraulic loads from a river in flood or from water flow on the landside during overtopping by reinforcing the grass root zone. However, after installation some time is needed to allow the roots to grow through the erosion control product into the soil below (Figure 22).

Experience has shown that vegetated dykes can withstand limited overtopping with little or no damage. Research utilizing actual vegetation on test embankments and experience with vegetated spillways has allowed quantification of the hydraulic loads required to generate failure on a vegetated embankment face [23]. The results of this research and experience have been used to develop simplified computational procedures for use in predicting an acceptable amount of overtopping flow during a major flood event. Allowing overtopping of dykes to the point of incipient failure of the slope protection is consistent with the U.S. Natural Resources Conservation Service approach of requiring passage of the design flood without breach of the embankment.

During overtopping flow the vegetation can provide protection against the initiation of concentrated erosion that leads to head cut development and breach. For larger flow rates, vegetation may delay breaching sufficiently to permit evacuation of downstream areas [24]. Vegetative cover is most viable as a protection method in humid climates that receive sufficient moisture to establish relatively dense, uniform turf grasses without supplemental irrigation. Good maintenance of the cover is essential to achieve significant protective benefits. Vegetation provides protection to an embankment in two functional ways: (1) protection of the soil surface by reduction of velocities and stresses at the embankment boundary as a result of the coverage provided by stems and leaves that lay down in the flow and blanket the surface; (2) the reinforcement of the underlying soil due to the presence of roots.

Figure 22. Examples of geosynthetics for erosion control. From left to right (**a–e**): geomat, steel mesh reinforced geomat, geogrid reinforced geomat, geoblanket and resulting vegetation growth.

Geosynthetic reinforced grass can resist significantly higher duration flow velocities compared to unreinforced grass; moreover the time before failure occurs is extended when vegetation is reinforced. A good grass cover can typically withstand flow velocities of 4.5, 3.2 and 2.8 m/s for durations of 1, 5 and 10 h respectively, while for grass reinforced by a 20 mm thick TRM these velocities are typically increased to 6.0, 5.5, and 5.0 m/s [25].

4.6. Geosynthetyics for Protection against Rodents Excavation

Beavers, nutria, badgers and other rodents or digging animals can sometimes cause significant damage on slopes of dykes. In isolated cases, such damage caused by these animals could lead to an increased risk of global slope failures with serious consequences of flooding for the communities living in the area. Cavities created by digging animals (Figure 23) can be of high risk for the stability of the dyke. The highest risk is beavers digging through the low permeability layers. An opening through these layers leads to increased flow through the dyke, ultimately causing erosion and water seeping through the embankment all the way to the landside. Most critical failures consist in partial or total collapse of the embankment by sliding.

Due to changing climate conditions, recent studies show an increase in the population of beavers, nutria and other rodents in vast regions of central Europe over the last 15 years, while the Netherlands' beaver population is expected to grow from 700 to 7000 by 2032 [26].

Safe and durable protection against rodent intrusion in dykes can be achieved by lining the dyke slopes with geomats reinforced with steel mesh (Figure 23). The strength of the steel mesh will act as an impenetrable barrier to the animals who will not be able to dig a hole through the steel mesh, while the geomat will provide the erosion protection function during flooding events.

Figure 23. Badger excavation at dyke toe and protection against rodent excavation with reinforced geomats (adapted from [26]).

An intervention against rodent intrusion was implemented in Germany along the Odra Dyke, District Sophienthal, Brandenburg, in 2013, in cooperation with the State Agency for Environment, Health and Consumer Protection Frankfurt/Oder and the Water and Dyke Association Oderbruch [26]. A reinforced geomat for beaver protection was installed in a 200 m long dyke stretch, near the community of Sophienthal. This project also shows the multi-functional use of geosynthetics in dykes: a geosynthetic clay liner (GCL) was placed for waterproofing of the riverside slope, a polyester geogrid was placed at the base of the embankment for basal reinforcement, and reinforced geomats were placed on the riverside and landside slopes for protection against erosion and rodents excavation.

4.7. Geosynthetics for River Training Works

Longitudinal dykes can be protected, particularly in curves, with groins built perpendicular to the river flow (Figure 24) [2]. Such structures must be resistant against high flow forces. The groins can be built using geosynthetic mattresses, geosynthetic tubes and geosynthetic containers. In addition, a filter geotextile underneath the groins is an essential component of the system that contributes to a long lasting training structure. The weight of the containers is typically chosen according to the hydraulic load and the installation done to avoid gaps in between the elements. These structures may also get an armor layer of steel or geosynthetic mattresses or armor stones (riprap) if necessary. The structures can also use cheap local material for the main structure fill with short transportation distances and thus reduced related impact.

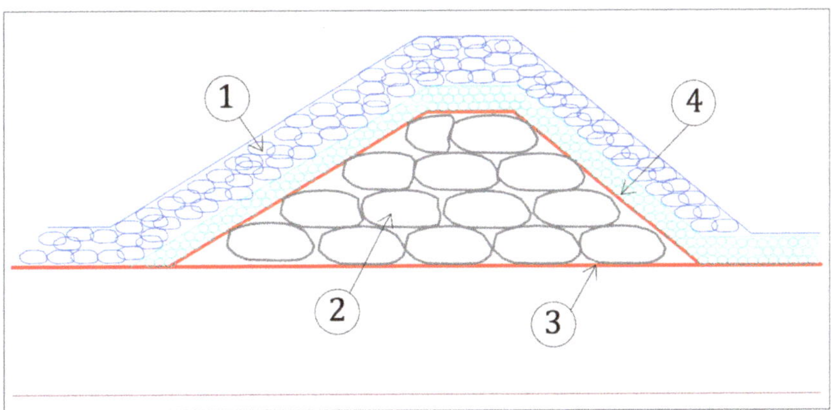

Figure 24. Groin construction with geosynthetics: (1) hard armor or geosynthetic containers; (2) geosynthetic containers for the core; (3) geotextile filter; (4) steel mattresses or armor stones (adapted from [2]).

4.8. Geosynthetics in Barrier Systems

To avoid failures due to water filtration inside and below a dyke, or due to erosion on the riverside slope, the most effective passive method is to separate water completely from the soil and thus prevent any interaction between the water and soil [1]. This can be done using barrier materials, which, also serve as protection for the underlying soil. Barrier layers can be provided by geosynthetic material, namely polymeric or bituminous geomembranes or geosynthetic clay liners (GCL). The advantage of these systems is the flexibility to adapt to deformations developing after the lining installation [1].

The use of geosynthetics also allows for the construction of a dyke out of higher permeability fill than would normally be adopted for dyke construction, in order to reduce seepage and internal erosion [12].

There is often a variety of locally available materials used to construct dykes, as local soils are the most cost effective construction material. However, the material is often very heterogeneous and its permeability can be variable. Many of these dykes have been subject to reconstruction, additional sealing and other improvement [3] as local weak points can lead to the failure of the entire structure.

Some potential applications of geosynthetic barriers in dykes are summarized in Figure 25.

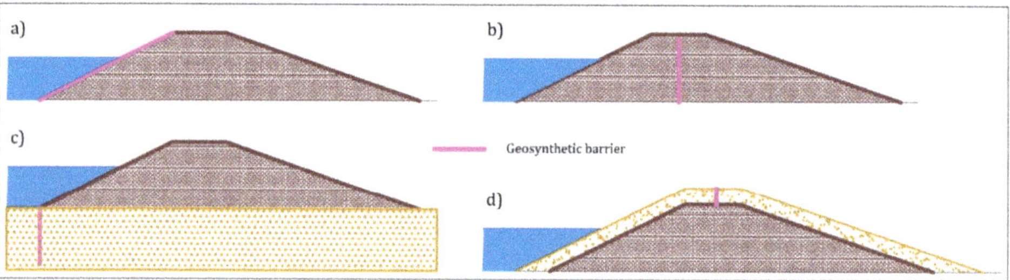

Figure 25. Some applications of geosynthetic barriers in dykes (adapted from [27]): (**a**) on the riverside slope; (**b**) vertically placed in the dyke body; (**c**) in the cut-off diaphragm below the dyke body; (**d**) vertically placed in the dyke expansion.

Near surface slips resulting from a quick drop in the water level on the riverside, and causing superficial sliding of the riverside slope, can leave the clay core unprotected against erosion. In the case of geomembranes or GCLs, placed on the riverside slope (Figure 25a), near-surface slips are also critical, but can be repaired more easily than along softened, partially eroded clay [8]. A specific requirement for barriers is a sufficient resistance against impact forces. Armor stones have to be laid on the barrier material with care [1]. In these cases a geomembrane requires protection provided by a nonwoven geotextile.

Brandl [8] also points out that geosynthetics have proved suitable as an alternative to clay cores in the body of a dyke.

Installing a barrier on the waterside of the dyke may not always result in the desired effect. An additional aspect to consider is the contact to the in situ soil layer at the upstream toe of the embankment. If the water is able to flow below the barrier and into the dyke, the effect of the lining will be negated. The seepage line in the embankment will ultimately reach nearly the same level as without the barrier. The only advantage is potentially to gain some time until this condition is reached, depending on the permeability of the embankment. If there is no soil layer with low hydraulic conductivity at the foot of the dyke, additional measures are necessary, e.g., cut-off diaphragm to a sufficient depth [1] (Figure 25c).

Cutoff walls can be constructed using interlocking geomembrane panels, which can be installed in slurry-filled trenches. The geomembrane panels are inserted with the aid of guide frames. The geomembrane panel joints are typically self-sealing with proprietary interlocking features. In some soil conditions, the geomembrane panels can be installed using a vibratory hammer and insertion plate, thus eliminating the need to excavate a slurry trench. The geomembrane for cut off walls are usually made of HDPE, with a thickness of 2–3 mm.

In case of dyke expansion, the expansion can be made of locally available soil with relatively high permeability; yet the expansion can be made impervious by placing a geosynthetic barrier vertically in the expansion on the dyke crest (Figure 25d).

Other impervious linings such as concrete slabs, concrete mattresses or asphalt layers are too stiff to adjust to subbase deformations. They will bridge indentations caused, e.g., by sub-erosion and thus no warning is given until a cavity under the lining has become too large and the whole structure collapses [1].

Significant experience on the use of GCLs as impermeable surface lining in dykes has been gained in Germany where following the Elbe River floods that took place between 2002 and the end of 2005, about 150 levee reconstruction projects are known to have been carried out in which about 700,000 m^2 of GCLs have been employed [10]. A schematic application of GCLs as an inland levee sealing system against flooding from rivers is illustrated in Figure 26. The dimensioning of GCLs in this case is covered in [28] and the installation of a geosynthetic clay liner generally follows [29].

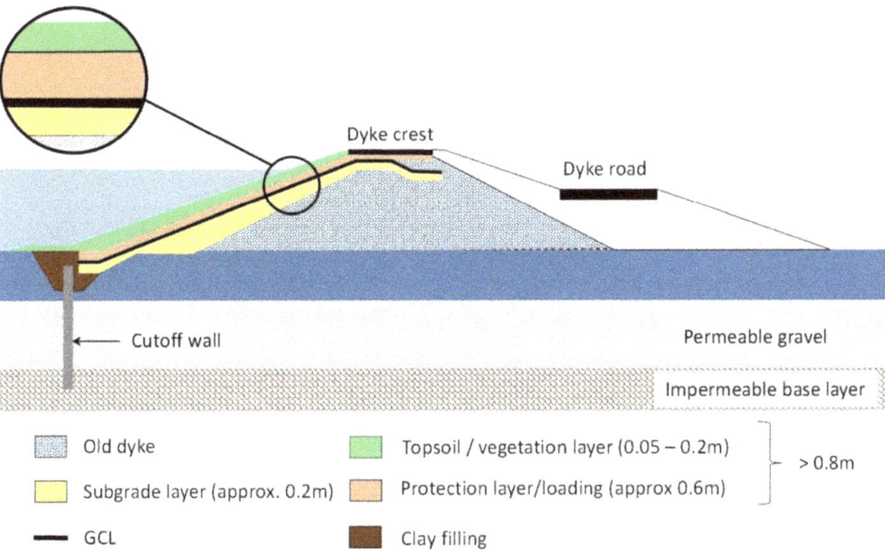

Figure 26. Application of a geosynthetic clay liners (GCL) in the course of dyke refurbishment works (adapted from [30,31]).

The hydraulic properties of GCLs and Geomembranes, compared with various lining materials are included in Table 1 [32].

Table 1. Properties and flow rates through lining materials including geosynthetic clay liners and geomembranes for an applied hydraulic head of 1 m for porous materials. The difference in pressure applied between both faces of the geomembranes and multicomponent GCLs is 100 kPa.

Material	Testing Conditions	Hydraulic Conductivity (ms^{-1})	Thickness (m)	Flow Rate $(m^3 m^{-2} d^{-1})$
Cement concrete	In the field [a]	10^{-10} [a]	0.1	9.5×10^{-5}
Roller compacted concrete		10^{-8} [a]	0.5	2.6×10^{-3}
Asphaltic concrete	In the field with excellent construction and quality control [a]	10^{-9} [a]	0.1	9.5×10^{-4}
Asphaltic concrete	In the field with ordinary construction and quality control [a]	10^{-8} [a]	0.1	9.5×10^{-3}
Compacted clay liner	With excellent construction and quality control [a]	10^{-9} [a]	1	1.7×10^{-4}
Compacted clay liner	With ordinary construction and quality control [a]	10^{-8} [a]	1	1.7×10^{-3}
Geosynthetic clay liners	As manufactured, confined and hydrated with low cation concentration solutions	10^{-11} [a]	0.01	8.7×10^{-5}
Multicomponent GCLs	As manufactured	Meaningless	0.01	$<2 \times 10^{-5}$ [b]
Geomembranes	As manufactured	Meaningless	≥ 0.001	$<10^{-6}$ [b]

[a] [33], [b] [34]. Adapted from [32].

5. Economic Advantages of Construction Using Geosynthetics

The use of geosynthetics in civil engineering applications often provides financial benefits by reducing the cost of imported materials, reducing the amount of waste, and generally provides more efficient use of resources compared with traditional solutions that use soil, concrete, and steel [35]. The cost, schedule, and risk reduction benefits provided by using, for example, geosynthetic reinforcement, have resulted in these products having been used for decades to improve stability of dykes. Based on a review, four types of cost savings have been identified [36]:

- Reduction of the quantity or need for select imported soil material,
- Easier and/or accelerated construction,
- Improved long-term performance, and
- Improved sustainability.

The questions of the reduction of the quantity or need for select imported soil material and improved long-term performance are discussed in the following subsections. Improved environmental benefits of the use of geosynthetics, contributing to sustainability, will be discussed in Section 6 of this paper.

5.1. Reduction of the Quantity of Soil Material

To reduce the quantity of natural soil materials required, geosynthetics often replace the given soil and rock materials at a material and installation cost that is less than that of the natural-material alternative. Furthermore, geosynthetics are often used in geotechnical systems and, due to improved performance efficiency, may decrease the volume of other geotechnical materials used in that system. In many cases the cost benefit is such that the use of geosynthetics is now the standard practice [36].

Normal flood embankments along rivers are not designed for overflow loads, except special designed overflow sections. Some technical code for flood protection dykes along rivers exclude the protection of the landside embankment from technical standard structures because of the high costs and poor experience with such structures [37]. Thus, overflow loads mostly lead to a very rapid and complete failure of the dyke structure forming a dyke breach with lengths from few meters to several hundreds of meters. A huge part of the damage that occurred due to dyke breaches in Germany in the past decades could have been avoided by focusing the application of overflow protection for the structures. In addition, [37] states that one major issue cited against the incorporation of overflow dyke structures are perceived high costs. According to them this issue can be addressed by using designs incorporating geosynthetics.

An example of how geosynthetics can economically contribute the construction of overflow dyke structures was presented in [13]. The USACE conducted extensive testing and pilot research programs to find alternative solutions for dyke overflow protection measures in order to reduce the costs and provide identical or better performance than traditional hard armoring systems such as rock riprap and concrete for hurricane and storm damage risk reduction systems. The study showed that a high-performance turf reinforcement mat (HP-TRM) solution is a more cost effective and environmentally friendly solution than traditional hard armor solutions.

The case study of the construction of dykes on embankments constructed on soft soils was presented by [36]. In the study dykes at Craney Island, Virginia, USA, construction, without reinforcement using geosynthetics would have required 8 to 10 times the volume of soil, as the fill would be pushed down into the soft soil below in order to build the structure above the surface. When reinforced embankments using geosynthetics were built on that project, less than one times the volume of soil was lost for one volume to construct the structure above the surface. Thus for very weak soils, geosynthetics used in reinforcing the embankment significantly reduced the amount of displacement of the foundation and thus the loss of material, resulting in a significant saving.

5.2. Reduction of the Need for Select Soil Material

The EU Regulation 305/2011 [38] lays down harmonized conditions for the marketing of construction products and particularly recommends predominantly using secondary materials as construction material. The demonstration of how the use of geosynthetics can improve the functionality of dykes constructed with fine-grained dredged materials as a replacement for standard dyke cover material was studied [3]. In this context geosynthetics contribute to reduce the cost of construction by allowing the use of inert waste products as part of the structural fill.

In many places in the world, gravel for the core and rock as armor is simply not available and concrete (for armor elements) is too expensive. In such cases, geotextile bags or containers offer an alternative solution. Local fill material is filled in geosynthetic containments to build the core of the river training structures. Using cheap local material for the major volume of the structure will result in lower costs with the same benefit [1].

More generally, a better use of residues and waste in engineering works is important to reduce the exploitation of natural material and to preserve the environment. The combination of geosynthetics with such residues may provide less expensive and concomitantly more environmentally friendly solutions [39].

5.3. Improved Long-Term Performance

The long term performance of geosynthetics in their application can both be discussed from the point of view of the product performance, and of the application in which the geosynthetics are installed. Indeed, geosynthetics, by their use, contribute to the increase of the lifetime of the application, as evidenced by the references mentioned in the following.

5.3.1. Durability of Geosynthetics

UV resistance is a key parameter for the durability of geomats and geoblankets for erosion control, especially in areas with lower chance of full vegetation, and for critical applications such as storm water management and flood protection [13]. Accelerated UV tests on geomats for a period of 3000 h and 6000 h have proven the long term durability of these geosynthetics for erosion control [13].

Reference [40] mentions the various ageing mechanisms that the GCLs can face while installed in dykes: stress during installation, surface resistance against abrasion due to rip-rap, washout of bentonite in relation with hydrodynamic action, freeze-thaw cycles, hydration-desiccation cycles combined with cation exchange that lead to significant increase in the hydraulic conductivity of the bentonite in the GCL, or root penetration. Proper design, and installation of GCL however allow to overcome those potential aging and failure mechanisms as emphasized by feedbacks from excavations at dykes. In particular a minimum 0.8 m thickness of cover soil is recommended. Fleischer and Heibaum [40] report on the excavation at three dykes of GCLs installed respectively in 1994, 2001 and 2004, thus 4 to 12 years after installation. Where the soil cover nature, thickness and slope were adapted, as well as the vegetation on top of the GCL and cover soil, good performance along time of the GCL was noticed.

Multicomponent GCLs are GCLs onto which is attached a film, coating, or membrane that decreases the hydraulic conductivity, protect the clay core, or both [41]; these geosynthetics could find some applications in dykes to solve some of the issues previously presented. Some hydraulic properties of multicomponent GCLs are given is Table 1.

5.3.2. Increased Life Span Thanks to the Use of Geosynthetics

Most of the applications in which geosynthetics are used are designed to perform at least equally to traditional design solutions. Part of the reason geosynthetic solutions have improved performance over traditional designs is that they work better than the geotechnical material they replace. The performance improvement is gained by using manufactured materials with known properties as compared to the relative high variability of soils and requirements for monitoring of the installation/compaction of soils to allow for their desired properties to be achieved in the field. In some applications geosynthetics also improve the performance of geotechnical material [36].

Haselsteiner et al. [37] mention that measures involving geosynthetics lead to a retardation of the flooding of the hinterland located behind the dyke and additionally a complete failure was avoided. Both effects result in a reduction of flood damage and in gaining more time for taking other flood protection measures. In spite of the fact that no design specification or standards were created, in comparison to commonly used overflow protection measures such as riprap or just flat embankment inclinations, geosynthetic overflow structures are an effective and efficient solution.

In addition, it was shown, among other conclusions, that HP-TRM maintained with engineered earth anchors (EEAs) can increase the safety factor of the armoring system against hydraulic forces and also provide slope surface stability [13].

5.3.3. Geosynthetics and Seismic Resilience

The good performance of geosynthetic-reinforced soil walls and slopes during earthquake has been documented widely in the literature [42]. In Japan, the greater seismic resistance of these structures compared to conventional retaining wall structures has led to their increasing use for new permanent structures and to replace conventional structures damaged in earthquakes. By extension, dykes constructed with geosynthetic reinforcement would be more resistant to seismic loading than those constructed from traditional materials and the geosynthetic reinforcement will provide additional tensile strength and resilience in the structure.

6. Mitigating Greenhouse Gases Emissions Using Geosynthetics

Two categories of actions are required to tackle climate change and its effects: (i) mitigation to reduce greenhouse gases (GHG) emissions and (ii) adaptation. Geosynthetics can make a contribution to mitigation by reducing GHG emissions from construction and operating infrastructure. In fact, the ecological advantages of construction methods using geosynthetics are well known, the use of geosynthetics can dramatically reduce emissions from soil that needs to be excavated, transported, and put in place [43].

The carbon footprint is a measure of total GHG emissions caused directly or indirectly by a person, organization, event, or product. The carbon footprint can include emissions over the entire life of a product or construction. Embodied carbon (EC) is an indicator of cumulative carbon emissions used in the solution adopted. Life-cycle assessment (LCA) is a tool for measuring the environmental impact of products or systems over their lifetime. It considers the extraction of raw materials, production, use, recycling, and disposal of waste [43].

The past few years have seen an improved mastery of techniques of LCA in the field of geosynthetics. The latest calculations [44–46] are indicative of the progress made in this field, in which an ever-more constructed standard approach is evolving by using EC values representative of geosynthetics [47] and by comparing the EC values for entire construction solutions. Thus, results are recognized and trusted when the conclusion indicates that solutions using geosynthetics significantly reduce environmental impact.

The sustainability of materials and processes is commonly assessed by calculating the carbon emissions (CO_2) generated that can be used as "short LCA" for the ecological evaluation [44]. Taking into account the extraction and production of the used construction materials, loading, transport and installation, the cumulated energy demand (CED) and CO_2 emissions are determined for each of the construction alternatives.

Although this is a simplification, the ease of calculation encourages comparisons between solutions and makes such assessments accessible, transparent, and repeatable so that the CO_2 emitted can more easily be counted towards industry, national, and international targets [44]. Some studies have incorporated other indicators such as cumulative energy demand, climate change, photochemical ozone formation, particulate formation, acidification, eutrophication, land competition, and water use. Such calculations were made for various applications [48].

The use of geosynthetics results in massive improvements to CO_2 savings as opposed to nearly all alternative civil engineering materials used [49]. In the particular context of dykes the example of the comparison of an external sealing for a river dyke on the Kinzig (southwest Germany) is given [50]. The use of a GCL is compared to the use of a compacted clay liner (CCL) with an average thickness of 0.625 m. The comparison turns out in favour of the GCL. The difference in the cumulated energy demand of the two sealing systems is, however, comparatively insignificant. A medium transport distance of 35 km (one-way) was assumed for the mineral barrier material, which greatly impacts the CED for the required sealing material of 45.000 tons (see Figure 27). For the GCL the main share in the CED is the polypropylene (PP), which at a surface weight of 0.69 kg/m^2 PP (including 6.2% overlapping) is a major factor. When comparing the two sealing systems, the transport distance for the mineral sealing material is the decisive parameter. If the place of extraction is on-site or very near to the place of installation, then the CCL—mostly because it has no energy content (feedstock)—can hardly be improved upon. In the case of the GCL, the main part of the CED is the energy content (feedstock) of the polypropylene (ca. 53%). The transport distance for the GCLs from the manufacturer's plant in Espelkamp to Offenburg (580 km) is, in comparison as regards the CED compared to the PP granulate material, of hardly decisive consequence (ca. 8.5%).

The covering soil which has to be put in position as weather protection for both barrier materials (here: d = 0.8 m), is 97 MJ/m^2 with an assumed average transport distance of 20 km for both cases studied, in particular when comparing these systems with other systems, of quite considerable consequence. The distribution concerning environmentally relevant CO_2 corresponds approximately to the CED, the GCL has a CO_2 emission of 4.0 kg/m^2, the CCL of 9.9 kg/m^2 and the covering soil is entered in the CO_2 balance sheet with 7.9 kg/m^2 (see Figure 28).

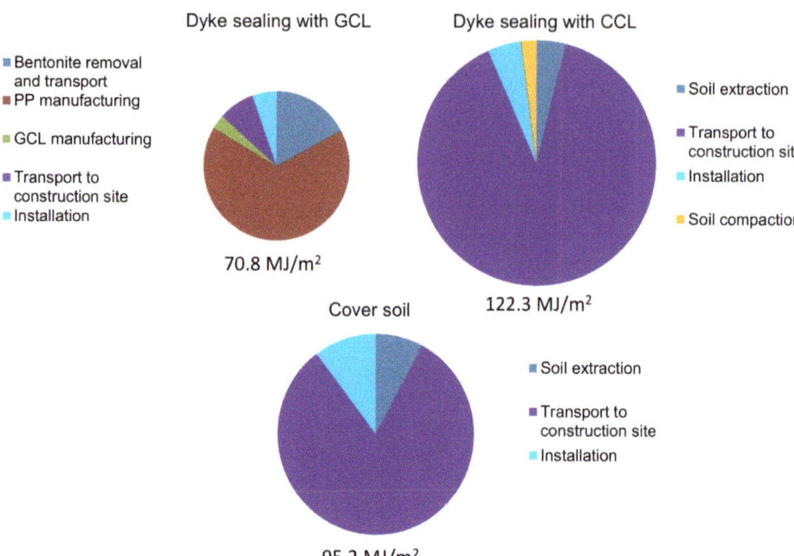

Figure 27. Comparison of cumulated energy demand for GCL and compacted clay liner (CCL) dyke sealing systems (adapted from [46]).

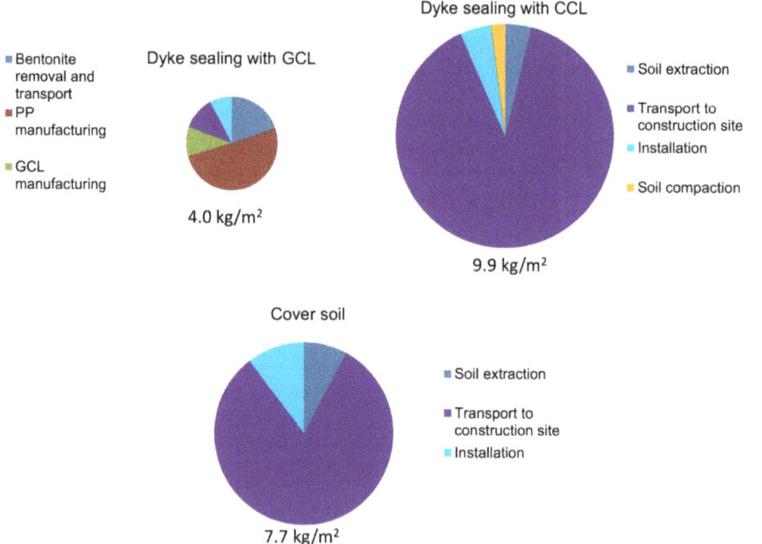

Figure 28. Comparison of CO_2 emissions for GCL and CCL dyke sealing systems (adapted from [46]).

In bank protection works, the carbon footprint of double twist wire mesh solutions is twice as little as riprap and can be improved more when locally available stones are used, which is usually the case as the diameter of stone used in the filling of the mesh is usually small. In this case the impact in terms of carbon footprint was 6 times less than for riprap for the case study presented [51].

7. The Minimum Energy Performance Concept

A key concept in sustainability of constructions is the minimum energy performance [52]: according to the EU Directive on the Energy Performance of Buildings, cost-optimal energy efficiency levels for buildings need to be ensured, where cost-optimal is the level of energy efficiency at which the total cost of the life cycle is minimized, taking into account the costs of construction, energy and maintenance [53].

Extending the concept from buildings to civil engineering constructions, such as river dykes, sound engineering practices would require the designers to understand how to use "living" and "inert" materials together by best combining both types of materials.

The challenge is to identify a system capable of providing the required resistance and to be able to incorporate the most appropriate solution, which would be that one defined by the cost-optimal energy efficiency levels; in civil engineering constructions this can be broadly defined as the minimum amount of intervention on the environment, which is required to solve the problem, ranging from the lowest level of no intervention up to the highest energy level, which may necessitate, as example, the construction of massive concrete structures.

Given the above outlined reduction of the quantity of soil material, reduction of the need for select soil material, improved long-term performance, durability, increased life span, and mitigating greenhouse gases emissions, Geosynthetics can be considered as one of the best ways to design according to the minimum energy performance approach, affording in most cases the cost-optimal energy efficiency level of the project.

8. Advantages and Disadvantages of the Use of Geosynthetics in Dyke Engineering

As discussed above, the use of geosynthetics in river dykes has many advantages, in terms of technical properties, ease of construction, economy, and environmental impact.

In this section the potential disadvantages for the use of geosynthetics in river dykes are discussed, based on recent research and experience from existing projects.

8.1. Geotextiles for Filtration

To the authors' knowledge there are no reported failures in literature of geotextile filters used in river dykes. The potential disadvantages of geotextile filters, compared to the traditional granular filters, lies in the risk of clogging and/or blinding due to fine particles being entrapped inside the geotextile and/or accumulated on its surface, to a level that the filtration capacity and the permeability are impaired. Recent results from excavation of geotextile filters have been reported [54].

It has to be noted that geotextile for filtration in earth dams has been one of the earliest applications: the first large earth dam using geosynthetic materials, Valcros dam, was built in 1970 in France. At that time the geotextiles were designed as a filter on the upstream slope between the rocks and the earth fill and on the downstream slope around the main drains [55]. The behavior of these geotextiles is fully documented because several excavations and investigations were carried out 6 years [56] and 21 years after installation [57]; the geotextiles at Valcros dam are still in operation without any impair factor to their filtration capacity after over 50 years.

Over 40 years of research into geotextile filters has led to the publication of several design criteria, which have been summarized in the ISO TR 18228 [58] series of standards for design of geosynthetics, including the filtration function.

Designing geotextile filters in river dykes according to this standard would prevent the potential risk of clogging and/or blinding, thus eliminating the only potential disadvantage for the use of a geotextile filter.

8.2. Geosynthetics for Drainage

Geosynthetic drains have very think profile (up to 20 mm) compared to granular drainage blankets (whose thickness is of the order of 0.5–1.5 m). Nevertheless the water flow inside geosynthetic drains is typically turbulent with much higher velocity than the filtration flow occurring in granular filters. Hence the flow rate of geosynthetic drains is very high even with their reduced thickness.

Recent research on geosynthetic drains has shown that the draining core can be subject to thickness reduction due to compressive creep over the entire design life, and due to the intrusion of the geotextile filters into the draining core, which would reduce the available flow capacity. This is a potential disadvantage compared to granular filters, which are not subject to the negative effects of long term overburden pressure.

A decrease in the available flow capacity due to chemical or biological clogging of the draining core has however been shown to not be of concern when fresh water is being transmitted, such as in the case of river dykes.

Research on geosynthetic drains has led to the development of test standards for the evaluation of their short and long term properties: water flow capacity in standard conditions (ISO 12958-1 [59]) and in operational conditions (ISO 12958-2 [60]), compressive creep testing (ISO 25619-1 [61]) and accelerated compressive creep testing with the Stepped Isothermal Method (SIM) (ASTM D7361 [62]).

These tests are the base for the design criteria for geosynthetic drains, which have been summarized in the ISO TR 18228 [58] series of standards for design of geosynthetics, including the drainage function.

Designing geosynthetic drains in river dykes according to this standard, taking into consideration the overburden pressure, would allow for the selection of a product able to provide the required flow rate at the end of the design life, with the prescribed factor of safety on the available drainage capacity.

Hence the potential disadvantage of geosynthetic drains compared to granular filters, due to compressive creep and geotextiles intrusion, can be easily overcome through sound engineering design.

8.3. Geosynthetics for Reinforcement and Stabilisation

To the authors' knowledge there is no potential disadvantage in the use of geosynthetics for reinforcement and stabilization in river dykes.

Obviously, the stability of reinforced or stabilized soil structures shall be designed taking into account all the potential limit states and failure mechanisms, including the water induced ones: for river dykes, the stability analyses in case of rapid draw-down shall always be carried out.

8.4. Geosynthetics for Surface Erosion Control

The erosion control function is complex, since the performance depends both on the geosynthetics that are used and on the establishment of the vegetation.

Given that the establishment of vegetation would require, in any case, the correct mixture of soils, the proper selection of seeds, the possible use of mulch, and the maintenance during the first growth season, the only potential disadvantage in the use of geosynthetics could be the difficulty to select a product that wouldn't impair the growth of vegetation while continuously providing erosion protection and root reinforcement.

The ISO TR 18228 [58] series of standards for design of geosynthetics, including the design of geosynthetics for the erosion control function, both on slopes and on river/channel banks, allows for the selection of the geosynthetics which can afford the required resistance to both the design water velocity and the design shear stresses applied by the stream.

8.5. Geosynthetics in Barrier Systems

The main barrier materials used in dykes engineering are GCLs. GCLs when used in river dykes are however subject to specific failure mechanisms. These are summarized as previously stated in [40] and include:

- Careful attention needs to be paid to panel overlaps as these are potential points of weakness in the barrier, so good construction quality assurance is critical.

- GCLs in dykes should not be exposed directly to hydraulic loads, as this could wash the bentonite out of the GCL over time. GCLs are therefore typically covered with a soil layer, minimum 800 mm thick.
- GCLs subject to freeze and thaw cycles can develop increased hydraulic conductivity; again the soil cover layer will usually prevent this failure mechanism.
- GCLs can be subject to root penetration from vegetation that establishes on the embankments. Sufficient cover depth, typically 0.8m, is usually specified to mitigate this risk.
- The bentonite in a GCL can be subject to cation exchange, reducing the swell capacity of the bentonite and thus the hydraulic conductivity. This effect is usually exacerbated by wet/dry cycling. As such designers should be aware that the GCL hydraulic conductivity is likely to increase over time and incorporate measures to deal with this in the design.

Von Maubeuge and Ehrenberg [63] discussed the benefits multicomponent GCLs could bring, by limiting bentonite erosion and wet dry cycles and also potential cation exchange with the soil covering the GCL, as could be expected in a dyke environment, thus further extending the design life of the material. The polymer layer in the multicomponent GCL will also form a barrier to root penetration, as a geomembrane would, which has been highlighted as a risk to the use of GCLs in dykes [40].

8.6. Environmental Benefits of Geosynthetics

Geosynthetics are designed to last: most geosynthetics can be supplied with certificates of conformance for an expected duration in excess of 100 years. It is essential that the design life of geosynthetics is at least equal to the design life of the structure they are incorporated in. As previously presented, the design life of geosynthetics used in geoenvironmental applications is subject to many factors. By taking these factors into consideration, and by using internationally accepted standards of specification, design and installation, their ability to function can be relied upon for well over the expected design life of these structures. Ref. [64] evaluated durability testing data collected over a 17 year period on HDPE geomembrane samples. Geomembranes used in applications such as river dykes will experience temperatures below the long-term average ambient temperatures. For a site with an annual average earth temperature of 15 degrees Celsius the expected life of a suitable specified and installed geomembrane would be well over the 100 years specified in CE marking. Strict quality control along the production chain but also during installation, for which procedures have been developed and are in use, is also a key factor.

The construction activities based on construction methods that do not involve geosynthetics have contributed significantly to global warming and associated geoenvironmental instabilities, such as soil erosion, hill/riverbank/coastal slope instability, ice/glacier melting, floods and sea-level rise, worldwide [65]. Carbon emissions associated to construction with geosynthetics has been shown to have a comparatively limited impact, so that in addition to bringing solutions that contribute to limiting the effects of floods, geosynthetics also, in a virtuous circle, contribute to limiting carbon emissions during construction of these measures. The use of geosynthetics also provides an increased durability and resilience to infrastructure, which also results in less carbon emissions long term due to the infrastructure being able to function for a longer time.

Finally, education, in particular through the Educate the Educators program of the IGS, is essential in offering undergraduate civil engineering students exposure to geosynthetics. It does this by providing Geotechnical Engineering university professors with the content and pedagogical tools needed [66]. The knowledge and knowhow thus capitalized over the years in terms of designing and building with geosynthetics can be used to the best and largest possible extent to ensure the longest duration of works and lowest impact to the environment.

9. Summary

In the context of climate change and the resulting expected increased frequency and magnitude of natural disasters, longitudinal dykes, constitute one of the most often used active structural methods to control the course of a river

Dykes are commonly made of different natural materials. They can also incorporate geosynthetics. Literature on the appropriate use of geosynthetics in dykes and their advantages in this context is sparse. The objective of this review paper was thus to compile a synthesis of the existing literature on this topic to address the many ways in which geosynthetics contribute to sustainable construction of dykes and thus contribute to water systems management.

A brief presentation of what geosynthetics are and the functions they can fulfil was given in Section 2 of this paper.

Section 3 gave a brief overview of dyke structure, when built from granular materials, and the typical failure mechanisms in earthen structures that are commonly encountered due to the action of water. Based on the review of these failure mechanisms, solutions incorporating geosynthetics, that contribute to increased resilience of dykes by increasing their strength and their longevity, are presented.

The main applications of geosynthetics in dykes were presented in Section 4. The advantages of using geotextiles filters, and the various locations in a dyke where they can be used were detailed. Insight was then given in the applications of geocomposites for drainage, that can replace granular materials efficiently to fulfil this function. The applications of reinforcement and stabilization were then addressed. Again the importance of the adaptation of the design and of using the right materials at the right location to optimize the performance was emphasized. The enlargement of dykes was discussed in this section.

Surface erosion control can be a major issue in dykes, both on the riverside and the landside. This topic is critical as overtopping of dykes during extreme intensity floods is thought to represent one of the main risks of failure of these structures. Various alternatives incorporating geosynthetics were presented.

The use of geosynthetics to ensure protection against rodents was also discussed.

Finally, details were given on the way geosynthetic barriers can be incorporated in dykes in order to separate water from soil in the structure and thus avoid failures due to water infiltration, was presented.

Through the presentation of the various potential uses of geosynthetics in dykes, emphasis was thus put on the many ways geosynthetics contribute to reduce the risk of failure of dykes and thus increase their resilience.

As further shown in Section 5, in addition to the technical benefits of construction with geosynthetics, the use of geosynthetics also results in economic advantages for a project. The paper has demonstrated how construction with geosynthetics can result in a reduction in the quantity of soil material required to form a dyke, and even of the type of fill required, making it possible to use dredged materials, coal combustion products or even residues. The increased life span of construction that results from the use of geosynthetics also contributes to the implementation of economies that can be significant.

In addition to the environmental benefits resulting from the reduction in quarrying of granular fill material, mitigation of greenhouse gas emissions during construction can be obtained and quantified as shown in Section 6 of this paper.

This review paper has illustrated the many possible contributions of geosynthetics to the construction of sustainable river management, not only by allowing the implementation of more economic construction methods, but also with increased durability, to be able to withstand the extreme stresses related to climate change that these structures are going to be expected to endure. Geosynthetics also bring a positive contribution to the reduction of greenhouse gas emissions and support a reduction in construction costs.

The advantages and disadvantages of the use of geosynthetics in water engineering in the context of dykes was finally discussed, showing that risks inherent in the use of these products can be prevented provided that proper design, specification and installation is ensured. Efforts in education made by the International Geosynthetics Society in the past years contributes to the effort of ensuring longevity of structures through the appropriate use of geosynthetics.

Author Contributions: Conceptualization, P.R., J.S., J.K. and N.T.; methodology, N.T.; software, P.R. and N.T.; validation, P.R., J.S., J.K. and N.T.; formal analysis, P.R., J.S., J.K. and N.T.; investigation, N.T., P.R., J.S. and J.K.; resources, N.T., P.R., J.S. and J.K.; data curation, N.T.; writing—original draft preparation, P.R., J.S., J.K. and N.T.; P.R., J.S., J.K. and N.T.; visualization, P.R. and N.T.; supervision, P.R. and N.T.; project administration, N.T.; funding acquisition, Not applicable. All authors have read and agreed to the published version of the manuscript.

Funding: This research received no external funding.

Institutional Review Board Statement: Not applicable.

Informed Consent Statement: Not applicable.

Data Availability Statement: Not applicable.

Conflicts of Interest: The authors declare no conflict of interest.

References

1. Heibaum, M. Geosynthetics for waterways and flood protection structures—Controlling the interaction of water and soil. *Geotext. Geomem.* **2014**, *42*, 374–393. [CrossRef]
2. Hsieh, C.; Heibaum, M. Geosynthetics for Canal and River Bank Erosion Control. In Proceedings of the Geoafrica 2017, Marrakech, Morocco, 8–11 October 2017; pp. 893–902.
3. Saathoff, F.; Cantré, S.; Sikora, Z. *South Baltic Guideline for the Application of Dredged Materials, Coal Combustion Products and Geosynthetics in Dyke Construction*; Saathoff, F., Cantré, S., Sikora, Z., Eds.; Universität Rostock: Rostock, Germany; p. 103.
4. IPCC Managing the Risks of Extreme Events and Disasters to Advance Climate Change Adaptation (Special Report (SREX), The Intergovernmental Panel on Climate Change). 2012. Available online: www.ipcc.ch (accessed on 26 November 2020).
5. IPCC Fifth Assessment Report (AR5). 2014. Available online: www.ipcc.ch (accessed on 26 November 2020).
6. Hov, O.; Cubasch, U.; Fischer, E.; Höppe, P.; Iversen, T.; Kvamstø, N.G.; Kundzewicz, Z.W.; Rezacova, D.; Rios, D.; Duarte Santos, F.; et al. *Extreme Events in Europe: Preparing for Climate Change Adaptation*; Norwegian Meteorological Institute: Blindern, Norway, 2013; p. 141.
7. DIN. *Anpassung an Den Klimawandel-Umgang Mit Unsicherheiten im Kontext Von Projektionen; Beiblatt 1: Sommerlicher Wärmeschutz Von Gebäuden-Ein Beispiel der Vulnerabilitätsanalyse Für Den Fall einer Temperaturerhöhung Von 2 °C Und Mögliche Maßnahmen Zur Anpassung an Die Folgen dieser Temperaturerhöhung*; DIN SPEC 35220 Beiblatt 1; DIN: Berlin, Germany, 2018.
8. Brandl, H. Geosynthetics applications for the mitigation of natural disasters and for environmental protection. *Geosyn. Int.* **2011**, *18*, 340–390. [CrossRef]
9. Kubetzek, T.; Steurnagel, J.; Ramm, H.; Saenger, N.; Zanke, U. Improvement of the general resistance of dikes against erosion during dike overflow. In Proceedings of the 4th International Symposium on Flood defence: Managing Flood Risk, Reliability and Vulnerability, Toronto, ON, Canada, 6–8 May 2008; pp. 1531–1536.
10. Heerten, G. 2010 Mitigation of flooding by improved dams and dykes. In Proceedings of the International Symposium, Exhibition, and Short Course on Geotechnical and Geosynthetics Engineering: Challenges and Opportunities on Climate Change, Bangkok, Thailand, 7–8 December 2010.
11. Kurakami, Y.; Nihei, Y. Resistance of Laminar Drain Reinforcement Levee against Overflow Erosion. *Water* **2019**, *11*, 1768. [CrossRef]
12. *International Levee Handbook*; CIRIA: London, UK, 2013; p. 1349.
13. Shahkolahi, A.; Loizeaux, D. Advanced technology for Cyclones Storm Damage Risk Reduction Systems and Flood Protection Levees. In Proceedings of the Stormwater 2018, Sydney, Australia, 8–12 October 2018; p. 11.
14. Heibaum, M.; Fourie, A.; Girard, H.; Karunaratne, G.P.; Lafleur, J.; Palmeira, E.M. Hydraulic applications of geosynthetics. In Proceedings of the 7th International Conference on Geosynthetics, Yokohama, Japan, 18–22 September 2006; Kuwano, J., Koseki, J., Eds.; Millpress Science Publishers: Rotterdam, The Netherlands, 2006; pp. 79–120.
15. CEN. *Geosynthetics–Part 1: Terms and definitions–Amendment 1*; EN ISO 10318-1:2015/AMD 1:2018; CEN: Brussels, Belgium, 2018.
16. Expertise Network for Flood Protection (ENW). *Fundamentals of Flood Protection*, English ed.; Ministry of Infrastructure and the Environment and the Expertise Network for Flood Protection: Hague, The Netherlands, 2017; p. 143.
17. APAT (Italian Agency for Environmental Protection). *Atlante Delle Opere di Protezione Fluviale (Atlas of River Protection Works), Manuali e Linee Guida*; APAT: Roma, Italy, 2004.
18. Giroud, J.P. Development of criteria for geotextile and granular filters. In Proceedings of the 9th International Conference on Geosynthetics, Guaruja, Brazil, 23–27 May 2010; pp. 45–64.

19. Palmeira, E. A review of some factors influencing the behaviour of nonwoven geotextile filters. *Soils Rocks* **2020**, *43*, 351–368. [CrossRef]

20. Rimoldi, P.; Fontana, F. The choice of geosynthetics for applications in dam engineering. In Proceedings of the Geoafrica 2017, Marrakech, Morocco, 8–11 October 2017; pp. 903–928.

21. *Design and Construction of Levees, Manual No. 1110-2-1913*; Department of the Army, U.S. Army Corps of Engineers: Washington, DC, USA, 2000.

22. Blond, E.; Boyle, S.; Ferrara, M.; Herlin, B.; Plusquellec, H.; Rimoldi, P.; Stark, T. *Applications of Geosynthetics to Irrigation, Drainage and Agriculture. Irrigation and Drainage*; John Wiley and Sons Ltd.: Hoboken, NJ, USA, 2018.

23. Temple, D.M.; Irwin, W. Allowable overtopping of earthen dams. In *Dam Safety 2006, Proceedings of the Association of State Dam Safety Officials Annual Conference, Boston, MA, USA, 10–14 September 2006*; ASDSO: Lexington, KY, USA, 2006.

24. Hepler, T.; Fiedler, B.; Vermeyen, T.; Dewey, B.; Wahl, T. Overtopping Protection for Dams—A Technical Manual Overview. In *Proceedings of the Dam Safety 2012-Association of State Dam Safety Officials*; Denver, CO, USA, 16–20 September 2012. Available online: https://www.usbr.gov/tsc/techreferences/hydraulics_lab/pubs/PAP/PAP-1093.pdf (accessed on 26 November 2020).

25. Morgan, R.P.C.; Rickson, R.J. *Slope Stabilization and Erosion Control: A Bioengineering Approach*; E & FN Spon Publisher: London, UK, 2011.

26. Di Pietro, P. Practical Applications with Geosynthetic Mats Reinforced with Steel Wire Meshes to Prevent Embankment Damage by Burrowing Large Rodents and Beavers. *J. Civ. Eng. Arch.* **2017**, *11*, 8–15. [CrossRef]

27. Cambiaghi, A.; Rimoldi, P. The use of geogrids in landslide control works: A case history from Valtellina (Northern Italy). In Proceedings of the International Conference on Slope Stability Engineering, Developments and Applications, Isle of Wight, UK, 15–18 April 1991; Thomas Telford: London, UK.

28. DWA. *Dichtungssysteme in Deichen. Deutsche Vereinigung Für Wasserwirtschaft, Abwasser Und Abfall DWA e. V.*; DWA: Hennef, Germany, 2005.

29. DGGT. *Empfehlungen Zur Anwendung Geosynthetischer Tondichtungsbahnen. German Geotechnical Society DGGT e.V.*; Ernst & Sohn: Berlin, Germany, 2002.

30. Haselsteiner., R.; Strobl, T. Constraints and Methods of Refurbishment Measures of Dikes. In Proceedings of the 3rd International Symposium on Integrated Water Resources Management, Bochum, Germany, 26–28 September 2006.

31. Werth, K.; Heerten, G.; Pries, J.-K.; Klompmaker, J. 20 years experience with GCLs in dams and dykes. In Proceedings of the GBR-C 2k10, 3rd International Symposium on Geosynthetic Clay Liners, Würzburg, Germany, 15–16 September 2010; p. 11.

32. Touze-Foltz, N. Healing the World: A geosynthetics solution. The 6th Giroud Lecture. In Proceedings of the 11th International Conference on Geosynthetics, Seoul, Korea, 16–21 September 2018; p. 59.

33. Giroud, J.-P.; Plusquellec, H. Water, canals and geosynthetics. In Proceedings of the Geoafrica Conference, Marrakech, Morocco, 8–11 October 2017; pp. 53–112.

34. Touze-Foltz, N.; Bannour, H.; Barral, C.; Stoltz, G. A review of the performance of geosynthetics for environmental protection. *Geotext. Geomem.* **2016**, *44*, 656–672. [CrossRef]

35. Jones, D.R.V. Using Geosynthetics for Sustainable development. In Proceedings of the GSI-Asia 2015 Geosynthetics Conference, Seoul, Korea, 24–26 June 2015; p. 4.

36. Christopher, B. Cost Savings by Using Geosynthetics in the Construction of Civil Work Projects. In Proceedings of the 10th International Conference on Geosynthetics, Berlin, Germany, 21–25 September 2014; Ziegler, M., Braü, G., Heerten, G., Laackma, K., Eds.; German Geotechnical Society: Essen, Germany, 2014.

37. Haselsteiner, R.; Strobl, T.; Heerten, G.; Werth, K. Overflow protection systems of flood embankments with geosynthetics. In Proceedings of the EuroGeo4, 4th European Geosynthetics Conference, Edinburgh, Scotland, 7–10 September 2008; p. 8.

38. Official Journal of the European Union. EU Regulation 305 of the European Parliament and of the Council of 9 March 2011, Laying Down Harmonised Conditions for the Marketing of Construction Products and Repealing Council Directive 89/106/EEC. 2011. Available online: https://eur-lex.europa.eu/LexUriServ/LexUriServ.do?uri=OJ:L:2011:088:0005:0043:EN:PDF (accessed on 27 November 2020).

39. Palmeira, E. Sustainability and innovation in Geotechnics: Contribution from Geosynthetics. Manuel Rocha Lecture. *Soils Rocks* **2016**, *39*, 113–135.

40. Fleischer, P.; Heibaum, M. Geosynthetic clay liners (GBR-C) for hydraulic structures. In Proceedings of the GBR-C 2k10: Third International Symposium on Geosynthetic Clay Liners, Würzburg, Germany, 15–16 September 2010; pp. 259–268.

41. Von Maubeuge, K.P.; Sreenivas, K.; Pohlmann, H. *The New Generation of Geosynthetic Clay Liners. Seminar Geosynthetics India' 11 and an Introductory Course on Geosynthetics, Chennai (Tamil Nadu), India, 8p.Empfehlungen Zur Anwendung Geosynthetischer Tondichtungsbahnen. German Geotechnical Society DGGT e.V.*; Ernst&Sohn: Berlin, Germany, 2002; 8p.

42. Koseki, J.; Barthurst, R.J.; Güler, E.; Kuwano, J.; Maugeri, M. Seismic stability of reinforced soil walls. In Proceedings of the 6th International Conference on Geosynthetics, Yokohama, Japan, 18–22 September 2006; Millpress Science Publishers: Rotterdam, The Netherlands, 2006; pp. 51–78.

43. Dixon, N.; Fowmes, G.; Frost, M. Global challenges, geosynthetic solutions and counting carbon. *Geosyn. Int.* **2016**, *24*, 451–464. [CrossRef]

44. Dixon, N.; Raja, J.; Fowmes, G.; Frost, M. Chapter 26 Sustainability aspects of using geotextiles. In *Geotexiles from Design to Applications*; Koerner, R.M., Ed.; Elsevier: Amsterdam, The Netherlands, 2016; pp. 577–596.

45. Damians, I.P.; Bathurst, R.J.; Adroguer, E.; Josa, A.; Lloret, A. Environmental assessment of earth retaining wall structures. *Environ. Geotech.* **2016**, *4*, 415–431. [CrossRef]
46. Damians, I.P.; Bathurst, R.J.; Adroguer, E.; Josa, A.; Lloret, A. Sustainability assessment of earth retaining wall structures. *Environ. Geotech.* **2016**, *5*, 187–203. [CrossRef]
47. Raja, J.; Dixon, N.; Fowmes, G.; Frost, M.; Assinder, P. Obtaining reliable embodied carbon values for geosynthetics. *Geosyn. Int.* **2015**, *22*, 393–401. [CrossRef]
48. Touze, N. Healing the World: A Geosynthetics Solution. *Geosyn. Int.* **2021**, *28*, 1–31.
49. Heerten, G. Reduction of Climate-damaging gases in geotechnical engineering by use of geosynthetics. *Geotext. Geomem.* **2012**, *30*, 43–49. [CrossRef]
50. Von Maubeuge, K.; Heerten, G.; Egloffstein, T.A. Reduction of Climate-damaging gases in geotechnical engineering by use of geosynthetics. In Proceedings of the 24th Annual GRI Conference Optimizing Sustainability with Geosynthetics, Dallas, TX, USA, 16 March 2011; pp. 58–71.
51. Sauli, G.; Pellizzari, L.; Vicari, M. Carbon footprint, capture and sequestration of double twist wire mesh solutions in river training works. In Proceedings of the 3rd IAHR Europe Congress, Porto, Portugal, 1 January 2014; p. 9.
52. Directive 2010/31/Eu of the European Parliament and of the Council of 19 May 2010 on the Energy Performance of Buildings. In Official Journal of the European Union L 153/13. 18 June 2010. Available online: https://eur-lex.europa.eu/LexUriServ/LexUriServ.do?uri=OJ:L:2010:153:0013:0035:en:PDF (accessed on 27 November 2020).
53. Arumägi, E.; Simson, R.; Kuusk, K.; Kalamees, T.; Kurnitski, J. *Analysis of Cost-Optimal Minimum Energy Efficiency Requirements for Buildings*; Tallinn University of Technology: Tallinn, Estonia, 2017.
54. Veylon, G.; Stoltz, G.; Mériaux, P.; Faure, Y.-H.; Touze-Foltz, N. Performance of geotextile filters after 18 years' service in drainage trenches. *Geotext. Geomemb.* **2016**, *44*, 515–533. [CrossRef]
55. International Standard Organization. *Design Using Geosynthetics—Part 1: General*; ISO/TR 18228-1:2020; International Standard Organization: Geneve, Switzerland, 2020.
56. *Geosynthetics-Method for Determining the Microbiological Resistance by a Soil Burial Test*; EN 12225; CEN: Bruxelles, Belgium, 2021.
57. Artières, O.; Oberreiter, K.; Aschauer, F. Geosynthetic systems for earth dams–35 years of experience. Active and passive defences against internal erosion. In Proceedings of the 7th ICOLD European Club Dam Symposium, Fresing, Germany, 17–19 September 2007; pp. 98–103.
58. Giroud, J.P.; Gourc, J.P.; Bally, P.; Delmas, P. Comportement d'un textile non tissé dans un barrage en terre. In Proceedings of the International Conference on the Use of Fabrics in Geotechnics, Paris, France, 20–22 April 1997; Amicale Anciens Elèves de l'E.N.P.C.: Paris, France, 1997; pp. 213–218.
59. International Standard Organization. *Geotextiles and Geotextile-Related Products—Determination of Water Flow Capacity in their Plane. Part 1–Index Test*; ISO 12958-1; International Standard Organization: Geneve, Switzerland, 2020.
60. International Standard Organization. *Geotextiles and Geotextile-Related Products—Determination of Water Flow Capacity in their Plane. Part 2–Performance Test*; ISO 12958-2; International Standard Organization: Geneva, Switzerland, 2020.
61. International Standard Organization. *Geosynthetics—Determination of Compression Behaviour—Part 1: Compressive Creep Properties*; ISO 25619-1; International Standard Organization: Geneve, Switzerland, 2021.
62. ASTM International. ASTM D7361. In *Standard Test Method for Accelerated Compressive Creep of Geosynthetic Materials Based on Time-Temperature Superposition Using the Stepped Isothermal Method*; ASTM International: West Conshohocken, PA, USA, 2018.
63. Von Maubeuge, K.P.; Ehrenberg, H. Investigation of Bentonite Mass per Unit Area Requirements for Geosynthetic Clay Liners. In Proceedings of the 10th International Conference on Geosynthetics, Berlin, Germany,, 21–25 September 2014; Ziegler, M., Braü, G., Heerten, G., Laackmann, K., Eds.; German Geotechnical Society: Essen, Germany, 2014.
64. Ewais, A.M.R.; Rowe, R.K.; Rimal, S.; Sangam, H. P 17-year elevated temperature study of HDPE geomembrane longevity in air, water and leachate. *Geosynth. Int.* **2018**, *25*, 524–544. [CrossRef]
65. Shukla, S.K. Geosynthetics and Ground Engineering: Sustainability considerations. *IJGGE* **2021**, *7*, 7–17.
66. Zornberg, J.G.; Touze, N.; Palmeira, E. Educate the Educators: An International Initiative on Geosynthetics Education. In *Proceedings of the Online International Conference on Geotechnical Engineering Education 2020 Athens, Greece 23–25 June 2020*; Marina, P., Michele, C., Margarida, P.L., Eds.; ISSMGE: London, UK, 2021; p. 11.

Article

Bottom Sediments from a Dam Reservoir as a Core in Embankments—Filtration and Stability: A Case Study

Karolina Koś [1], Andrzej Gruchot [2,*] and Eugeniusz Zawisza [2]

[1] Department of Rural Building, University of Agriculture in Kraków, al. Mickiewicza 24/28, 30-059 Kraków, Poland; karolina.kos@urk.edu.pl

[2] Department of Hydraulic Engineering and Geotechnics, University of Agriculture in Kraków, al. Mickiewicza 24/28, 30-059 Kraków, Poland; kiwig@urk.edu.pl

* Correspondence: rmgrucho@cyf-kr.edu.pl; Tel.: +48-12-662-4136

Abstract: A possibility of using bottom sediments from dam reservoirs in earth structures was considered. Sediments from the Rzeszow reservoir (Poland) were used as research material, which, according to geotechnical standards, were classified as low permeable silt with high organic content. As fine, cohesive soil with a low coefficient of permeability, the sediments can be used in sealing elements of hydraulic engineering embankments. In order to verify the suitability of the sediments, stability and filtration calculations were carried out for embankments with a sealing in the form of a core made of the sediments. It was stated that by using a core made of sediments, the volume of seepage on the downstream side during continuous or variable backwater was significantly lower in relation to an embankment without a core, and the phreatic line did not extend to the downstream slope. It is estimated that, in the case of a planned dredging in Rzeszow Reservoir, the amount of dredged sediment would exceed 1.5 million m^3, and therefore, the possibility of their economic use is essential. The search for materials that could replace natural soil in earthen structures is an important issue from both the ecological and economic points of view.

Keywords: bottom sediments; soil core; levee; stability analysis

Citation: Koś, K.; Gruchot, A.; Zawisza, E. Bottom Sediments from a Dam Reservoir as a Core in Embankments—Filtration and Stability: A Case Study. *Sustainability* **2021**, *13*, 1221. https://doi.org/10.3390/su13031221

Received: 19 November 2020
Accepted: 13 January 2021
Published: 25 January 2021

Publisher's Note: MDPI stays neutral with regard to jurisdictional claims in published maps and institutional affiliations.

1. Introduction

Hydraulic engineering embankments are earth structures temporarily or permanently damming water. They fulfill a variety of functions as levees, embankments of rivers, causeways, and so forth. These structures have a trapezoidal shape, and depending on the class or function, they may include sealing and drainage elements within the body and the subsoil (Figure 1).

Granular materials with the greatest particle size distribution should be used to form embankments. In the construction of the embankments of rivers and water reservoirs, their proper tightness should be ensured to prevent excessive water filtration through the body of the earth structure and water leaks, which would pose a threat to the stability of the structure and surrounding area [1–3]. The soil used for forming the embankments should therefore be resistant to water. In the case of using permeable materials, the use of additional sealing in the form of a screen or a core should be considered. Depending on the place of incorporation into the embankment, the soil should be characterized by different requirements. For the construction of the outer part of an earth-fill dam, the so-called static part, stony soil (screen, residual clay soil, and cobbles), coarse-grain soil (gravel and gravel–sand), and fine non-cohesive soil (sand) can be used. Cohesive soils (clay, silt, gravel sand mixed with clay, and gravel mixed with clay) are incorporated into the dam seal and are designed to reduce the filtration through the dam body.

Due to difficulties in obtaining suitable natural soils for incorporation, whose resources are limited and non-renewable, research continues on materials that could replace them [5,6]; industrial waste is a typical example [7,8]. In this situation, the use of bottom sediment from dam reservoirs is also worth considering [9]. One of the main problems related to the performance of reservoirs is protection from sediment deposition [10]. Since the process of silting occurs in every dam reservoir, the extraction of sediments is one of the ways to recover lost volume and to restore proper functioning of the reservoir. In Poland, water reservoirs in the south, especially in the Carpathians, get silted up the fastest [11].

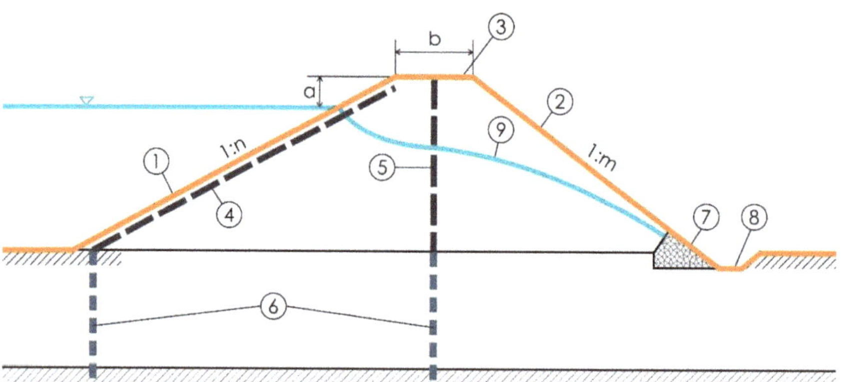

Figure 1. Schematic diagram of the cross-section of a levee, including the main elements [4]. (1)—upstream slope, (2)—downstream slope, (3)—crest, (4)—tight screen, (5)—tight core, (6)—subsoil ground sealing, (7)—drain, (8)—drainage ditch, (9)—depression curve in the case of a homogeneous embankment (without core). (a)—safety reserve, (b)—levee crest width, (1:n)—upstream slope inclination, (1:m)—downstream slope inclination.

The silting of water reservoirs as a result of the sedimentation of material carried by tributaries causes a decrease in their capacity. The intensity of the silting process depends on the surface and topography of the catchment area, its use, types of soil, as well as the precipitation intensity and duration. The type and location of the reservoir, as well as the hydrological regime of the river, also play an important role. Therefore, considering reservoir water management, it is important to determine the rate of the siltation process, and thus, the amount of mineral and organic material deposited in the reservoir. It is believed that if the capacity is reduced by 80%, the reservoir loses its retention function [12].

According to the Act [13], bottom sediments are classified as waste and should be disposed of when they are extracted. However, if they are moved within surface waters for purposes related to water management or reducing the effects of floods, drought, or remediation, these sediments are not dangerous [14]. While considering the potential use of bottom sediments, another important aspect is their pollution. The chemical composition of the sediments depends on the type of soils in the catchment area and their agricultural use, as well as the type of industry present in the area [15–17]. Therefore, identification of the chemical composition of bottom sediments using chemical and ecotoxicological analyses is important, not only to assess the degradation of the water reservoir, but also to determine the potential applications of the extracted sediments [18,19]. There is also research carried out on anaerobic methane oxidation processes, which occur in sediments that create habitats in freshwater dam reservoirs of small volumes; these processes play a significant part in global warming [20].

Bottom sediments that show a neutral or alkaline reaction and high content of fine fractions can be used to improve physicochemical properties of light and acidic soils [21]. The possibilities of the agricultural use of sediments are mainly connected with a low content of heavy metals, but also an appropriate content of available forms of magnesium, potassium, and phosphorus, which prove their fertility. Research on the potential uses of bottom sediments, including agricultural, was carried out by Fonseca et al. [22], Canet et al. [23], Baran et al. [24], and others. Using the example of sediments from the Rożnów reservoir located in southern Poland, Tarnawski et al. [25] indicated that there is a great possibility of using sediments for environmental purposes (agriculture, reclamation). In order to fully assess their potential usage, it is necessary to analyze the ecological risk connected to their extraction and possible environmental pollution with heavy metals, PAHs (polycyclic aromatic hydrocarbon), PCBs (polychlorinated biphenyl) [25]. These materials can also be used in the reclamation of mine and energy waste landfills, especially the ones with fly ashes [26]. Sediments in dam reservoirs, if they meet geotechnical requirements, can be a full-value material that can be used in earth constructions [9,27,28]. These materials should not contain more than 3% of organic parts and not more than 5% of water-soluble parts [2]. In plastic sealing elements of hydrotechnical embankments, mainly cohesive soils should be used, which contain more than 25% of particles with a diameter smaller than 0.01 mm, and the permeability coefficient of such soils should be less than 10^{-7} m·s^{-1}. There is also research on bottom sediments stabilized by the addition of bentonite [28,29] and using bottom sediments in brick production [30].

There has also been research conducted on the characteristics of sediments deposited in the bottom of lakes and seas, but from the point of view of their geological structures [31–33]. For example, tests on the sediments from Lake Bergsee in southern Germany have shown that they are much weaker and highly-compressive compared to the sediments from Scandinavian lakes [31]. Whereas Ballas et al. [33] showed that sediments in the Romanian part of the Black Sea are highly plastic, clayey materials that have high compressibility, low shear strength, and moderate sensitivity. They also noticed a slight increase in the shear strength of sediments where the precipitation of iron sulphides and calcium carbonates occurred, which cemented this material. Ballas et al. [33] clearly indicate that it is the cementation process, connected with the precipitation of chemical compounds, that causes the seabed to be stable.

The purpose of the work was to analyze the filtration output and stability of an earth embankment with a seal made of fine-grained bottom sediments in the form of a central core and a cut-off wall in the ground. The calculations were also made for the embankment without these seals. The results of the tests on the geotechnical parameters of the bottom sediments from a water reservoir in Rzeszów (Poland) were used in the calculations. The calculations were carried out to assess the usability of bottom sediments from that reservoir for earthworks, in particular, in hydrotechnical constructions. The purpose of the work resulted in the need to look for materials, other than natural mineral soils, that could potentially be used as a sealing material in hydrotechnical earth embankments. As shown above, there are a number of studies describing problems related to the chemical pollution of bottom sediments from dam reservoirs and possible attempts to use them in agriculture. However, there are very few studies related to the use of sediments in earthworks; moreover, the amount of deposited material is significant, so their economic potential is great.

2. Materials and Methods

2.1. Study Area

The water reservoir in Rzeszów (Podkarpackie Voivodeship, Rzeszów, Poland) was created in 1973 as a result of building a barrage in 63.7 km of the Wisłok River (Figure 2). Initially, the reservoir capacity was 1.8 million m^3, and its area was 68.2 hectares. The catchment area of the reservoir is 2060.7 km^2, with a total catchment area of the Wisłok River of 3528.2 km^2. As a result of the silting process, the reservoir is currently about 60% silted up, so its original functions cannot be fully realized. It is estimated that approximately 1.5 million m^3 of bottom sediments may be extracted from the reservoir. This intensive silting process is mainly the result of the way the catchment area is used, where agricultural land is over 70% (arable land—53.9%, orchards—1.8%, meadows and pastures—14.3%) [34]. The remaining part of the catchment area is forest land. In the catchment area, for the most part, there are poorly permeable, brown and silty soils, as well as alluvial soils with a mechanical composition of heavy clays and clayey silts [34].

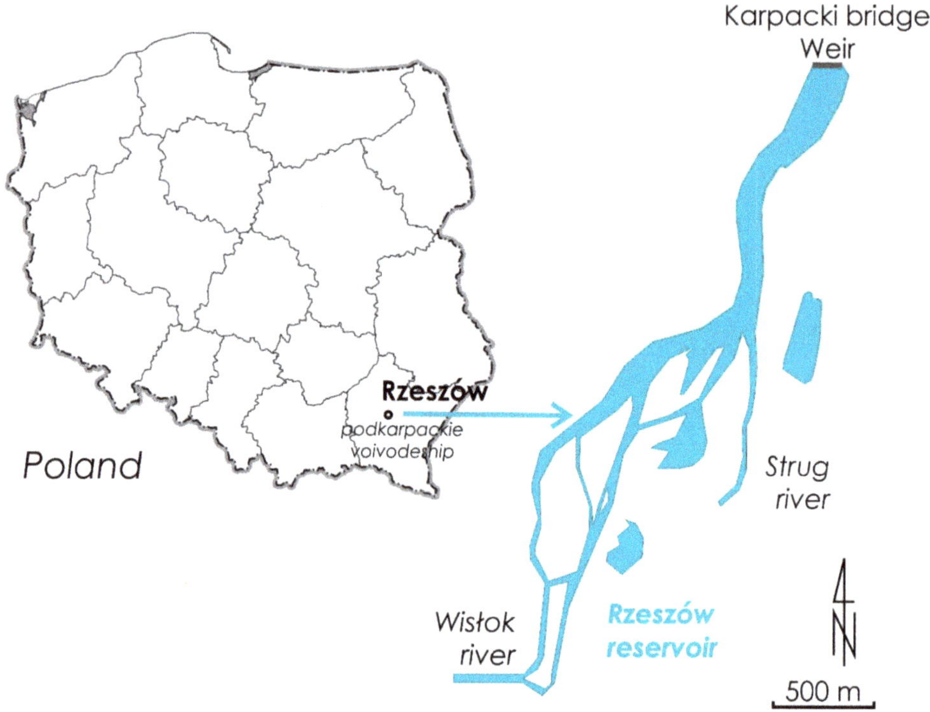

Figure 2. Location and view of the Rzeszów reservoir.

2.2. Geotechnical Parameters

Bottom sediment tests were carried out on an averaged material, which was a mixture of all samples taken from the bottom of the reservoir. The scope of the research included grain size composition, specific density, consistency limits, compaction parameters, shear strength, coefficient of permeability, and organic matter content. The grain size composition was determined by the areometric method [35,36], and the specific density by the pycnometer method [37]. Consistency limits, that is, the liquid limit, was determined by the Casagrande method, and the plastic limit, by the standard thread rolling method [38]. Compaction parameters, that is, the optimum moisture content and the maximum dry density, were determined in the Proctor apparatus at the compaction energy of 0.59 J·cm^{-3} [39]. Shear strength parameters, the angle of internal friction and cohesion, were determined in a direct shear apparatus [40]. Samples measuring 60×60 mm were formed directly in the apparatus box at the optimum moisture content to achieve the compaction index $I_S = 1$. The samples were subjected to normal stress of 50, 100, and 200 kPa, and then sheared at a rate of 0.1 mm·min^{-1}. The coefficient of permeability was determined in an oedometer, on samples (diameter—6 cm, height—2 cm) formed at optimum moisture content and a compaction index of $I_S = 1$. The organic matter content was determined by two methods: oxidation and loss on ignition [41].

2.3. Stability and Filtration Calculations

Two methods were used to calculate slope stability: the analytical Bishop's limit equilibrium method and the numerical finite element method. The Bishop's method is a limit equilibrium method, where the potential landslide mass is divided into blocks with vertical walls, for which the resisting and sliding forces are analyzed. The factor of safety refers to the strength parameters of the soil, so it is determined how much the values of the angle of internal friction and cohesion should decrease for the slope to lose stability. It is assumed that, along the potential slip surface, the soil is plastic and meets the Mohr–Coulomb limit condition. The calculations were made using the GeoStudio package—the SLOPE/W software was used for stability, and SEEP/W for filtration [42].

Z_Soil.PC software was used to calculate the stability and filtration, by the finite element method [43]. As a model of soil, the elastic-perfectly plastic Mohr-Coulomb model was used and the method of reducing the shear strength parameters was used to determine the factor of safety. The method of reducing shear strength is based on reducing the values of the angle of internal friction (φ) and cohesion (c) in subsequent calculation steps, maintaining the ratio $FS = \frac{\tan \varphi}{\tan \varphi^F} = \frac{c}{c^F}$ that determines the factor of safety (FS). The values φ^F and c^F are the limit values of the angle of internal friction and cohesion; when they are exceeded, the shearing process starts. The method allows for the determination of the slip surface where the balance between shear stress and shear strength was obtained the fastest. Reducing or increasing the parameters allows for the determination of consecutive failure surfaces and tracking of the destruction process [44]. The stability analysis was carried out for the case of flat deformation (two-dimensional analysis).

Filtration calculations were carried out in the conditions of steady and non-steady filtration. In the case of steady-state filtration, the filtration pressure distribution depends only on the coefficient of permeability of individual materials, and the values of the coefficients themselves determine, in turn, the flow rate—the parameters describing the flow in the unsaturated zone are of secondary importance [45]. In the case of unsteady filtration in the unsaturated zone, a model that takes into account the changes in conductivity related to the presence of the air should be used. The specificity of water flow through the unsaturated zone is related to the time and space variable pressure distribution, which depends directly on the degree to which the pores are filled with water, that is, the volume moisture content. In the unsaturated zone, the pore pressure is lower than atmospheric pressure and has a negative value. The van Genuchten model was used to describe unsteady filtration, which illustrates the relationship between conductivity and the degree of saturation, as well as the relationship between the degree of saturation and the value of negative water pressure [46]. According to van Genuchten [47], the relative permeability coefficient is described by the following equation:

$$k_w(h) = k_s \cdot \frac{\left\{1 - \left(\alpha \cdot h^{(n-1)}\right) \cdot \left[1 + (\alpha \cdot h^n)^{-m}\right]\right\}^2}{\left[(1 + \alpha \cdot h)^n\right]^{\frac{m}{2}}} \tag{1}$$

The value of the permeability coefficient is a function of suction pressure, which, in turn, depends on the volumetric moisture content of the soil:

$$\theta = \theta_r + \frac{(\theta_s - \theta_r)}{\left[1 + (\alpha \cdot h)^n\right]^m}, \tag{2}$$

where k_s (m·s^{-1}) is the saturated permeability coefficient, k_w (m·s^{-1}) is the unsaturated coefficient of permeability, θ (−) is the volumetric moisture content, θ_s (−) is the saturated volumetric moisture content, θ_r (−) is the residual volumetric moisture content, h (cm) is the pressure head in cm, the water column, is α (cm^{-1}), and the n and m constants are ($m = 1 - \frac{1}{n}$).

It was assumed that the bottom sediments and soil in the body of the dam and in the subsoil had the same vertical and horizontal hydraulic parameters. The coefficient of permeability was determined based on the authors' tests, and the parameters of the van Genuchten equation were estimated in the RETC program, based on the geotechnical characteristics of the soils (grain size composition, bulk density) [48–50]. The values of the Young modulus (E) and Poisson ratio (v) do not have an influence on the factor of safety [44], so for each material, the default values were assumed to be E = 100 MPa, ν = 0.3, and the angle of dilatancy, ψ = 0°.

The calculations were made for the assumed embankment structures made of medium sand without seals and with a seal in the form of a core made of bottom sediments from the Rzeszów reservoir. The seal was in the form of a centrally located core width of 1 m in the upper part, enlarged at the base of the embankment, prolonged in the ground in the form of a vertical wall, maintaining the unsealed zone below the wall with a thickness equal to the height of the dam, according to the relevant requirements [51]. The foundation of the embankments was assumed to be on the ground built with permeable medium sands laying on an impermeable layer of clay. Calculations using the Bishop's method were made for two calculation cases: without backwater and with continuous backwater; and using the FEM method, for three calculation cases: without backwater, with continuous backwater, and with variable backwater (dependent on the flood wave). It was assumed that during the passage of a flood wave, the water level rises to the maximum backwater level for two days (t = 2 d), remains at a constant level for three days (t = 5 d), after which the water level falls for two days (t = 7 d). The crest width of the embankments was assumed to be 3 m, with a slopes gradient of 1:2, heights of 4, 6, and 8 m, and position of the water level 1.5 m below the crest (Figure 3). According to the regulations on technical conditions to be met by hydrotechnical structures and their location, which are enforced in Poland [52], the minimum safety factor—regardless of the building class or with detailed tests on the parameters of soil in the body and ground under the basic load system—is FS = 1.3.

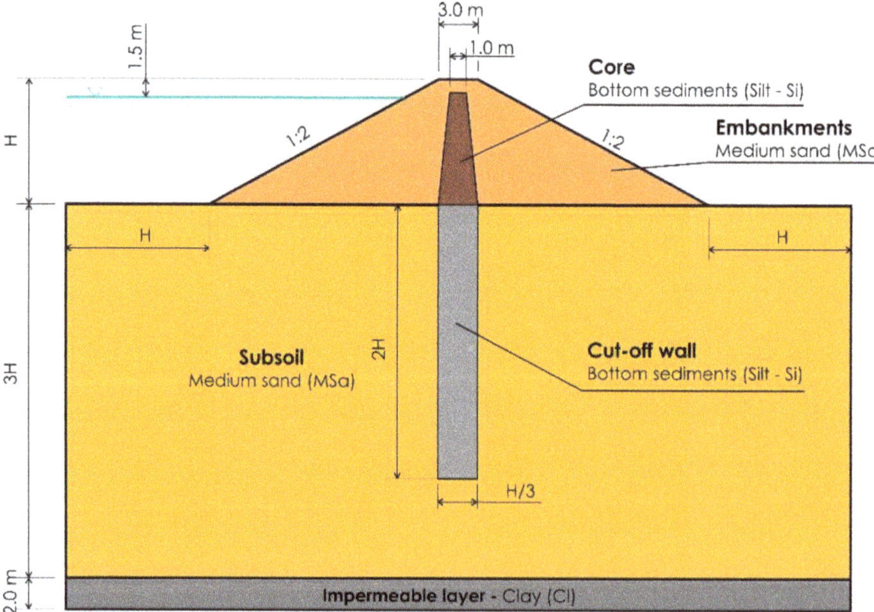

Figure 3. Diagram for stability calculations of the embankment made with medium sand and sealed with a core made of bottom sediments.

3. Results

3.1. Characteristic of Bottom Sediments

Bottom sediments taken from several places of the Rzeszów reservoir were classified as well-grained coarse silts or coarse clayey silts (Figure 4). The grain size composition was dominated by the silt fraction, which was from about 86% to 92%; the sand fraction was from 1% to 2%, and the clay fraction, from 6% to almost 13%. The specific density was from 2.58 to 2.62 g·cm^{-3}.

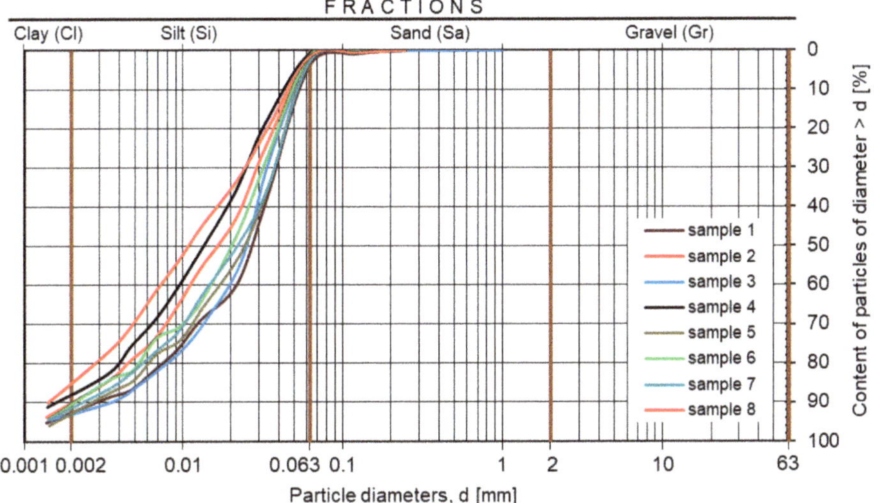

Figure 4. Grain size composition of the bottom sediments from the Rzeszów reservoir.

The high content of silt is related to the location of the reservoir (in the downstream part of the Wisłok river), the predominance of silty soils in the catchment area, and the presence of the Besko reservoir in the upstream part of that river, where coarser fractions are accumulated [53]. In addition, this reservoir was built in the part where Wisłok is a lowland river, so the water flow velocities are low.

The other geotechnical parameters of the sediments were determined for the averaged sample. It was assumed that, during extraction, the sediments would be mixed and, as such, built into earthen embankments.

The material was characterized by an increased content of organic parts; the loss on ignition amounted to nearly 5%, while organic content was over 3%, which should qualify it as a low-organic soil. Sediments are a low permeable material, and the filtration rate decreased from 1×10^{-8} to about 1×10^{-9} m·s^{-1}, with an increase in the compaction index from 0.9 to 1. The liquid limit of the Rzeszów reservoir sediment was above 40%, while the plasticity limit was about 27%, so the material was characterized by a high plasticity index of close to 14%. The maximum dry density was 1.4 g·cm^{-3} at the optimum moisture content of 27%. Sediments at the optimum moisture content were characterized by relatively high values of the angle of internal friction and cohesion. The angle of internal friction increased from 23° to almost 34° and cohesion from 33 to about 36 kPa, respectively [53].

The values of the geotechnical parameters for stability and the filtration calculations were adopted based on the test results (Table 1). The geotechnical parameters of the medium sand, which was in the body of the hydrotechnical embankment, and the clay (impermeable layer) were adopted based on the authors' own tests.

As fine, cohesive soil with a low coefficient of permeability, the Rzeszow reservoir bottom sediments meet most of the Polish criteria [2,54] for sealing elements in hydraulic engineering embankments (Table 2). They were characterized by an appropriate coefficient of permeability and fine particle content. In the case of two parameters—the coefficient of uniformity and organic content—a value slightly lower than required was obtained (the coefficient of uniformity was 13.3, with the desired value of at least 15, and the organic content was 3.3%, with a limit value of 2% or 3%, depending on the source). Therefore, bottom sediment could be used as a construction material in the sealing elements in hydraulic engineering embankments [53].

Table 1. Geotechnical parameters of bottom sediments from the Rzeszów reservoir and soils adopted for stability calculations.

Parameter	Value for Material:		
	Bottom Sediments—Silt (Si), $I_s = 0.95$	Medium Sand (MSa), $I_D = 0.8$	Clay (Cl), $I_L = 0.2$
Angle of internal friction [°]	22.3	35	18
Cohesion [kPa]	42.7	0	30
Unit weight [kN·m^{-3}]	20.1	18.6	21
Coefficient of permeability [m·s^{-1}]	3.4×10^{-9}	1×10^{-4}	1×10^{-10}
Voids ratio [–]	0.843	0.565	0.422

where: IS—compaction index [–], ID—degree of compaction [–], IL—plasticity index [–].

Table 2. Geotechnical parameters of bottom sediments from the Rzeszów reservoir in comparison with the requirements for sealing elements in hydraulic engineering structures.

Parameter	Requirements	Value for the Bottom Sediments
Coefficient of uniformity, C_u	>15	13.3
Clay fraction content, Cl [%]	<30	8.5
Content of particles smaller than 0.005 mm, [%]	$10 \div 25$	19
Content of particles smaller than 0.01 mm, [%]	>25	30
Organic matter content, [%]	$<2 \div 3$	3.33
Coefficient of permeability, [m·s^{-1}]	$<10^{-8}$	1.05×10^{-8} ($I_S = 0.9$) 1.2×10^{-9} ($I_S = 1$)

Gwóźdź [9] has also researched the possibility of using bottom sediments for earth constructions. His research has shown that the sediments from the Rożnów reservoir (Poland) are suitable for mineral sealing layers in landfills. Their coefficient of permeability ranged from 1.1×10^{-9} to 4.7×10^{-10} m·s^{-1}, with an increase in compaction from IS = 0.95 to 1 [9], and the plasticity index was PI = $15 \div 65$, which also indicates the usefulness of this material. The value of compaction index (IS) expresses the ratio of the dry density of soil for moulded samples in laboratory tests to the value of the maximum dry density determined in a Proctor's method. The value of compaction index (IS) expresses the ratio of the dry density of soil for molded samples in laboratory tests (or from the field test) to the value of the maximum dry density determined in a Proctor's method. Whereas the plasticity index (PI) is a measure of the plasticity of a soil and is the size of the range of the moisture contents at which the soil remains plastic.

3.2. Filtration Calculations

In each calculation case in the GeoSlope software for an embankment without a core, seepage occurred on the downstream slope. The phreatic line decreased as a result of using seals in the form of a core made of bottom sediments from the Rzeszow reservoir. Figure 5 shows exemplary pore water pressure values in the body of the embankment and its base with and without the seal. The course of phreatic lines and a significant reduction in water pressure indicate a restriction of the water flow.

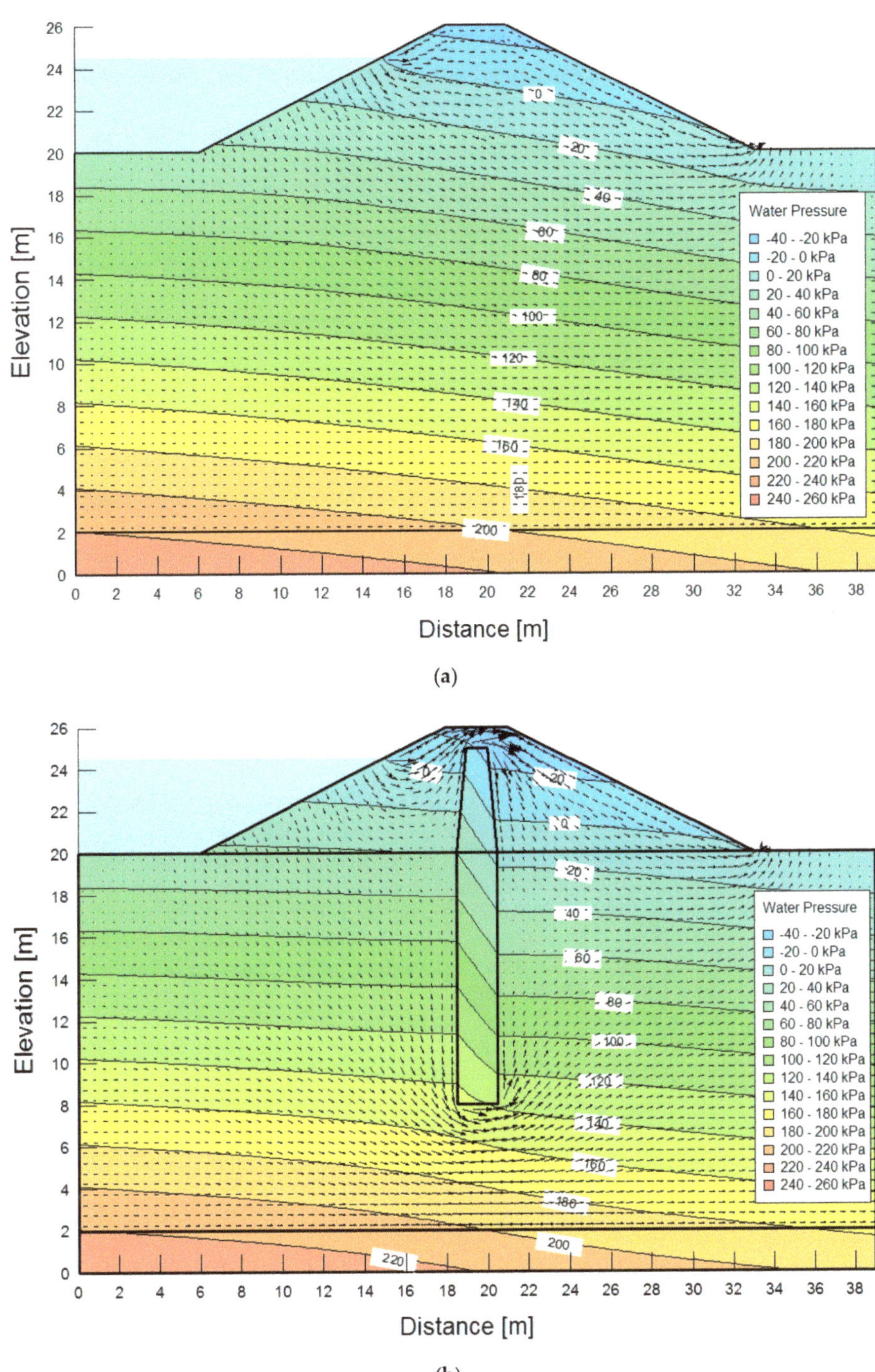

Figure 5. Location of the phreatic line and water pressure distribution in soil pores—embankment with a height of 6 m without a core (**a**) and with a core (**b**) (pressure in kPa).

Calculations of the seepage through the body and base of the embankment using the finite element method showed that the value of the seepage for an embankment without a core was large and ranged from 7.4 to 32.1 $m^3 \cdot (d \cdot m)^{-1}$, depending on the height of the embankment (Table 3). Considerably lower values were obtained for an embankment with a core—from 4.4 to more than 13.2 $m^3 \cdot (d \cdot m)^{-1}$; there was thus an approximately two-fold reduction in the size of seepage.

Table 3. Results of filtration calculations.

Conditions	Embankment	Height of embankment [m]								
		4			6			8		
		Height of maximum water level [m]								
		2.5			4.5			6.5		
		Maximum leakage Q [$m^3 \times (d \cdot m)^{-1}$] for:								
		body	subsoil	total	body	subsoil	total	body	subsoil	total
With continuous backwater	Without a core	1.28	6.13	7.41	3.36	15.92	19.38	7.28	24.82	32.1
	With a core	0.66	3.79	4.45	1.2	8.4	9.59	1.15	12.09	13.24
With variable backwater	Without a core	1.29	6.06	7.35	2.9	16.01	18.91	7.56	24.81	32.36
	With a core	0.79	3.81	4.6	1.28	8.25	9.52	0.88	12.02	12.9

The analysis of seepages for the embankment with and without sealing, and with continuous and variable backwater, shows significant differences in the obtained values (Table 3). Seepages through the downstream slope and the subsoil for an embankment without a core were high and ranged from over 7 to 32 $m^3 \cdot (d \cdot m)^{-1}$, depending on the height of the embankment. The use of seals in the form of a core made of bottom sediments resulted in a lower phreatic line, a significant reduction in seepages, which amounted to more than 4 to 13 $m^3 \cdot (d \cdot m)^{-1}$.

Detailed analysis of seepage through the embankment showed that the differences in its values between the continuous and variable backwater were small (Table 3). These values did not exceed 0.5 and 0.3 $m^3 \cdot (d \cdot m)^{-1}$ adequately with and without the sealing. The same was found for the base of the embankment, but here the differences did not exceed 0.15 $m^3 \cdot (d \cdot m)^{-1}$. On the other hand, there was a significant reduction of the seepage through the body of the embankment and its base with or without a sealing. The seepage through the body decreased by two to six times with an increase of the embankment height from 4 to 8 m. On the other hand, the seepage through the subsoil decreased two times as a result of using sealing, regardless of the height of the embankment. In conclusion, it should be clearly stated that the use of bottom sediments from the Rzeszów reservoir to seal the body of the embankment and its base significantly reduced the size of seepage.

3.3. Stability Calculations

3.3.1. Bishop's Method

The results of stability calculations of the slopes of the analyzed embankments, depending on the backwater (with and without continuous backwater) and the type of embankment (with and without a seal) are shown in Table 4. For the case without backwater, the factor of safety was 1.4 at any height of the embankment, with or without a seal. This follows from the assumption that the slope material is composed of medium sand, for which zero cohesion was assumed in the calculations. For each embankment height, the smallest factor of safety was obtained for the slip surfaces, which were at low depths, located parallel to the edge of the slope, with a surface landslide occurring in each case. The factor of safety was higher than that required (FS $\geq$ 1.3), so, in the absence of backwater, medium sands with the adopted parameters can be used for the construction of embankment slopes with the selected gradient to a height of 8 m.

Table 4. The results of stability calculations using the Bishop's method.

Height of Embankments [m]	Slope	Conditions			
		Without Backwater		With Continuous Backwater	
		Embankment			
		Without a Core	With a Core	Without a Core	With a Core
		Factor of Safety, FS [–]			
4	Upstream	1.403	1.403	1.476	1.451
	Downstream			1.174	1.379
6	Upstream	1.403	1.403	1.611	1.477
	Downstream			0.958	1.321
8	Upstream	1.402	1.402	1.67	1.522
	Downstream			0.866	1.255

The calculation results for the case with continuous backwater show significant differences between the factor of safety of the upstream and downstream slopes, depending on the height of the embankment and whether the seal was used or not (Table 4). These values for upstream slopes increased from 1.48 to 1.67 (embankment without a core), and from 1.45 to 1.52 (embankment with a core), as the height of the embankment increased from 4 to 8 m. For downstream slopes, there was an inverse relationship, and the stability factor decreased from 1.17 to 0.96 and 0.87 (embankment without a core) and from 1.38 to 1.32 and 1.26 (embankment with a core), with the height of the embankment increasing from 4 to 6 and 8 m. As can be seen, the downstream slope of the embankment without a core with heights of 6 and 8 m was unstable, and at a height of 4 m, the factor of safety was less than required.

Based on calculations using the Bishop's method, it can be said that in conditions without backwater, the use of a core made of bottom sediment from the Rzeszow reservoir had no effect on the stability of the embankment. In the case of continuous backwater, an embankment without a core retained stability only at a height of 4 m, wherein the value of the stability factor was less than that required (FS = 1.17). For embankments with a height of 6 and 8 m, the factor of safety of downstream slopes was smaller than 1, so the slope was unstable.

3.3.2. FEM Method

Calculations for the case without backwater indicate that stability is maintained for both an embankment with and without a core at each adopted embankment height. The factor of safety ranged from 1.8 to 1.6 and did not show variability, depending upon the type of the embankment (Table 5). As in the case of the Bishop's method, this results from the parameters of the material (cohesion value of zero), so the yield zones are shallow and parallel to the slope surface.

Calculations for the case with continuous backwater (Table 5, Figures 6 and 7) also indicate that the stability of both embankments was maintained. The potential slip curves for the embankment without a core and with a core, with a height of 8 m, occurred on the downstream side, whereas, in the case of an embankment with a core with heights of 4 and 6 m, the soil yielded as a result of displacement caused by the reduction of the strength parameters. The factor of safety decreased from 1.7 to 1.2 for the embankment without a core and from 1.9 to 1.7 for the embankment with a core, as the height of the embankment increased from 4 to 8 m, respectively. Therefore, the use of a core resulted in a 1.2-fold (on average) increase in the value of the factor of safety.

Table 5. Results of embankment stability calculations using the FEM method.

		Height of the Embankment [m]		
Conditions	Embankment	4	6	8
		Factor of safety, FS [−]		
Without backwater	Without a core	1.83	1.68	1.63
	With a core	1.76	1.68	1.63
With continuous backwater	Without a core	1.72	1.43	1.22
	With a core	1.88	1.81	1.67
With variable backwater — At maximum backwater (t = 5 d)	Without a core	1.72	1.43	1.22
	With a core	1.89	1.81	1.68
With variable backwater — After the drop of the flood wave (t = 7 d)	Without a core	2.12	1.97	1.88
	With a core	2.07	1.98	1.74

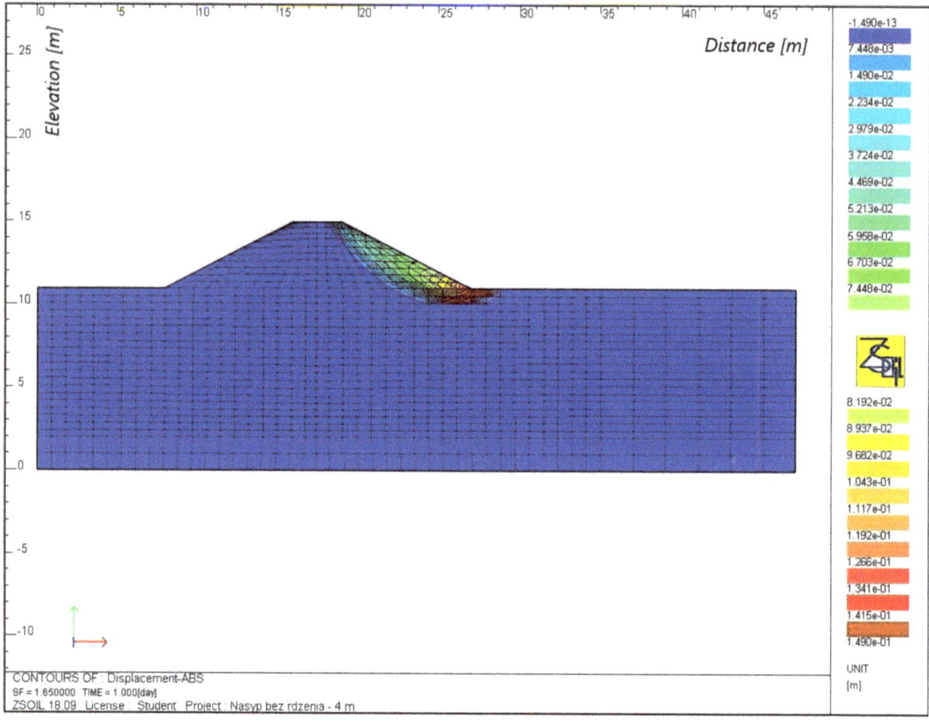

(a)

Figure 6. *Cont.*

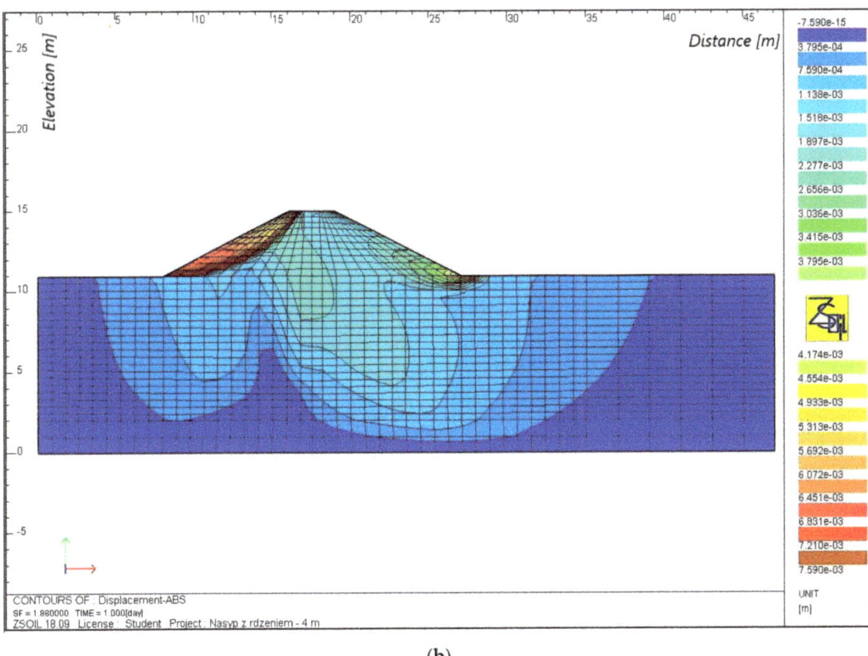

(b)

Figure 6. Movement isolines for FEM grid nodes for an embankment with a height of 4 m, without a core (**a**) and with a core (**b**), with continuous backwater.

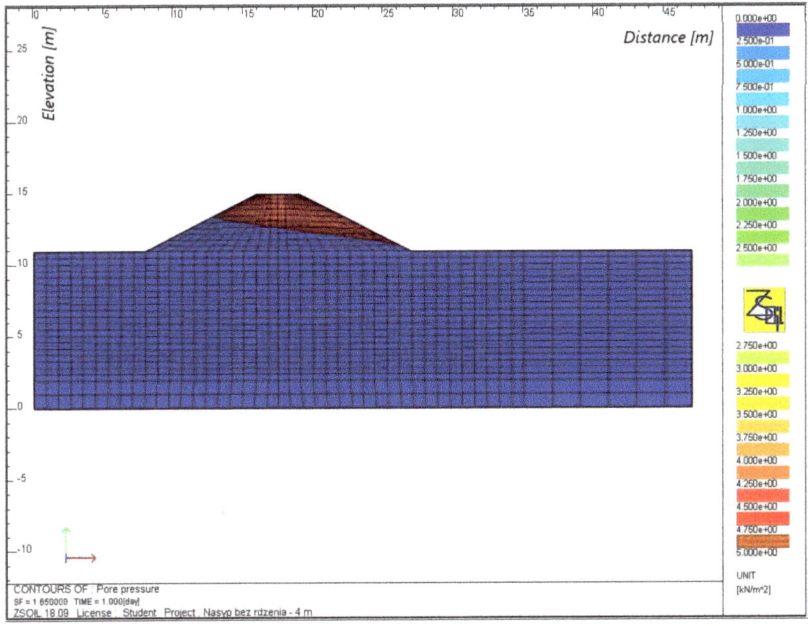

(a)

Figure 7. *Cont.*

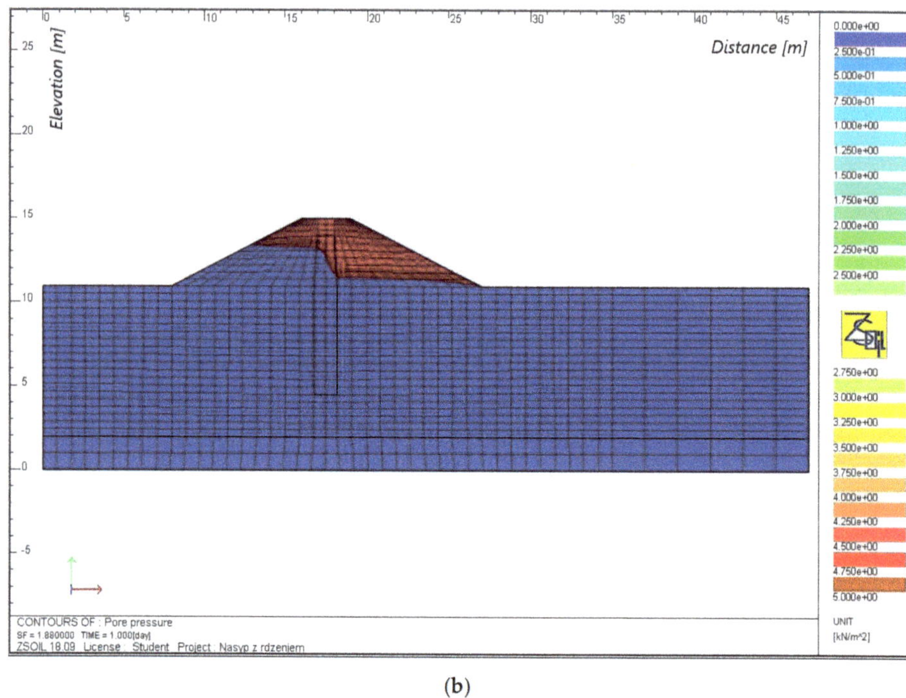

(**b**)

Figure 7. The phreatic line for an embankment with a height of 4 m, without a core (**a**) and with a core (**b**), with continuous backwater.

Calculations for the case with a passing of a flood wave (Table 5, Figure 8) indicate that both embankments maintained stability. For the calculations carried out at a maximum backwater (t = 5 d), the potential slip curves occurred as in the case with continuous backwater, with the exception of an embankment with a core, with a height of 4 m, where the movement occurred on both sides of the embankment. The results of the calculations after the water level drops indicate that the potential slip curves for an embankment with a core appear on the upstream slope, and for an embankment without a core, on both sides of the embankment. After 3 days of maximum backwater (t = 5 d), there was a steady filtration; hence, the factors of safety at the maximum backwater (t = 5 d) were the same as for the case with continuous backwater, while in the descending phase (t = 7 d) they decreased from about 2.1 to 1.9 (embankment without a core) and from 2.1 to 1.7 (embankment with a core), as the height of the embankment increased from 4 to 8 m. The size of seepage on the downstream side was slightly lower than for the conditions of continuous backwater (Table 5).

In the case of calculations simulating the passage of a flood wave, the stability of the embankment was similar as in the case of continuous backwater, while in the descending phase, the values of the stability factor for upstream slopes in each calculation case met the requirement of FS $\geq$ 1.3 [52]. Seepage on the downstream side of the embankment was similar as in the case of continuous backwater, so the use of a core had a corresponding effect on the stability conditions of these embankments.

Based on these results. it can be stated that in conditions without backwater, both an embankment without a core and an embankment with a core maintain stability. The resulting stability factors were high and comparable for the two types of embankments adopted; the use of a core therefore did not affect the conditions of stability. In the case of calculations with continuous backwater, lower values of the factor of safety were obtained, which, however, guaranteed the stability of the slopes and were higher than the value required by [52] for the Classes III and IV of embankments.

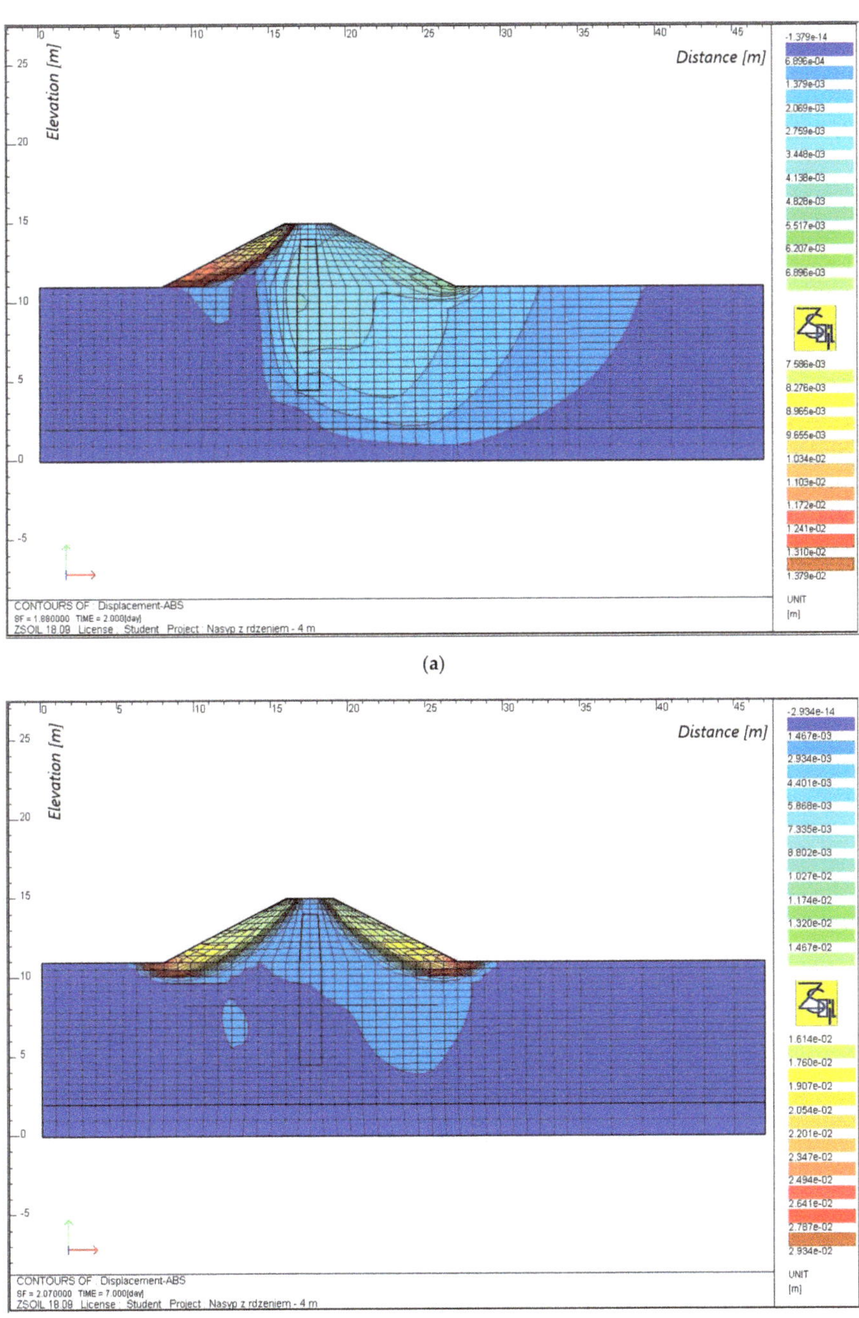

(a)

(b)

Figure 8. Movement isolines for FEM grid nodes at maximum backwater (**a**) and after the falling of the wave (**b**), for an embankment with a height of 4 m, with a core made of sediment at the passing of a flood wave.

3.4. Summary and Discussion of the Results

The values of the factor of safety of slopes that were determined using numerical methods were generally consistent with those obtained using the limit equilibrium method. In all calculation cases, the values of the factor of safety that were obtained using the analytical method were smaller than those obtained using the numerical method; it can thus be stated that this method provides a greater safety margin. It was a non-cohesive slope, so the critical failure surface in the Bishop's method was shallow, and the factor of safety was lower. The biggest differences occurred in the case of an embankment without a core with heights of 6 and 8 m with continuous backwater; according to the analytical method, the slopes in this embankment were unstable, and in the numerical method, the factors of safety were 1.4 and 1.2 respectively. These differences arose from different assumptions made in the above methods. In the analytical method, the loss of stability for the assumed slip curve occurs if the forces maintaining the balance of the rigid body are smaller than the sliding forces (a component of the weight of the given block, tangential to the slip surface, is greater than the friction and cohesion of the soil), while in the numerical method, the determination of the yielding point results from the analysis of stress distribution in the given soil medium. The advantage of the Bishop's method is the speed and simplicity of the calculations, and in the case of simple structures, such as hydraulic engineering embankments in the technical Classes III and IV, it may be successfully used to determine stability conditions. On the other hand, finite element calculations do not require the assumption of the shape of the slip surface in advance, and allow the determination of the development of plastic strain with the progressive reduction of strength parameters, until the formation of a plastic strain zone, which determines the course and shape of the potential slip surface.

Dams of reservoirs are an obstacle to mineral material transported by rivers. This causes the development of the landing processes of these reservoirs. The high degree of overloading of the Rzeszów reservoir prompted the authors to carry out research aimed at recognizing the geotechnical properties of the sediments deposited in the reservoir. In the available literature, there are few publications on research related to the possibility of using bottom sediments for earth construction purposes. Research on the properties of bottom sediments is most often carried out in the aspect of assessing the content and migration of heavy metals [55,56] or organic pollutants [57,58]. On the other hand, the potential use of bottom sediments is seen in agriculture for the fertilization and reclamation of post-industrial areas [22–24,59]. Therefore, the problem presented in the article is important from the point of view of environmental engineering. The authors carried out a series of tests and numerical calculations in terms of the use of bottom sediments for earth construction purposes. The performed calculations indicate the potential possibilities of using bottom sediments in applications related to hydrotechnical construction.

4. Conclusions

Based on the suitability analysis of bottom sediments from the Rzeszow reservoir for civil engineering, it was concluded that this material can be used for sealing elements in hydraulic engineering embankments. The obtained values of geotechnical parameters of the sediments show that this material fulfills most of the criteria required to use them in sealing layers. The calculations of stability and filtration through the embankments in this regard confirmed the usefulness of the analyzed sediments. An embankment made of medium sand, with a core made of bottom sediments maintained stability at a height of 8 m; the values of the stability factor in each calculation case were greater than that required by regulation. By using a core made of bottom sediment, the volume of seepage on the downstream side during continuous or variable backwater was significantly lower in relation to an embankment without a core, and the phreatic line did not extend to the downstream slope.

It is estimated that in the case of a planned dredging in the Rzeszow reservoir, the amount of dredged sediments could exceed 1.5 million m^3; therefore, the possibility of their economic use, as indicated above, is essential. The tests and calculations carried out allowed the determination of the possible use of bottom sediments in civil engineering. As a result, the sediments could provide an alternative to natural soil, whose resources are limited and non-renewable. The search for materials that could replace natural soil in earthen structures is a very important issue from both the ecological and economic points of view.

Author Contributions: Conceptualization, K.K., A.G. and E.Z.; methodology, K.K.; software, K.K.; validation, A.G. and K.K.; formal analysis, K.K. and A.G.; investigation, A.G. and K.K.; resources, K.K.; data curation, K.K.; writing—original draft preparation, K.K. and A.G.; writing—review and editing, A.G. and K.K.; visualization, K.K. and A.G.; supervision, A.G.; project administration, A.G.; funding acquisition, A.G. All authors have read and agreed to the published version of the manuscript.

Funding: This research received no external funding.

Institutional Review Board Statement: Not applicable.

Informed Consent Statement: Not applicable.

Conflicts of Interest: The authors declare no conflict of interest.

References

1. Sobczak, J. *Zapory z Materiałów Miejscowych*; PWN: Warszawa, Poland, 1975.
2. Flores-Berrones, R.; López-Acosta, P. Internal Erosion due to Water Flow through Earth Dams and Earth Structures. In *Soil Erosion*; Godone, D., Stanchi, S., Eds.; IntechOpen: Rijeka, Croatia, 2011; pp. 283–306. [CrossRef]
3. U.S. Society on Dams. *Materials for Embankment Dams; Prepared by the USSD Committee on Materials for Embankment Dams*; U.S. Society on Dams: Denver, CO, USA, 2011.
4. Bednarczyk, S.; Jarzębińska, T.; Mackiewicz, S.; Wołoszyn, E. *Vademecum ochrony przeciwpowodziowej*; Krajowy Zarząd Gospodarki Wodnej: Gdańsk, Poland, 2006.
5. Park, J.; Son, Y.; Noh, S.; Bong, T. The suitability evaluation of dredged soil from reservoirs as embankment material. *J. Environ. Manag.* **2016**, *183*, 443–452. [CrossRef]
6. Ipsita, P.; Surabhi, J.; Sarat, K.; Jayabalan, R. Characterization of red mud as a structural fill and embankment material using bioremediation. *Int. Biodeterior. Biodegrad.* **2017**, *119*, 368–376. [CrossRef]
7. Zydroń, T.; Gruchot, A. Influence of Moisture and Compaction on Shear Strength and Stability of Embankments from Ash-Slag Mixture. (In Polish: Wpływ wilgotności i zagęszczenia na wytrzymałość na ścinanie popioło-żużli i stateczność budowanych z nich nasypów). *Annu. Set Environ. Prot. Rocznik Ochrony Środowiska* **2014**, *16*, 498–518.
8. Gruchot, A. *Utilisation of Coal Mining Wastes and Fuel Ashes for Engineering Purposes as a Factor of Environmental Development and Protection. (In Polish: Utylizacja Odpadów Powęglowych i Poenergetycznych do Celów Inżynierskich Jako Czynnik Kształtowania i Ochrony środowiska)*; Zeszyty Naukowe (Rozprawy), 410; Uniwersytet Rolniczy im. Hugona Kołłątaja: Kraków, Poland, 2016; p. 533.
9. Gwóźdź, R. Geotechnic properties of sediment deposited in the Rożnowskie lake and the possibility of using it in municipal waste disposal site soil construction. (In Polish: Właściwości geotechniczne osadów zdeponowanych w jeziorze rożnowskim oraz możliwości ich wykorzystania do budowy przesłon mineralnych w składowiskach odpadów komunalnych). *Czasopismo Techniczne-Środowisko* **2008**, *1*, 13–23.
10. Dysarz, T.; Wicher-Dysarz, J. Analysis of Flow Conditions in the Stare Miasto Reservoir Taking into Account Sediment Settling Properties. *Annu. Set Environ. Prot. Rocznik Ochrony Środowiska* **2013**, *15*, 584–605.
11. Łajczak, A. *Studium nad Zamulaniem Wybranych Zbiorników Zaporowych w Dorzeczu Wisły*; Monografie Komitetu Gospodarki Wodnej PAN; Oficyna Wydawnicza PWN, Zeszyt 8: Warszawa, Poland, 1995.
12. Batuca, G.D.; Jordaan, M.J. *Silting and Desilting of Reservoirs*; A.A. Balkema: Rotterdam, The Netherlands, 2000.
13. Journal of Laws. Dz.U. z 2014 r., poz. Rozporządzenie Ministra Środowiska z dnia 9 grudnia 2014 r. w sprawie katalogu odpadów. 1923. Available online: https://dziennikustaw.gov.pl/DU/rok/2014/pozycja/1923 (accessed on 10 September 2020).
14. Journal of Laws. Dz.U. z 2013 r., poz. 21. Ustawa z dnia 14 grudnia 2012 r. o odpadach z późniejszymi zmianami. Available online: https://dziennikustaw.gov.pl/DU/rok/2013/pozycja/21 (accessed on 10 September 2020).
15. Sojka, M.; Siepak, M.; Gnojska, E. Assessment of heavy metal concentration in bottom sediments of Stare Miasto pre-dam reservoir on the Powa River. (In Polish: Ocena zawartości metali ciężkich w osadach dennych wstępnej części zbiornika retencyjnego Stare Miasto na rzece Powie). *Annu. Set Environ. Prot. Rocznik Ochrona Środowiska* **2013**, *15*, 1916–1928.
16. Zhang, Y.; Zhang, X.; Bi, Z.; Yu, Y.; Shi, P.; Ren, L.; Shan, Z. The impact of land use changes and erosion process on heavy metal distribution in the hilly area of the Loess Plateau, China. *Sci. Total Environ.* **2020**, *718*. [CrossRef] [PubMed]
17. Augustyniak, R.; Grochowska, J.; Łopata, M.; Parszuto, K.; Tandyrak, R. Characteristics of bottom sediments in polish lakes with different trophic status. In *Polish River Basins and Lakes—Part I. The Handbook of Environmental Chemistry*; Korzeniewska, E., Harnisz, M., Eds.; Springer: Cham, Switzerland, 2020; p. 86. [CrossRef]

18. Baran, A.; Tarnawski, M.; Kaczmarski, M. Assessment of agricultural utilization of bottom sediments from the Besko reservoir. *Czasopismo Techniczne* **2011**, *108*, 3–11.
19. Szara, M.; Baran, A.; Klimkowicz-Pawlas, A.; Tarnawski, M. Ecotoxicological characteristics and ecological risk assessment of trace elements in the bottom sediments of the Rożnów reservoir (Poland). *Ecotoxicology* **2020**, *29*, 45–57. [CrossRef] [PubMed]
20. Szal, D.; Gruca-Rokosz, R. Anaerobic oxidation of methane in freshwater sediments of Rzeszów reservoir. *Water* **2020**, *12*, 398. [CrossRef]
21. Wiśniowska-Kielian, B.; Niemiec, M. Effect of bottom sediment addition to the substratum on the quality of produced maize biomass. *Ecol. Chem. Eng.* **2007**, *14*, 581–589.
22. Fonseca, R.M.; Barriga, F.; Fyfe, W.S. Reversing desertification by using same reservoir sediments as agriculture soils. *Episodes* **1998**, *21*, 218–224. [CrossRef]
23. Canet, R.; Chaves, C.; Pomares, R.; Alibach, R. Agricultural use of sediments from the Albufera Lake (eastern Spain). *Agric. Ecosyst. Environ.* **2003**, *95*, 29–36. [CrossRef]
24. Baran, A.; Jasiewicz, C.; Tarnawski, M. Effect of bottom desposit on trace element content in light soil. *Ecol. Chem. Eng.* **2010**, *17*, 1553–1561.
25. Tarnawski, M.; Baran, A.; Koniarz, T.; Wyrębek, M.; Grela, J.; Piszczek, M.; Koroluk, A. The possibilities of the environmental use of bottom sediments from the silted inlet zone of the Rożnów Reservoir. *Geol. Geophys. Environ.* **2017**, *43*, 335–344. [CrossRef]
26. Pelczar, J.; Loska, K.; Melaniuk, E. Wpływ nawożenia osadem dennym na aktywność enzymatyczną zwałowiska odpadów węgla kamiennego. *Arch. Ochr. Śr.* **1998**, *24*, 93–101.
27. Koś, K. Charakterystyka geotechniczna osadów dennych cofki Zbiornika Czorsztyńskiego i możliwości ich wykorzystania do celów budownictwa ziemnego. *Inżynieria Morska i Geotechnika* **2013**, *2*, 135–143.
28. Bounouara, Z.; Malab, S.; Mekerta, B.; Benaissa, A.; Bourokba, S.A. Treatment of Dredged Sediments of Bouhanifia Dam for Their Valorization in Passive Barrier of Landfill. *Geotech. Geol. Eng.* **2020**, *38*, 3997–4011. [CrossRef]
29. Koś, K.; Zawisza, E. Stabilization of bottom sediments from Rzeszowski Reservoir. *Ann. Warsaw Univ. Life Sci.—SGGW Land Reclam.* **2015**, *47*, 127–137. [CrossRef]
30. Chiang, K.-Y.; Chien, K.-L.; Hwang, S.-J. Study on the characteristics of building bricks produced from reservoir sediment. *J. Hazard. Mater.* **2008**, *159*, 499–504. [CrossRef]
31. Becker, A.; Bucher, F.; Davenport, C.A.; Flisch, A. Geotechnical characteristics of post-glacial organic sediments in Lake Bergsee, southern Black Forest, Germany. *Eng. Geol.* **2004**, *74*, 91–102. [CrossRef]
32. Kataoka, S.; Yamashita, S.; Kawaguchi, T.; Suzuki, T. The Soil Properties of Lake-Bottom Sediments in the Lake Baikal Gas Hydrate Province. *Soils Found.* **2009**, *49*, 757–775. [CrossRef]
33. Ballas, G.; Garziglia, S.; Sultan, N.; Pelleter, E.; Toucanne, S.; Marsset, T.; Riboulot, V.; Ker, S. Influence of early diagenesis on geotechnical properties of clay sediments (Romania, Black Sea). *Eng. Geol.* **2018**, *240*, 175–188. [CrossRef]
34. Madeyski, M.; Michalec, B.; Tarnawski, M. *Silting of Small Water Reservoirs and Quality of Sediments. (In Polish: Zamulanie Małych Zbiorników Wodnych i Jakość Osadów Dennych)*; Infrastuktura i Ekologia Terenów Wiejskich (Infrastructure And Ecology Of Rural Areas); Polska Akademia Nauk: Kraków, Poland, 2008.
35. Polski Komitet Normalizacyjny. *PN-EN ISO 14688-2:2006. Badania Geotechniczne. Oznaczanie i Klasyfikowanie Gruntów. Część 2: Zasady Klasyfikowania*; Polski Komitet Normalizacyjny: Warszawa, Poland, 2006.
36. Polski Komitet Normalizacyjny. *PN-EN ISO 17892-4:2017-01. Rozpoznanie i Badania Geotechniczne. Badania Laboratoryjne Gruntów. Część 4: Badanie Uziarnienia Gruntów*; Polski Komitet Normalizacyjny: Warszawa, Poland, 2017.
37. Polski Komitet Normalizacyjny. *PN-EN ISO 17892-3:2016-03. Rozpoznanie i Badania Geotechniczne. Badania Laboratoryjne Gruntów. Część 3: Badanie Gęstości Właściwej*; Polski Komitet Normalizacyjny: Warszawa, Poland, 2016.
38. Polski Komitet Normalizacyjny. *PN-EN ISO 17892-12:2018-08. Rozpoznanie i Badania Geotechniczne. Badania Laboratoryjne Gruntów. Część 12: Oznaczanie Granic Płynności i Plastyczności*; Polski Komitet Normalizacyjny: Warszawa, Poland, 2018.
39. Polski Komitet Normalizacyjny. *PN-EN 13286-2:2010. Mieszanki Niezwiązane i Związane Hydraulicznie. Część 2: Metody Badań Laboratoryjnych Gęstości na Sucho i Zawartości Wody. Zagęszczanie Metodą Proktora*; Polski Komitet Normalizacyjny: Warszawa, Poland, 2010.
40. Polski Komitet Normalizacyjny. *PN-EN ISO 17892-10:2019-01. Rozpoznanie i Badania Geotechniczne. Badania Laboratoryjne Gruntów. Część 10: Badania w Aparacie Bezpośredniego Ścinania*; Polski Komitet Normalizacyjny: Warszawa, Poland, 2019.
41. Polski Komitet Normalizacyjny. *PN-B-04481:1988. Grunty Budowlane. Badania Próbek Gruntu*; Polski Komitet Normalizacyjny: Warszawa, Poland, 1988.
42. GeoStudio. *Stability Modeling with GeoStudio*; GEO-SLOPE International LTD.: Calgary, AB, Canada, 2020.
43. Truty, A.; Podles, K. *User Developments in ZSoil®*; ZSoil®.PC 200101 Report; Zace Services Ltd., Software Engineering: Lausanne, Switzerland, 2020.
44. Cała, M.; Flisiak, J.; Tajduś, A. Numerical methods of slope stability analysis (In Polish: Numeryczne metody analizy stateczności skarp i zboczy). In *Materiały Sympozjum Warsztaty Górnicze z Cyklu "Zagrożenia Naturalne w Górnictwie", Bełchatów*; Polska Akademia Nauk. Instytut Gospodarki Surowcami Mineralnymi i Energią: Bełchatów, Poland, 2004; pp. 34–50.
45. Baran, P.; Grodecki, M. Strength and strain parameters of colliery spoils in the light of laboratory tests and numerical analyses. (In Polish: Parametry wytrzymałościowe i odkształceniowe odpadów powęglowych w świetle badań laboratoryjnych i analiz numerycznych). *Wydawnictwo Politechniki Śląskiej Zeszyty Naukowe Budownictwo* **2003**, *98*, 15–22.

46. Żurek, A.; Czop, M. Modelling of water flow and precipitation chemistry transformation during infiltration process in the lysimeter experiment. (In Polish: Modelowanie warunków przepływu i przekształceń składu chemicznego wód opadowych w trakcie procesu infiltracji, na przykładzie doświadczenia lizymetrycznego). *Biuletyn Państwowego Instytutu Geologicznego Hydrogeologia* **2010**, *442*, 181–188.

47. Van Genuchten, M.T. A closed form equation for predicting the hydraulic conductivity of unsaturated soils. *Am. Soc. Soil Sci.* **1980**, *44*, 892–898. [CrossRef]

48. Arya, L.M.; Paris, J.F. A physico-empirical model to predict the soil moisture characteristic from particle-size distribution and bulk density data. *Soil Sci. Soc. Am. J.* **1981**, *45*, 1023–1030. [CrossRef]

49. Aubertin, M.; Mbonimpa, M.; Bussire, B.; Chapuis, R.P. A model to predict the water retention curve from basic geotechnical properties. *Can. Geotech. J.* **2003**, *40*, 1104–1122. [CrossRef]

50. Yang, X.; You, X. Estimating Parameters of Van Genuchten Model for Soil Water Retention Curve by Intelligent Algorithms. *Appl. Math. Inf. Sci.* **2013**, *7*, 1977–1983. [CrossRef]

51. Borys, M.; Zawisza, E.; Gruchot, A.; Chmielowski, K. River embankments. In *Open Channel Hydraulics, River Hydraulic Structures and Fluvial Geomorphology: For Engineers, Geomorphologists and Physical Geographers*; Radecki-Pawlik, A., Pagliara, S., Hradecký, J., Eds.; CRC Press: Florida, USA, 2018; pp. 133–167.

52. Journal of Laws. Dz.U. z 2007 r., nr 86, poz. 579. Rozporządzenie Ministra Środowiska z dnia 20 kwietnia 2007 r. w sprawie warunków technicznych, jakim powinny odpowiadać budowle hydrotechniczne i ich usytuowanie. 2007. Available online: https://dziennikustaw.gov.pl/DU/2007/s/86/579 (accessed on 10 September 2020).

53. Koś, K.; Zawisza, E. Geotechnical characteristics of bottom sediments from Rzeszowski reservoir. (In Polish: Charakterystyka geotechniczna osadów dennych zbiornika Rzeszowskiego). *J. Civil Eng. Environ. Archit. Czasopismo Inżynierii Lądowej Środowiska i Architektury* **2015**, *62*, 195–208. [CrossRef]

54. Polski Komitet Normalizacyjny. *PN-B-12095:1997. Urządzenia Wodno-Melioracyjne. Nasypy. Wymagania i Badania Przy Odbiorze*; Polski Komitet Normalizacyjny: Warszawa, Poland, 2016.

55. Szarek-Gwiazda, E.; Czaplicka-Kotas, A.; Szalińska, E. Background Concentrations of Nickel in the Sediments of the Carpathian Dam Reservoirs (Southern Poland). *Clean Soil Air Water* **2011**, *39*, 368–375. [CrossRef]

56. Kulbat, E.; Sokołowska, A. Methods of Assessment of Metal Contamination in Bottom Sediments (Case Study: Straszyn Lake, Poland). *Arch. Environ. Contam. Toxicol.* **2019**, *77*, 605–618. [CrossRef] [PubMed]

57. Mazurkiewicz, J.; Mazur, A.; Mazur, R.; Chmielowski, K.; Czekała, W.; Janczak, D. The Process of Microbiological Remediation of the Polluted Słoneczko Reservoir in Poland: For Reduction of Water Pollution and Nutrients Management. *Water* **2020**, *12*, 3002. [CrossRef]

58. Ziemińska-Stolarska, A.; Imbierowicz, E.; Jaskulski, M.; Szmidt, A. Assessment of the Chemical State of Bottom Sediments in the Eutrophied Dam Reservoir in Poland. *Int. J. Environ. Res. Public Health* **2020**, *17*, 3424. [CrossRef] [PubMed]

59. Fonseca, R.; Barriga, F.J.A.S.; Fyfe, W.S. Suitability for Agricultural Use of Sediments from the Maranhão Reservoir, Portugal. In *Optimization of Plant Nutrition*; Developments in Plant and Soil Sciences, 53; Fragoso, M.A.C., Van Beusichem, M.L., Houwers, A., Eds.; Springer: Dordrecht, The Netherlands, 1993. [CrossRef]

Article

Artificial Neural Network Optimized with a Genetic Algorithm for Seasonal Groundwater Table Depth Prediction in Uttar Pradesh, India

Kusum Pandey [1], Shiv Kumar [2], Anurag Malik [3,*] and Alban Kuriqi [4,*

[1] Department of Soil and Water Engineering, Punjab Agricultural University, Ludhiana 141004, Punjab, India; khushipandey166@gmail.com
[2] Department of Irrigation and Drainage Engineering, College of Technology, G.B. Pant University of Agriculture & Technology, Pantnagar 263145, Uttarakhand, India; shivacte@gmail.com
[3] Punjab Agricultural University, Regional Research Station, Bathinda 151001, Punjab, India
[4] CERIS, Instituto Superior Técnico, Universidade de Lisboa, Av. Rovisco Pais 1, 1049-001 Lisbon, Portugal
* Correspondence: amalik19@pau.edu (A.M.); alban.kuriqi@tecnico.ulisboa.pt (A.K.)

Received: 8 September 2020; Accepted: 24 October 2020; Published: 27 October 2020

Abstract: Accurate information about groundwater level prediction is crucial for effective planning and management of groundwater resources. In the present study, the Artificial Neural Network (ANN), optimized with a Genetic Algorithm (GA-ANN), was employed for seasonal groundwater table depth (GWTD) prediction in the area between the Ganga and Hindon rivers located in Uttar Pradesh State, India. A total of 18 models for both seasons (nine for the pre-monsoon and nine for the post-monsoon) have been formulated by using groundwater recharge (GW_R), groundwater discharge (GW_D), and previous groundwater level data from a 21-year period (1994–2014). The hybrid GA-ANN models' predictive ability was evaluated against the traditional GA models based on statistical indicators and visual inspection. The results appraisal indicates that the hybrid GA-ANN models outperformed the GA models for predicting the seasonal GWTD in the study region. Overall, the hybrid GA-ANN-8 model with an 8-9-1 structure (i.e., 8: inputs, 9: neurons in the hidden layer, and 1: output) was nominated optimal for predicting the GWTD during pre- and post-monsoon seasons. Additionally, it was noted that the maximum number of input variables in the hybrid GA-ANN approach improved the prediction accuracy. In conclusion, the proposed hybrid GA-ANN model's findings could be readily transferable or implemented in other parts of the world, specifically those with similar geology and hydrogeology conditions for sustainable planning and groundwater resources management.

Keywords: artificial intelligence; Ganga; groundwater; Hindon; statistical indicators; Uttar Pradesh

---

1. Introduction

Groundwater is one of the most vital natural resources. It promotes healthy human life, economic growth, and environmental sustainability. It becomes a reliable source of water in all climatic regions of the world [1]. Due to rapid population growth, industrial development, agricultural activities, and increased domestic use, most of the world's countries will face a freshwater shortage problem [2]. The spatial-temporal variation, discrepancies of groundwater resources, and increased groundwater dependence have also impacted groundwater levels [3,4]. The physical-based model requires explicit knowledge about the study region's physical properties (characterization and quantification), boundary conditions, and big dataset; usually, these aspects are very laborious, costly, and time-consuming [1,5]. To overcome these difficulties, the machine learning-based model has proved the ability to solve large complex problems, including rainfall-runoff modeling [6–8], hydrometeorological drought prediction [9–11], evaporation estimation [12–15], simulation of evapotranspiration [16–18], and prediction of groundwater level [19–26]. Recently, Deb et al. [27] and Den and Kiem [28] explored the contribution of surface water (SW) and groundwater (GW) in a rainfall-runoff relationship simulation in two heterogeneous, semi-arid catchments located in southeast Australia (SEA) under different climatic conditions by using physical-based models. They found highly improved runoff simulation during dry conditions linked with SW–GW in study catchments.

In a related context, extensive application of the Artificial Neural Network (ANN) has been found in water resources engineering, such as modeling of groundwater discharge [29–35], groundwater quality prediction [36–39], and aquifer parameter estimation [40–42]. Also, the hybrid and straightforward versions of ANN have been widely utilized for modeling groundwater fluctuation in confined aquifers, unconfined aquifers, leaky or semi-confined aquifers, and multi-layered aquifer systems [43–52]. Furthermore, different machine learning models optimized with nature-inspired algorithms have been proposed for attaining higher reliability in water resource problems [53–55].

The genetic algorithm (GA) is a stochastic approach capable of solving complex multi-dimensional problems for finding the optimal global solution. In the last 25 years, GA has been successfully explored in the diverse field of engineering [56–61]. Likewise, Dash et al. [62] integrated the ANN with GA (GA-ANN) to predict the groundwater level in the lower Mahanadi river basin of Orissa State, India. The hybrid GA-ANN model results were compared with standalone ANN, optimized with back-propagation, Levenberg–Marquardt, and Bayesian regularization algorithms. They found a better performance with the GA-ANN model of medium and high groundwater level prediction than the other models. Supreetha et al. [63] investigated the GA-ANN model's capability to forecast the groundwater fluctuation in the Udupi District, Karnataka (India). The monthly rainfall and water level records of 10 years were used as input. The results of the analysis revealed that the GA-ANN model outperformed the ANN model. Hosseini and Nakhaei [64] employed the back-propagation network (BP) coupled with GA (GA-BP) in the Shabestar Plain of Iran to predict monthly groundwater levels using hydrometeorological inputs. They found that the GA-BP method performed superiorly to the simple BP method. However, other studies also support the feasibility of the hybrid GA-ANN model for groundwater management problems [65–67].

Consequently, Wibowo et al. [68] proposed hybrid multiple back-propagation neural networks with GA (M-BPNN-GA) to resolve the limitation of a Seasonal Auto-Regressive Integrated Moving Average (SARIMA), Auto-Regressive Integrated Moving Average (ARIMA), Back-Propagation Neural Network (BPNN), and Exogenous Input Nonlinear Auto-Regressive Network (NARX) for predicting the groundwater level in the Dungun River, Malaysia. They found that the M-BPNN-GA improved significantly over the ARIMA/ SARIMA, BPNN, and NARX techniques. Li et al. [69] used the particle swarm optimization (PSO), GA with back-propagation (GA-BP), artificial bee colony with back-propagation (ABC-BP), and BP to project the groundwater level in overexploited arid regions of northwest China. They found a promising performance of the ABC-BP method with double hidden nodes in long-term groundwater fluctuation prediction.

To this end, for socio-economic development in the Gangetic plain of eastern Uttar Pradesh, the groundwater plays a dynamic role. Although agriculture is the primary income source in Uttar Pradesh, farmers are mostly dependent on the groundwater for non-monsoon season irrigation, as stated by the Central Groundwater Board (CGWB) [70]. With the tremendous increase in population growth, this region faces higher freshwater demand, leading to groundwater scarcity during the non-monsoon season. According to the authors' knowledge and reported literature so far, no study has been conducted to examine the potential of a hybrid machine learning model—that is, a genetic algorithm integrated with an artificial neural network (GA-ANN) for predicting the seasonal groundwater table depth fluctuations in the area between the Ganga and Hindon rivers by using various hydrogeological variables. The outcomes of the hybrid GA-ANN models were compared with the traditional GA models based on eight statistical indicators (coefficient of determination (R^2), coefficient of efficiency (CE), correlation coefficient (r), mean absolute deviation (MAD), root mean square error (RMSE), coefficient of variation of error residuals (CVRE), absolute prediction error (APE), and performance index (PI)) and through visual interpretation for optimal utilization of groundwater resources in the study area.

2. Materials and Methods

2.1. Description of the Study Area

The study region is situated between the Hindon River and the Ganga River, consisting of an alluvial cover of the Gangetic plain covering approximately 563,647 hectares with varying elevations from 215 m to 273 m above MSL (mean sea level). The study region lies between the latitude of 28°66′ N and 29°92′ N and longitude of 77°46′ E and 78°02′ E, consisting of 24 blocks: 3 blocks of Saharanpur district (Baliakheri, Nagal, Deoband), 6 blocks of Muzaffarnagar district (Charthawal, Purkaji, Muzaffarnagar, Shahpur, Janpath, Khatauli), 12 blocks of Meerut district (Sardhana, Rohta, Daurala, Mawana, Meerut, Janikurd, Saroorpur, Paricchitgarh, Hastinapur, Rajpura, Kharkhoda, Macchra) and 3 blocks of Ghaziabad district (Muradnagar, Razapur, Bhojpur) of Uttar Pradesh. Figure 1 illustrates the location map study area with all the blocks. The entire study region's climate is sub-tropical monsoon with annual rainfall ranging from 933 mm to 1204 mm.

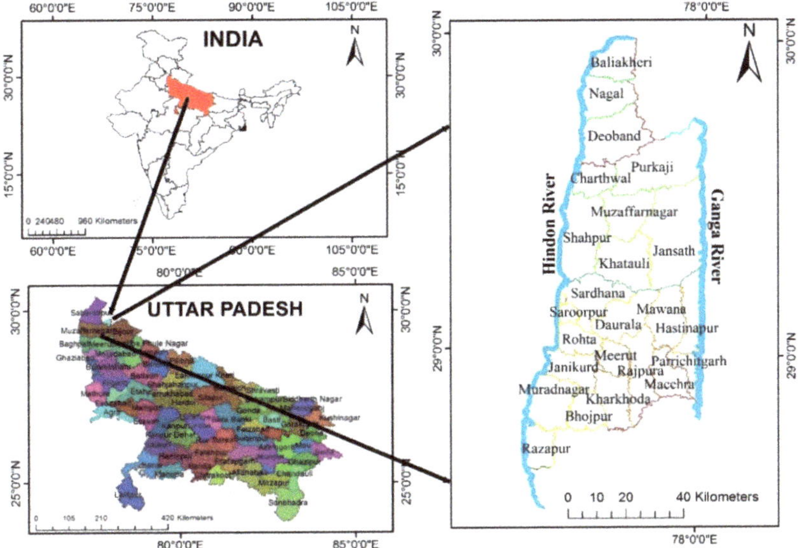

Figure 1. Location Map of the Study Area.

2.2. Hydrogeology of Study Area and Data Acquisition

The study region covers alluvium plain, consisting of sand, silt, clay, and minerals like sodium carbonate, sodium chloride, and sodium sulfate with calcium and magnesium, and detrital traces in varying proportions. Typically, the deposit of the sand bed contains the groundwater in the area. The abstractions of groundwater are utilized for irrigation, drinking, industrial, and domestic purposes in the study area. The groundwater abstraction rate is 942 m^3/h, while the demand is estimated to be between 5069 and 12,672 m^3/h [70]. This study location has heterogeneous types of aquifers, which are divided into three categories:

- Aquifer type group I is composed of different types of basalt rocks, like weathered, dense, and vesicular. Groundwater occurs under water table conditions 30 m or lower than ground level. The tube well discharge varies from 97 to 227 m^3/h for drawdown between 2.68 m and 0.68 m.

- Aquifer type group II is mainly sandstone, siltstone, limestone, and schist. The piezometric head of the horizontal flowing wells lies between 6.63 and 8.92 m above the groundwater level. In the non-flowing wells, it ranges between 1.55 and 11.34 m below the ground level. Most tube wells constructed in this region register a free flow, ranging from 80 to 210 m^3/h. In the non-flowing wells, the discharge head varies from 2 to 8 m.

- Aquifer type group III is composed of gravel, pebbles, grit, sand, clay, etc. Quaternary aquifers belong to this group. Groundwater occurs under unconfined conditions in surface or near-surface aquifers. Water depth varies from 0.2 to 9.7 m.

For this analysis, the daily rainfall data of 21 years (1994–2014) were obtained from the Nazarat district headquarters, while the groundwater table depth (GWTD) data measured at 38 observation wells (Dabthawa, Daurala, Hastinapur, Ganeshpur, Janikurd, Atrada, Kaili, Macchra, Mawana, Pilona, Meerut, Khatki, Paricchitgarh, Sathla, Incholi, Mau Khas, Rajpura, Salawana, Sardhana, Sarrorpur, Surana, Razapur, Charthwal, Kutsera, Jansath, Kakrauli, Barla, Sohanjini, Deoband, Rajupur, Kapoori Govindpur, Nagal, Rohta, Khatauli, Lakhnaur, Amrala, Muradnagar, and Muzaffarnagar) for the pre- and post-monsoon seasons were collected from the Groundwater Department of Uttar Pradesh (India) for the same period. Figure 2 shows the location of observation wells in the study area. The pre-monsoon, post-monsoon, monsoon, and non-monsoon seasons are March–May, October–December, June–September, and January–February, respectively. Figure 3a,b demonstrates the variation of groundwater table depth measured at 38 observation wells during the pre- and post-monsoon seasons. The study region's statistical data were taken from the corresponding district statistical departments regarding the number of minor irrigation structures, the area taken by minor and major crops, area irrigated by minor irrigation structures, and the human population and livestock.

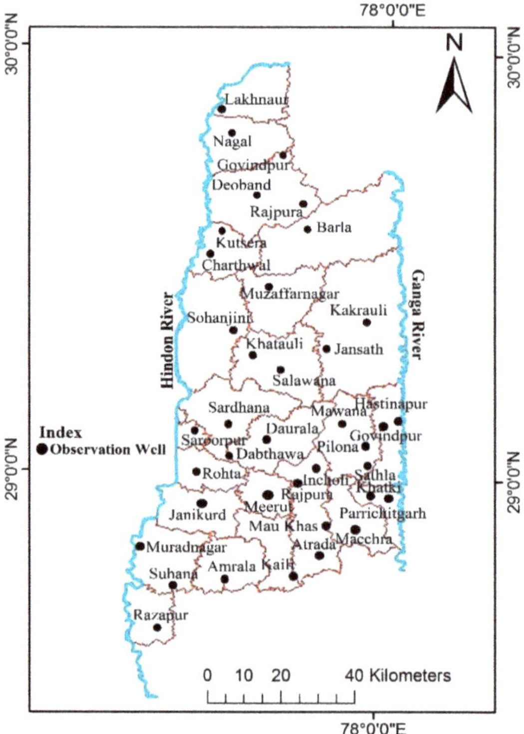

Figure 2. Study Area with 38 Observation Wells.

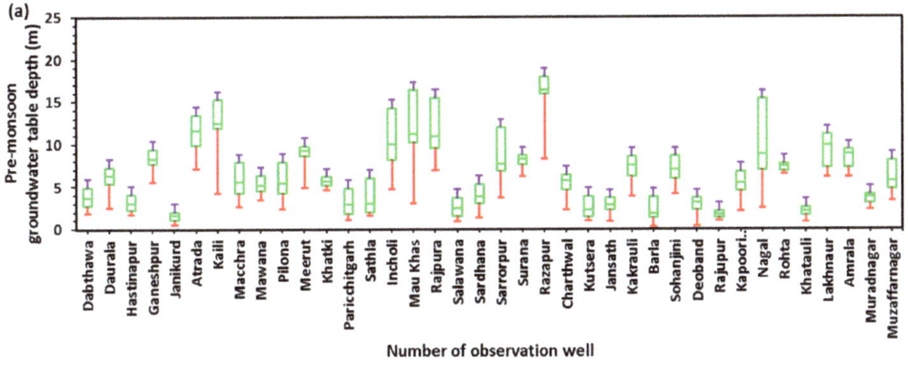

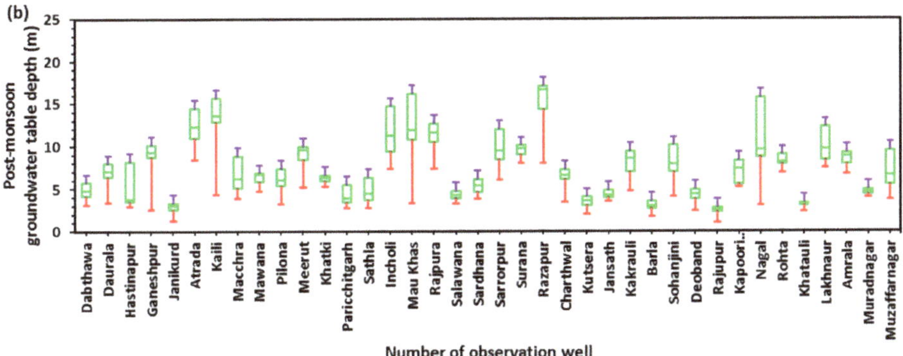

Figure 3. The Fluctuation of Groundwater Table Depth at 38 Observation Wells in (**a**) Pre-Monsoon, and (**b**) Post-Monsoon Seasons, 1994–2014.

2.3. Genetic Algorithm (GA)

The GA is a robust optimization metaheuristic algorithm, driven by natural and biological selection based on Darwin's survival theory [71,72]. The GA has no hypotheses like linearity, stationary, or uniformity, and does not depend on any specific conceptual phenomenon. It involves chromosomes, population set, fitness function, mutation, and selection steps. This provides a set of solutions, named populations, that are governed by chromosomes. The solutions are taken from one population and used to originate a new population, based on the idea that the newly developed population will be better than the older population.

Furthermore, solutions are chosen to develop new solutions (offspring) as per the fitness function. The above procedure will be repeated until the number of offspring in the final population is the same as that equal to the number of parents in the initial population. Two genetic operators are used in these processes: crossover and mutation. In this study, double point crossover and Gaussian mutation operators were used with a crossover and mutation probability of 0.01. Figure 4 shows the general process chart of the genetic algorithm. The necessary steps of the GA are outlined below:

1. Start: Generate chromosomes by random population.
2. Fitness: Determine the fitness function in the populations of every chromosome.
3. New Population: Develop the new population by following the steps that follow until completing the new population.

    (a)   Selection: Based on their fitness, identify two parent chromosomes from a population.

(b)   Crossover: Cross the parents to create a new spring (children), with the possibility of a crossover. When there is no crossover, offspring are the exact duplicate of the parents.

(c)   Mutation: Mutate new offspring at each locus, with the likelihood of mutation.

(d)   Accepting: In the new population, locate new offspring.

4.   Replace: For the further running of the algorithm, use the newly created population.

5.   Test: When the ended conditions are encountered, the current population's best outcome will stop and return.

6.   Loop: Switch back to Step 2.

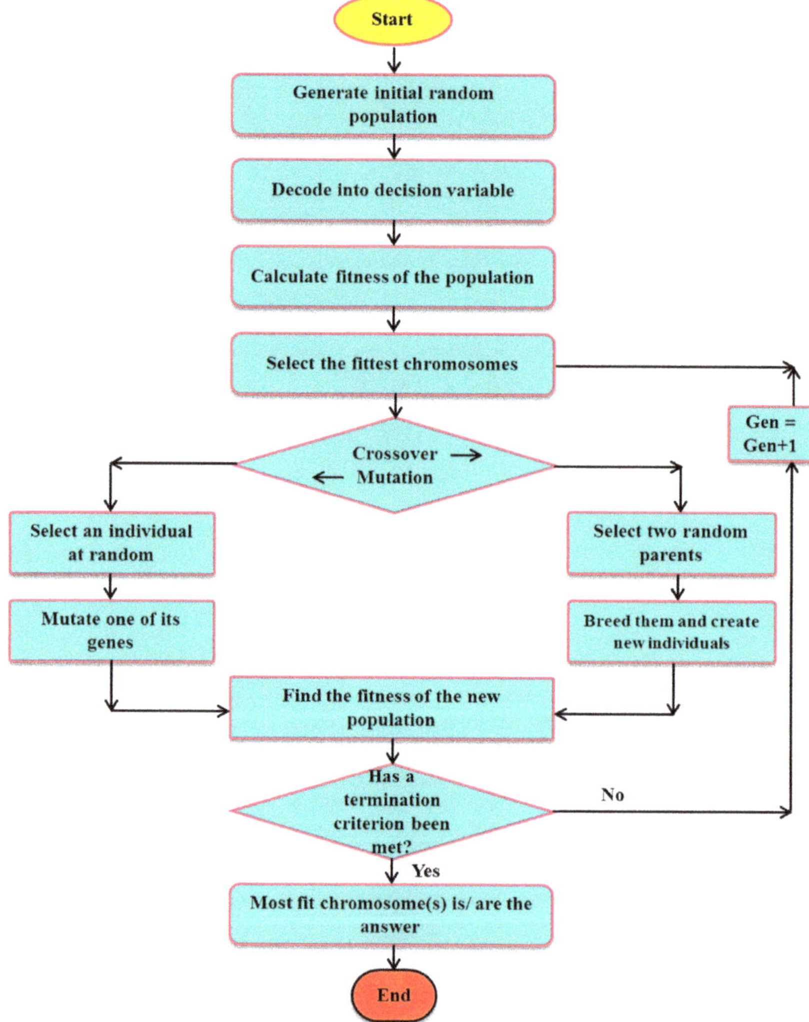

Figure 4. Flow Chart of the Genetic Algorithm (GA) Model.

2.4. Hybrid Genetic Algorithm-Artificial Neural Network (GA-ANN)

In this research, the Hybrid GA-ANN model was developed by incorporating the ANN into a single topology coupled with the GA. The single ANN model suffers from certain drawbacks, such as getting trapped through local minima and slow learning rates. Therefore, optimization algorithms such as the GA with ANN can significantly improve ANN efficiency [73–75] over the aforementioned weaknesses. The integrated GA-ANN strategy fulfills the goal based on two steps:

- The GA technique is used to improve the topology of the ANN and its variables.
- The optimal response is obtained using ANN.

In this study, the GA method was chosen to maximize the optimal number of hidden neurons, weights, and bias values for the ANN models. The GA variables, like crossover likelihood, selection method, mutation rate, size of the population, and the generation numbers, were calculated based on a hit and trial procedure; the details of GA parameters are summarized in Table 1. The flow diagram of the proposed hybrid GA-ANN technique is depicted in Figure 5.

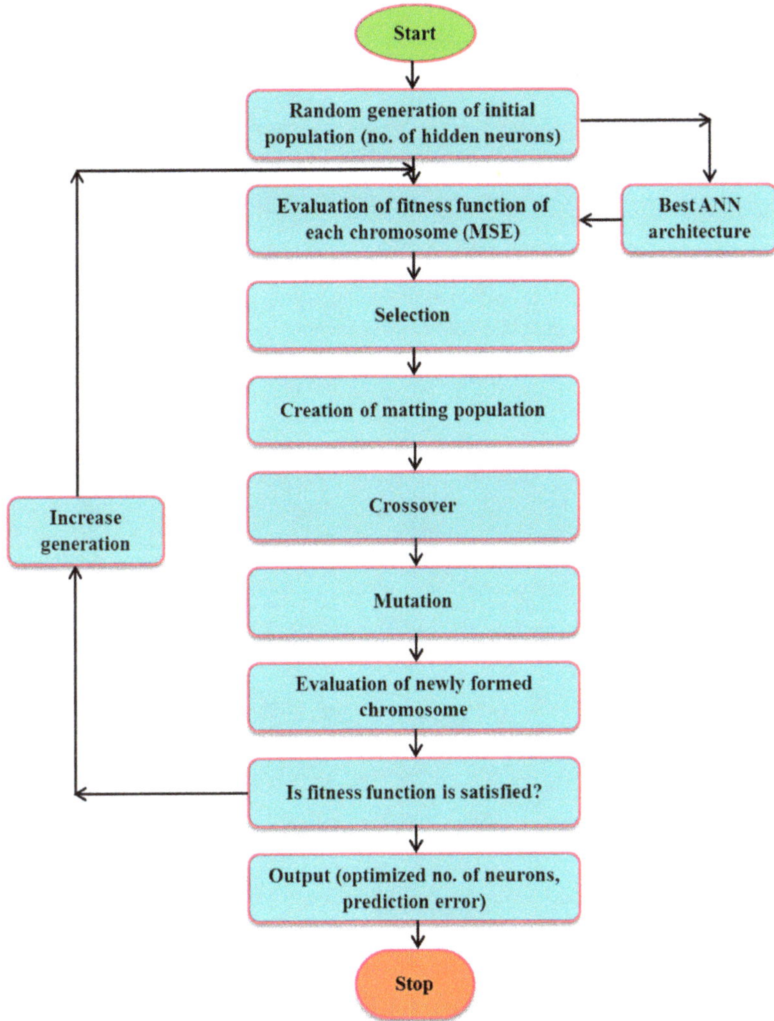

Figure 5. Flow Chart of the Hybrid GA-ANN Model.

Table 1. GA parameters for the optimization of the Artificial Neural Network (ANN) model.

GA Parameter	Type/Value
Population size	50
Population type	Double vector
Population initial range	[3 × 1double]
Selection Mechanism	Roulette wheel
Basis of chromosome selection	Fitness function (MSE)
Crossover type	Double
Crossover Probability	0.8–1.0
Mutation type	Gaussian
Mutation Probability	0.001–0.01
Elite count	2
Migration direction	Forward
Migration interval	20
Time limit	Infinite
Stall Generation limit	Inf
Maximum number of generations	100
Termination Criteria	0.001 m^2
Display	Iteration

2.5. Determination of the Parameters the ANN Model

Architecture: Sets the relationship between a series of inputs and the desired outputs. The ANN with a single hidden node structure has been used to forecast the seasonal GWTD in the study area.

Training algorithm: Training is a process in which iterative modification and optimization techniques are adjusted to update the ANN model parameters such as connection weights and bias values. After several iterations, the training will stop or converge to a specified minimum error rate. In this study, GA optimization techniques were employed to reduce the error between the target and the predictors.

Activation function: Used to convert the input signal to output. In this study, a linear transfer function in the output layer and a logistic sigmoid transfer function in the hidden node were used for ANN models. The functional limitations of the sigmoid logistic factor range between 0 and 1.

Learning rate: The trained mechanism's efficiency is highly vulnerable to the selection of the learning rate. A non-conventional GA optimization method was used to evaluate the favorable learning rate.

Hidden neuron optimization: In general, a hit and trial procedure was utilized to determine the neurons' best-hidden numbers. Few researchers showed the utility of GA for optimizing hidden neurons [76,77]. Hence, in this study, the GA optimization method was utilized to calculate the hidden neurons' optimum numbers.

Error function: Denoted by E, the means square error used for the optimization of the weights and described by Equation (1):

$$E = \frac{1}{n}\left(\sum\nolimits_{i=1}^{n}\left(a_i - p_i\right)^2\right) \tag{1}$$

where a_i is the actual value, p_i is the predicted value, and n is the number of observations.

Weight optimization: In this study, the learning of error correction was used to develop a channel to attain favorable connection weights by reducing the risk of error between the network's actual performance of a neuron and the response targeted from that neuron. The initial range of weights chosen was from 0 to 1 and was then prioritized using the GA technique.

2.6. Development of GA-ANN and GA Models for GWTD Prediction

In this research, a total of 18 models (9 for pre-monsoon and 9 for post-monsoon) were developed with different input parameters (groundwater discharge, groundwater recharge, and antecedent water table depth), as listed in Table 2, for predicting the seasonal GWTD in the study region. The total available data were separated into two classes: (i) training data included from 1994 to 2008 (70%), and (ii) testing data obtained from 2009 to 2014 (30%). In both the seasons, the number of observations varied from 548 to 570 for the training, and 206 to 228 for the testing. Figure 6 illustrates the length of data utilized for the training and testing of the GA and GA-ANN models. The entire ANN and GA modeling exercises were carried out using the MATLAB R2013a software.

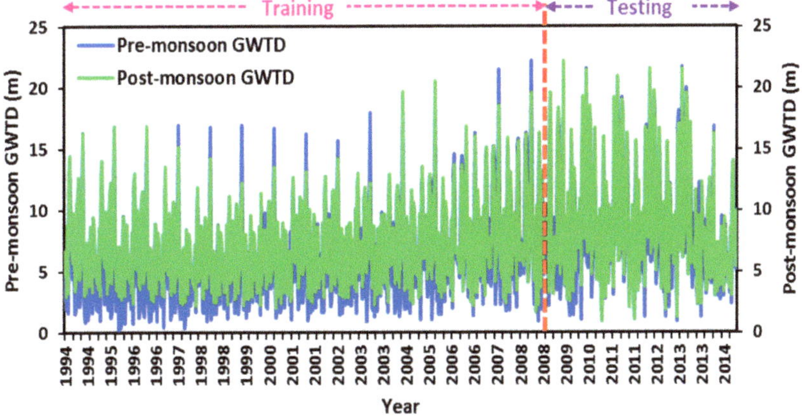

Figure 6. Time-Series Plot of Pre- and Post-Monsoon GWTD in the Study Region.

In Table 2, the groundwater recharge (GW_R) and discharge (GW_D) were computed using Equations (2) and (3) given by the Ministry of Water Resources, India (MWRI) [78]:

$$GW_R = R_r + R_c + R_i + R_g + R_s + R_t \tag{2}$$

$$GW_D = E_s + O_g + E_t + T_p \tag{3}$$

where R_r is recharge due to rainfall, R_c is recharge due to seepage from canals, R_i is recharge due to return flow of irrigation water, R_g is groundwater inflow into the area, R_s is influent seepage from rivers, R_t is recharge due to seepage from tanks and ponds, E_s is effluent seepage to rivers, O_g is groundwater outflow from the area, E_t is evapotranspiration loss from groundwater reservoir, and T_p is groundwater pumpage through wells. The groundwater recharge due to rainfall was estimated using Equation (3) given by [79]:

$$R_r = 3.47(P - 38)^{0.4} \tag{4}$$

In which, R_r is the rainfall penetration (cm), and P is the annual rainfall (cm). The groundwater recharge data due to canal seepage were obtained from the Canal Irrigation Department of Uttar Pradesh's divisional offices. Recharge due to percolation of irrigation water applied to the field took 35% of the total water applied for irrigation, as suggested by Agricultural Refinance and Development Corporation (ARDC) and CGWB, India [70,80]. Seepage from tanks and ponds was calculated based on the Groundwater Department of Uttar Pradesh's norms. The year-wise data of the number of minor irrigation structures present in the study area were collected from the districts' District Statistical Officer. Then, pumpage was determined by multiplying the number of minor irrigation structures by their respective unit drafts [70,80]. In the study region, groundwater withdrawal for industrial habits was taken as 1% of the pumpage through minor irrigation units [70,80]. The groundwater requirement for domestic purposes was taken as 135 L/day/capita. The total groundwater requirement for livestock consumption was assumed to be 10% of the total water requirement, as livestock consumes a significant portion of surface water resources.

Table 2. Parameters of nine developed models for pre- and post-monsoon seasons.

Model	Output	Input
Pre-monsoon season		
1	WT_{pr}	$F[GW_{R(ms,n-1)} + (Nms,n-1\ to\ n),\ GW_{D(ms,n-1)} + (Nms,n-1\ to\ n),\ WT_{pr,n-1}]$
2	WT_{pr}	$F[GW_{R(ms,n-1)} + (Nms,n-1\ to\ n),\ GW_{D(ms,n-1)} + (Nms,n-1\ to\ n),\ WT_{pr,n-1},\ WT_{ps,n-1}]$
3	WT_{pr}	$F[GW_{R(ms,n-1)} + (Nms,n-1\ to\ n),\ GW_{D(ms,n-1)} + (Nms,n-1\ to\ n),\ \Delta WT_{(pr,n-1-ps,n-1)}]$
4	WT_{pr}	$F[GW_{R(Nms,n-1\ to\ n) + (ms,n-1) + (Nms,n-2\ to\ n-1) + (ms,n-2)},\ GW_{D(Nms,n-1\ to\ n) + (ms,n-1) + (Nms,n-2\ to\ n-1) + (ms,n-2)},\ WT_{pr,n-2}]$
5	WT_{pr}	$F[GW_{R(Nms,n-1\ to\ n) + (ms,n-1) + (Nms,n-2\ to\ n-1) + (ms,n-2)},\ GW_{D(Nms,n-1\ to\ n) + (ms,n-1) + (Nms,n-2\ to\ n-1) + (ms,n-2)},\ WT_{pr,n-2},\ WT_{ps,n-2},\ WT_{pr,n-1},\ WT_{ps,n-1}]$
6	WT_{pr}	$F[GW_{R(Nms,n-1\ to\ n) + (ms,n-1) + (Nms,n-2\ to\ n-1) + (ms,n-2)},\ GW_{D(Nms,n-1\ to\ n) + (ms,n-1) + (Nms,n-2\ to\ n-1) + (ms,n-2)},\ \Delta WT_{(pr,n-2-ps,n-2)},\ \Delta WT_{(ps,n-2-pr,n-1)},\ \Delta WT_{(pr,n-1-ps,n-1)}]$
7	WT_{pr}	$F[GW_{R(Nms,n-1\ to\ n) + (ms,n-1) + (Nms,n-2\ to\ n-1) + (ms,n-2) + (Nms,n-3\ to\ n-2) + (ms,n-3)},\ GW_{D(Nms,n-1\ to\ n) + (ms,n-1) + (Nms,n-2\ to\ n-1) + (ms,n-2) + (Nms,n-3\ to\ n-2) + (ms,n-3)},\ WT_{pr,n-3}]$
8	WT_{pr}	$F[GW_{R(Nms,n-1\ to\ n) + (ms,n-1) + (Nms,n-2\ to\ n-1) + (ms,n-2) + (Nms,n-3\ to\ n-2) + (ms,n-3)},\ GW_{D(Nms,n-1\ to\ n) + (ms,n-1) + (Nms,n-2\ to\ n-1) + (ms,n-2) + (Nms,n-3\ to\ n-2) + (ms,n-3)},\ WT_{pr,n-3},\ WT_{ps,n-3},\ WT_{pr,n-2},\ WT_{ps,n-2},\ WT_{pr,n-1},\ WT_{ps,n-1}]$
9	WT_{pr}	$F[GW_{R(Nms,n-1\ to\ n) + (ms,n-1) + (Nms,n-2\ to\ n-1) + (ms,n-2) + (Nms,n-3\ to\ n-2) + (ms,n-3)},\ GW_{D(Nms,n-1\ to\ n) + (ms,n-1) + (Nms,n-2\ to\ n-1) + (ms,n-2) + (Nms,n-3\ to\ n-2) + (ms,n-3)},\ \Delta WT_{(pr,n-3-ps,n-3)},\ \Delta WT_{(ps,n-3-pr,n-2)},\ \Delta WT_{(pr,n-2-ps,n-2)},\ \Delta WT_{(ps,n-2-pr,n-1)},\ \Delta WT_{(pr,n-1-ps,n-1)}]$
Post-monsoon season		
1	WT_{ps}	$F[GW_{R(ms,n)} + (Nms,n-1\ to\ n),\ GW_{D(ms,n)} + (Nms,n-1\ to\ n),\ WT_{ps,n-1}]$
2	WT_{ps}	$F[GW_{R(ms,n)} + (Nms,n-1\ to\ n),\ GW_{D(ms,n)} + (Nms,n-1\ to\ n),\ WT_{ps,n-1},\ WT_{pr,n}]$
3	WT_{ps}	$F[GW_{R(ms,n)} + (Nms,n-1\ to\ n),\ GW_{D(ms,n)} + (Nms,n-1\ to\ n),\ \Delta WT_{(ps,n-1-pr,n)}]$
4	WT_{ps}	$F[GW_{R(ms,n) + (Nms,n-1\ to\ n) + (ms,n-1) + (Nms,n-1\ to\ n-2)},\ GW_{D(ms,n) + (Nms,n-1\ to\ n) + (ms,n-1) + (Nms,n-1\ to\ n-2)},\ WT_{ps,n-2}]$
5	WT_{ps}	$F[GW_{R(ms,n) + (Nms,n-1\ to\ n) + (ms,n-1) + (Nms,n-1\ to\ n-2)},\ GW_{D(ms,n) + (Nms,n-1\ to\ n) + (ms,n-1) + (Nms,n-1\ to\ n-2)},\ WT_{ps,n-2},\ WT_{pr,n-1},\ WT_{ps,n-1},\ WT_{pr,n}]$
6	WT_{ps}	$F[GW_{R(ms,n) + (Nms,n-1\ to\ n) + (ms,n-1) + (Nms,n-1\ to\ n-2)},\ GW_{D(ms,n) + (Nms,n-1\ to\ n) + (ms,n-1) + (Nms,n-1\ to\ n-2)},\ \Delta WT_{(ps,n-2-pr,n-1)},\ \Delta WT_{(pr,n-1-ps,n-1)},\ \Delta WT_{(ps,n-1-pr,n)}]$
7	WT_{ps}	$F[GW_{R(ms,n) + (Nms,n-1\ to\ n) + (ms,n-1) + (Nms,n-1\ to\ n-2) + (ms,n-2) + (Nms,n-3\ to\ n-2)},\ GW_{D(ms,n) + (Nms,n-1\ to\ n) + (ms,n-1) + (Nms,n-1\ to\ n-2) + (ms,n-2) + (Nms,n-3\ to\ n-2)},\ WT_{ps,n-3}]$
8	WT_{ps}	$F[GW_{R(ms,n) + (Nms,n-1\ to\ n) + (ms,n-1) + (Nms,n-1\ to\ n-2) + (ms,n-2) + (Nms,n-3\ to\ n-2)},\ GW_{D(ms,n) + (Nms,n-1\ to\ n) + (ms,n-1) + (Nms,n-1\ to\ n-2) + (ms,n-2) + (Nms,n-3\ to\ n-2)},\ WT_{ps,n-3},\ WT_{pr,n-2},\ WT_{ps,n-2},\ WT_{pr,n-1},\ WT_{ps,n-1},\ WT_{pr,n}]$
9	WT_{ps}	$F[GW_{R(ms,n) + (Nms,n-1\ to\ n) + (ms,n-1) + (Nms,n-1\ to\ n-2) + (ms,n-2) + (Nms,n-3\ to\ n-2)},\ GW_{D(ms,n) + (Nms,n-1\ to\ n) + (ms,n-1) + (Nms,n-1\ to\ n-2) + (ms,n-2) + (Nms,n-3\ to\ n-2)},\ \Delta WT_{(ps,n-3-pr,n-2)},\ \Delta WT_{(pr,n-2-ps,n-2)},\ \Delta WT_{(ps,n-2-pr,n-1)},\ \Delta WT_{(pr,n-1-ps,n-1)},\ \Delta WT_{(ps,n-1-pr,n)}]$

WT is the groundwater table depth, GW_R is the groundwater recharge, GW_D is the groundwater discharge, Pr is the pre-monsoon, Ps is the post-monsoon, n is the number of years, ms is the monsoon season, Nms in the non-monsoon season, and Δ is the change in storage.

2.7. Statistical Indicators

The predictive efficacy of the formulated GA-ANN and GA models was evaluated based on eight statistical measures: coefficient of determination (R^2), coefficient of efficiency (CE), correlation coefficient (r), mean absolute deviation (MAD), root mean square error (RMSE), coefficient of variation of error residuals (CVRE), absolute prediction error (APE) and performance index (PI). These R^2, CE, r, MAD, RMSE, CVRE, APE, and PI, were computed using Equations (5)–(12) as given by [10,81]:

$$R^2 = 1 - \frac{\sum_{i=1}^{n}(a_i - P_i)^2}{\sum_{i=1}^{n}(a_i - a_{avg})^2} \tag{5}$$

$$CE = \frac{\sum_{i=1}^{n}(a_i - a_{avg})^2 - \sum_{i=1}^{n}(a_i - p_i)^2}{\sum_{i=1}^{n}(a_i - a_{avg})^2} \tag{6}$$

$$r = \sum_{i=1}^{n} \frac{(a_i - a_{avg})(p_i - p_{avg})}{(p_i - p_{avg})^2 * (a_i - a_{avg})^2} \tag{7}$$

$$MAD = \frac{1}{n}\sum_{i=1}^{n} p_i - a_i \tag{8}$$

$$RMSE = \sqrt{\frac{1}{n}\left(\sum_{i=1}^{n}(a_i - p_i)^2\right)} \tag{9}$$

$$CVRE = \frac{1}{a_{avg}}\sqrt{\frac{\sum_{i=1}^{n}(a_i - p_i)^2}{n}} \tag{10}$$

$$APE = \frac{\sum_{i=1}^{n} a_i - p_i}{\sum_{i=1}^{n} a_i} \tag{11}$$

$$PI = \frac{\sqrt{\sum_{i=1}^{n}(a_i - p_i)^2}}{\sum_{i=1}^{n} a_i} \tag{12}$$

where a_i is the actual value of GWTD, p_i is the predicted value of GWTD, n is the number of observations, a_{avg} is the average of actual GWTD values, and P_{avg} is the average of predicted GWTD values.

3. Results and Discussion

3.1. Prediction of GWTD Using Traditional GA Method

Firstly, the GA technique was optimized by computing the minimum value of the root mean square error (RMSE) to build the GWTD prediction models. The generation limit values, population size, and the number of binary variables with their lower and upper limits were established by the number of variables in the models. Table 3 shows the GA parameters' values for the nine pre-monsoon season and nine post-monsoon season models, respectively. It was noted from Table 3 that the minimum RMSE was 2.26 for GA-5 with a population size and generation limit of 150 and 200, respectively, for the pre-monsoon season, while the minimum RMSE was 2.73 for post-monsoon season GA-8, with a population size of 100 and generation limit of 150.

Table 4 displays the values of performance or statistical indicators in the pre-monsoon season. The values of performance indicators for the GA-5 model were found to be better during the pre-monsoon season. For this model, in the testing period, the maximum values of coefficient of determination (R^2), coefficient of efficiency (CE), and correlation coefficient (r) were 0.42, 0.33, and 0.65, respectively, while the minimum values of mean absolute deviation (MAD), root mean square error (RMSE), coefficient of variation of error residuals (CVRE), absolute prediction error (APE), and performance index (PI) were 2.14, 5.11, 0.43, 0.23, and 0.03, respectively. Therefore, GA-5 was selected to forecast the pre-monsoon GWTD in the study area. For the testing data set, the observed and predicted GWTD values by GA-1 to GA-9 models during the pre-monsoon season are illustrated in Figure 7, which shows that the predicted values of GWTD in the pre-monsoon season were not in reasonable consistency with the observed GWTD values. From the 228 (38 × 6 (nodes × years)) expected values of GWTD in the pre-monsoon season, only 113 values were ensured a 10 % variation during the testing period.

Table 3. Value of GA model parameters for pre- and post-monsoon seasons.

Model	Population Size	Generation Limit	Minimum RMSE	Generation at Minimum RMSE
Pre-monsoon				
GA-1	50	60	3.95	45
GA-2	50	60	2.42	43
GA-3	100	120	3.18	85
GA-4	150	200	4.44	120
GA-5	150	200	2.26	120
GA-6	150	200	4.57	120
GA-7	100	150	4.65	120
GA-8	150	200	4.70	120
GA-9	150	200	4.01	135
Post-monsoon				
GA-1	100	120	4.41	65
GA-2	100	120	4.54	100
GA-3	150	200	3.45	100
GA-4	100	150	4.03	86
GA-5	100	150	5.89	84
GA-6	100	150	5.49	95
GA-7	150	200	5.56	135
GA-8	150	200	2.73	150
GA-9	150	200	5.31	150

Table 4. Performance indicators of GA models during pre-monsoon season.

Model	Period	R^2	CE	r	MAD	MSE	CVRE	APE	PI
GA-1	Training	0.26	−0.76	0.03	1.01	19.60	0.14	0.14	1.45
GA-1	Testing	0.39	−0.31	0.62	2.63	15.60	0.65	0.33	0.05
GA-2	Training	0.40	−6.19	0.55	5.02	11.23	0.12	0.72	1.28
GA-2	Testing	0.39	0.17	0.62	2.68	5.86	0.48	0.29	0.03
GA-3	Training	0.11	−1.16	0.39	6.01	25.77	0.91	0.08	0.03
GA-3	Testing	0.02	−0.14	0.14	3.9	10.11	0.56	0.42	0.04
GA-4	Training	0.40	0.35	0.03	0.52	9.33	0.75	0.07	0.02
GA-4	Testing	0.39	0.15	0.62	2.67	19.71	0.48	0.29	0.07
GA-5	Training	0.54	0.43	0.60	0.52	4.75	0.10	0.07	0.01
GA-5	Testing	0.42	0.33	0.65	2.14	5.11	0.43	0.23	0.03
GA-6	Training	0.23	0.42	0.54	0.71	7.06	0.40	0.1	0.02
GA-6	Testing	0.15	0.14	0.39	3.65	20.88	0.49	0.39	0.03
GA-7	Training	0.47	0.32	0.50	0.71	8.3	0.43	0.13	0.02
GA-7	Testing	0.36	0.11	0.60	2.34	21.62	0.50	0.25	0.03
GA-8	Training	0.29	0.05	0.40	0.64	11.54	0.52	0.09	0.02
GA-8	Testing	0.34	0.08	0.58	2.57	22.1	0.51	0.28	0.04
GA-9	Training	0.44	0.07	0.05	1.69	11.07	0.47	0.24	0.02
GA-9	Testing	0.39	0.16	0.62	3.29	16.08	0.45	0.36	0.03

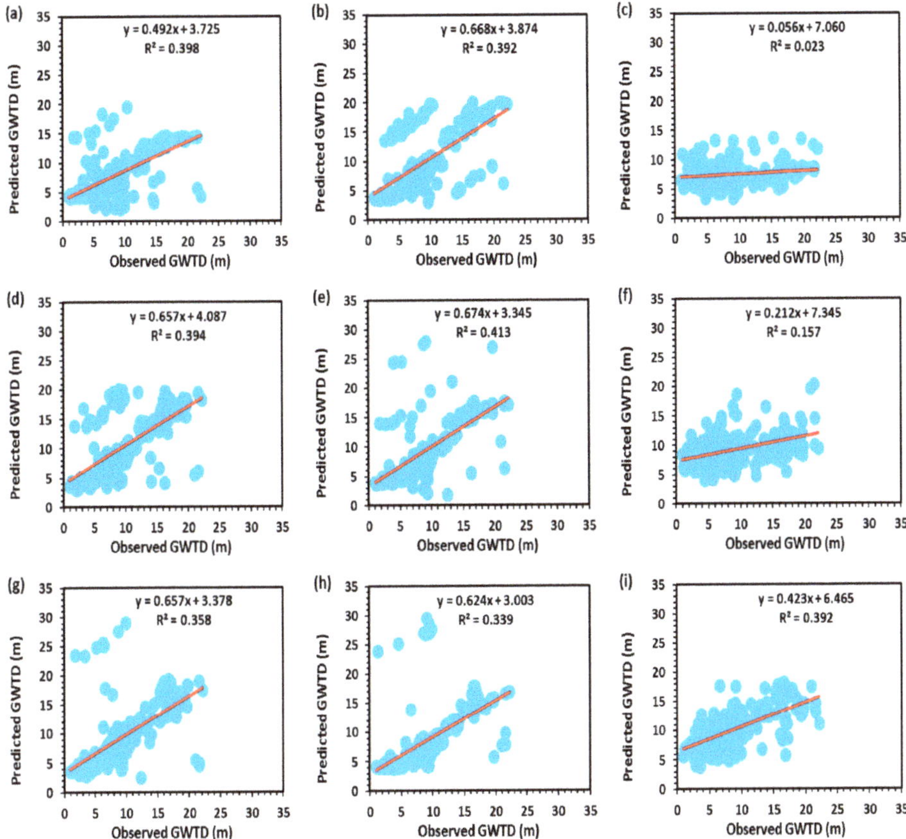

Figure 7. Scatter plot of observed and predicted GWTD by (**a**) GA-1, (**b**) GA-2, (**c**) GA-3, (**d**) GA-4, (**e**) GA-5, (**f**) GA-6, (**g**) GA-7, (**h**) GA-8, and (**i**) GA-9 models in testing period for pre-monsoon season.

Similarly, the values of performance indicators during the post-monsoon season are given in Table 5. In the post-monsoon season, the GA-8 model produced the highest values of R^2, CE, and r at 0.47, 0.68, and 0.31, respectively, while the values of MAD, MSE, CVRE, APE, and PI were the lowest at 1.87, 7.45, 0.47, 0.22 and 0.03, respectively. The GA-8 model was elected best to predict GWTD for the post-monsoon season in the study area. Figure 8 demonstrates the comparison among the observed and predicted values of GWTD in the post-monsoon season with the testing dataset. It can be seen in Figure 8 that the GA-8 model has less scattering than the other models. Based on the assessment, it can be concluded that the GA model had the potential ability to recognize the trend of groundwater table depth data during both seasons. However, GA models were not able to predict the GWTD accurately in the study region during both seasons.

Table 5. Performance indicators of GA models during post-monsoon season.

Model	Period	R^2	CE	r	MAD	MSE	CVRE	APE	PI
GA-1	Training	0.43	0.47	0.40	0.75	10.81	0.57	0.13	0.02
	Testing	0.40	0.19	0.63	2.56	19.45	0.51	0.29	0.03
GA-2	Training	0.42	0.23	0.40	0.67	11.71	0.59	0.11	0.02
	Testing	0.39	0.14	0.36	2.52	20.61	0.53	0.29	0.03
GA-3	Training	0.13	−0.29	0.08	0.33	19.78	0.78	0.04	0.03
	Testing	0.01	−0.24	0.32	4.11	11.9	0.63	0.48	0.05
GA-4	Training	0.33	0.19	0.04	0.57	12.68	0.67	0.09	0.02
	Testing	0.39	0.07	0.36	2.38	16.24	0.55	0.27	0.03
GA-5	Training	0.27	0.08	0.04	0.80	13.93	0.53	0.13	0.03
	Testing	0.29	−0.44	0.54	2.74	34.69	0.68	0.32	0.04
GA-6	Training	0.08	−0.30	0.2	0.77	20.44	0.85	0.13	0.03
	Testing	0.01	−0.24	0.1	4.18	30.14	0.63	0.48	0.03
GA-7	Training	0.41	0.32	0.05	0.62	10.31	0.52	0.10	0.02
	Testing	0.39	−0.31	0.36	2.82	30.91	0.65	0.33	0.04
GA-8	Training	0.56	0.51	0.47	0.27	7.42	0.46	0.04	0.02
	Testing	0.47	0.31	0.68	1.87	7.45	0.47	0.22	0.03
GA-9	Training	0.06	−0.32	0.06	0.9	20.12	0.77	0.15	0.03
	Testing	0.07	−0.19	0.26	3.91	28.19	0.62	0.46	0.06

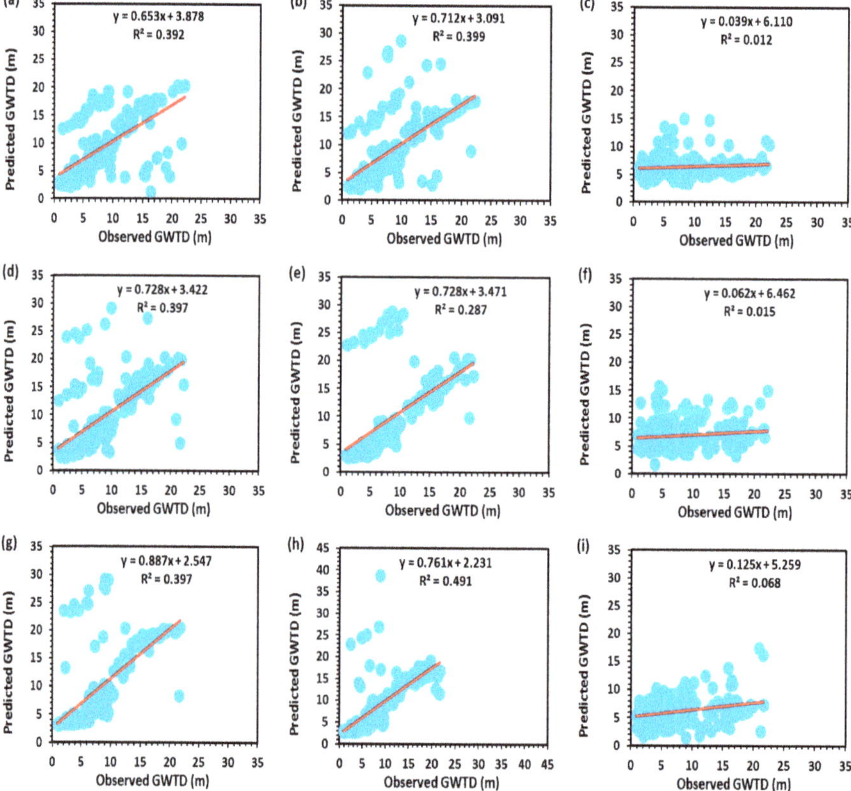

Figure 8. Scatter diagram of observed and predicted GWTD by (**a**) GA-1, (**b**) GA-2, (**c**) GA-3, (**d**) GA-4, (**e**) GA-5, (**f**) GA-6, (**g**) GA-7, (**h**) GA-8, and (**i**) GA-9 models in testing period for post-monsoon season.

3.2. Prediction of GWTD Using GA-ANN Models

The training and testing results based on the effect of population size on the mean square error (MSE) for the post-monsoon and pre-monsoon seasons of all GA-ANN models are listed in Table 6. It was observed in Table 6 that the minimum MSE values for different GA-ANN models were obtained from a population size of 50 as compared with the MSE from the population sizes of 100 and 200 for all models. Hence, the population size of 50 was selected as optimal for GA-ANN development to predict seasonal GWTD in the study region. The optimal number of generations, optimal population size, and respective MSE values for all GA-ANN models for pre- and post-monsoon seasons are summarized in Table 7. In a single-layered ANN structure, the number of neurons was enhanced for all the developed GA-ANN models using MATLAB R2013a software. The methodology of optimizing the GA was used to maximize the number of neurons per model. The optimal numbers of neurons for each GA-ANN model corresponding to the minimum mean square error (MSE) are given in Table 8.

Table 6. Effect of population size on GA-ANN models for pre- and post-monsoon seasons.

Model	Data Set	Population Size = 50		Population Size = 100		Population Size = 200	
		MSE		MSE		MSE	
		Pre-Monsoon	Post-Monsoon	Pre-Monsoon	Post-Monsoon	Pre-Monsoon	Post-Monsoon
GA-ANN-1	Training	0.20	0.20	0.50	0.20	0.50	0.30
	Testing	0.30	0.10	0.30	0.30	0.30	0.20
GA-ANN-2	Training	0.60	0.30	0.60	0.60	0.50	0.60
	Testing	0.40	0.20	0.50	0.40	0.40	0.50
GA-ANN-3	Training	0.15	0.14	0.15	0.15	0.14	0.18
	Testing	0.12	0.12	0.13	0.14	0.13	0.16
GA-ANN-4	Training	0.60	0.30	0.60	0.30	0.80	0.20
	Testing	0.40	0.20	0.50	0.25	0.50	0.30
GA-ANN-5	Training	0.30	0.40	0.30	0.40	0.60	0.60
	Testing	0.10	0.10	0.20	0.20	0.40	0.20
GA-ANN-6	Training	0.12	0.11	0.12	0.11	0.14	0.15
	Testing	0.10	0.90	0.11	0.12	0.12	0.13
GA-ANN-7	Training	0.40	0.40	0.50	0.40	0.50	0.60
	Testing	0.30	0.30	0.40	0.60	0.40	0.20
GA-ANN-8	Training	0.30	0.30	0.30	0.30	0.30	0.30
	Testing	0.20	0.20	0.50	0.60	0.40	0.50
GA-ANN-9	Training	0.10	0.20	0.11	0.30	0.13	0.25
	Testing	0.06	0.10	0.09	0.20	0.11	0.10

Table 7. Optimal population and generation for developed GA-ANN models during pre- and post-monsoon seasons.

Model	Optimal Population Size		Optimal Generation		MSE	
	Pre-Monsoon	Post-Monsoon	Pre-Monsoon	Post-Monsoon	Pre-Monsoon	Post-Monsoon
GA-ANN-1	49	49	35	45	0.02	0.03
GA-ANN-2	35	37	30	35	0.04	0.02
GA-ANN-3	45	50	40	45	0.19	0.46
GA-ANN-4	42	42	30	30	0.08	0.06
GA-ANN-5	39	47	35	35	0.05	0.05
GA-ANN-6	47	48	45	45	0.72	0.62
GA-ANN-7	47	32	30	25	0.02	0.02
GA-ANN-8	42	47	35	40	0.06	0.03
GA-ANN-9	49	48	45	45	0.94	0.95

Table 8. Structure of different developed GA-ANN models.

Pre-Monsoon			Post-Monsoon		
Model	Neuron in the Hidden Layer	Mean Square Error	Model	Neuron in the Hidden Layer	Mean Square Error
GA-ANN-1	4	0.02	GA-ANN-1	7	0.03
GA-ANN-2	6	0.04	GA-ANN-2	7	0.02
GA-ANN-3	5	0.19	GA-ANN-3	8	0.46
GA-ANN-4	6	0.08	GA-ANN-4	11	0.06
GA-ANN-5	7	0.05	GA-ANN-5	8	0.05
GA-ANN-6	8	0.72	GA-ANN-6	12	0.62
GA-ANN-7	6	0.02	GA-ANN-7	7	0.02
GA-ANN-8	9	0.06	GA-ANN-8	9	0.03
GA-ANN-9	8	0.94	GA-ANN-9	8	0.95

Finally, the value of performance indicators of hybrid GA-ANN models for pre-monsoon season during training and testing periods are listed in Table 9, which indicates that the performance of GA-ANN-8 was better than other GA-ANN models. The values of R^2, CE, and r for the GA-ANN-8 were found to be 0.91, 0.91, and 0.96, respectively, in the training period. In testing, the values of these variables were 0.94, 0.94, and 0.97, respectively. The values of MAD, RMSE, CVRE, APE, and PI were 0.45, 0.22, 0.12, 0.01, and 0.03, respectively, during the training period, while in the testing period were 0.48, 0.17, 0.11, 0.03, and 0.02, respectively. The GA-ANN-8 model was chosen as the best to predict the pre-monsoon GWTD in the study area. The observed and predicted values of GWTD in pre-monsoon by the GA-ANN models for the testing dataset are plotted in Figure 9. It was noted from Figure 9 that the expected values of pre-monsoon season GWTD were in better agreement with the measured (observed) values of GWTD during the testing period.

Table 9. Performance indicators of GA-ANN models in the pre-monsoon season.

Model	Structure	Dataset	R^2	CE	r	MAD	RMSE	CVRE	APE	PI
GA-ANN-1	3-4-1	Training	0.93	0.88	0.96	0.56	0.25	0.14	0.11	0.04
		Testing	0.92	0.89	0.95	0.59	0.28	0.13	0.10	0.03
GA-ANN-2	4-6-1	Training	0.90	0.87	0.95	0.57	0.30	0.15	0.12	0.05
		Testing	0.89	0.84	0.94	0.65	0.59	0.16	0.13	0.04
GA-ANN-3	3-5-1	Training	0.89	0.82	0.94	0.85	0.85	0.19	0.14	0.06
		Testing	0.54	0.46	0.73	0.75	0.96	0.31	0.16	0.07
GA-ANN-4	3-6-1	Training	0.80	0.77	0.79	0.62	0.78	0.19	0.12	0.06
		Testing	0.83	0.79	0.89	0.54	0.76	0.17	0.09	0.08
GA-ANN-5	6-7-1	Training	0.86	0.89	0.93	0.59	0.76	0.18	0.02	0.03
		Testing	0.85	0.82	0.92	0.49	0.74	0.14	0.04	0.05
GA-ANN-6	5-8-1	Training	0.83	0.82	0.92	0.65	0.79	0.21	0.10	0.06
		Testing	0.75	0.80	0.87	0.56	0.80	0.15	0.10	0.09
GA-ANN-7	3-6-1	Training	0.86	0.90	0.92	0.49	0.31	0.15	0.11	0.07
		Testing	0.84	0.82	0.91	0.61	0.42	0.13	0.10	0.06
GA-ANN-8	8-9-1	Training	0.91	0.91	0.96	0.45	0.22	0.12	0.01	0.03
		Testing	0.94	0.94	0.97	0.48	0.17	0.11	0.03	0.02
GA-ANN-9	7-8-1	Training	0.75	0.59	0.87	0.84	0.75	0.21	0.15	0.08
		Testing	0.62	0.45	0.79	0.87	0.98	0.23	0.19	0.04

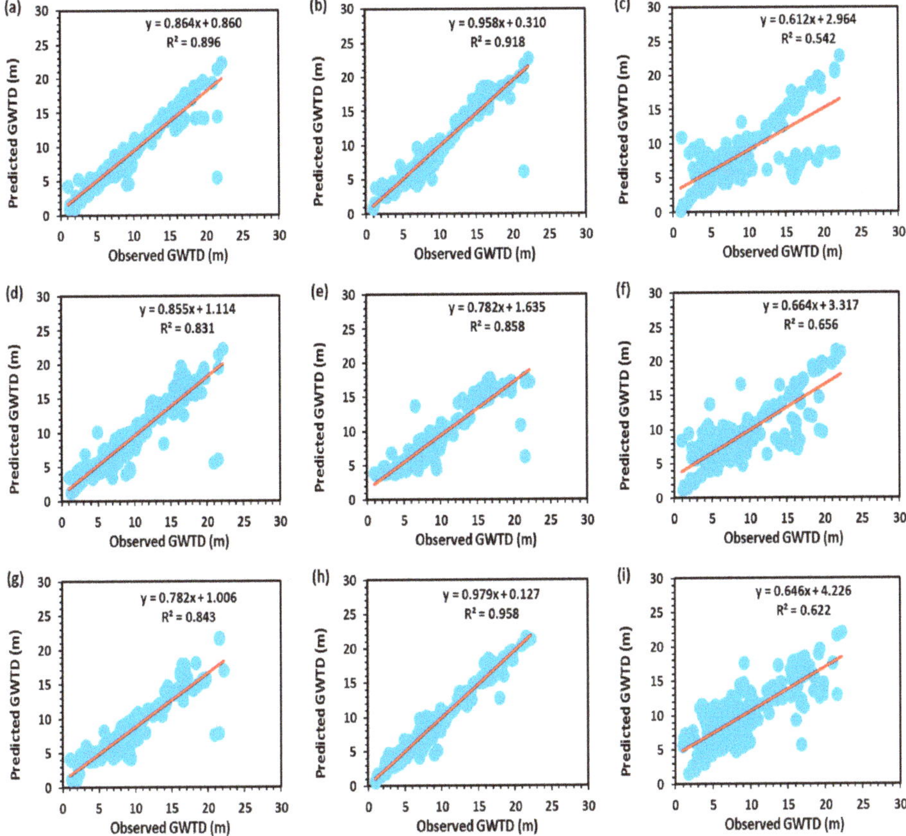

Figure 9. Scatter diagram of observed and predicted GWTD by (**a**) GA-ANN-1, (**b**) GA-ANN-2, (**c**) GA-ANN-3, (**d**) GA-ANN-4, (**e**) GA-ANN-5, (**f**) GA-ANN-6, (**g**) GA-ANN-7, (**h**) GA-ANN-8, and (**i**) GA-ANN-9 models in the testing period for pre-monsoon season.

Similarly, for the post-monsoon season, the performance indicator values of hybrid GA-ANN models are summarized in Table 10 for both the periods and found that the GA-ANN-8 performed significantly better than other GA-ANN models. The values of R^2, CE, and r for the GA-ANN-8 were obtained as 0.89, 0.90, and 0.94, respectively, during the training period and 0.95, 0.96, and 0.97, respectively, during the testing period. While the values of MAD, RMSE, CVRE, APE, and PI for GA-ANN-8 were found to be 0.56, 0.31, 0.15, 0.11, and 0.03, respectively, in the training, and 0.45, 0.42, 0.13, 0.10, and 0.01, respectively, in the testing. The observed and predicted GWTD values yielded by GA-ANN-1 to GA-ANN-9 models for post-monsoon throughout the testing period are illustrated in Figure 10. It was found that the expected value of GWTD in post-monsoon had a better association with the observed values of GWTD in the testing period. The reason for the better performance of the GA-ANN-8 model may be the development of the model using annual data from three years, including the values of previous groundwater table depth for both seasons. However, in the GA-ANN-1 and GA-ANN-5 models, only annual data from one and two years were used. Therefore, the GA-ANN-8 model was nominated as the best model to predict the post-monsoon GWTD in the study area.

Table 10. Performance indicators of GA-ANN models in the post-monsoon season.

Model	Structure	Dataset	R^2	CE	r	MAD	RMSE	CVRE	APE	PI
GA-ANN-1	3-7-1	Training	0.89	0.87	0.94	0.58	0.52	0.17	0.14	0.04
		Testing	0.92	0.93	0.96	0.53	0.54	0.15	0.11	0.03
GA-ANN-2	4-7-1	Training	0.87	0.85	0.93	0.64	0.75	0.21	0.17	0.06
		Testing	0.90	0.87	0.95	0.59	0.56	0.18	0.12	0.04
GA-ANN-3	3-8-1	Training	0.88	0.86	0.94	0.60	0.58	0.19	0.16	0.05
		Testing	0.77	0.75	0.88	0.53	0.69	0.30	0.15	0.05
GA-ANN-4	3-6-1	Training	0.74	0.69	0.86	0.85	0.88	0.21	0.13	0.05
		Testing	0.89	0.85	0.89	0.84	0.79	0.14	0.14	0.04
GA-ANN-5	6-8-1	Training	0.86	0.81	0.93	0.60	0.79	0.19	0.12	0.04
		Testing	0.92	0.93	0.96	0.53	0.55	0.12	0.11	0.02
GA-ANN-6	5-12-1	Training	0.60	0.62	0.77	0.72	1.15	0.56	0.15	0.08
		Testing	0.50	0.57	0.71	0.69	1.35	0.86	0.86	0.09
GA-ANN-7	3-7-1	Training	0.86	0.82	0.93	0.83	0.35	0.19	0.11	0.06
		Testing	0.88	0.87	0.94	0.80	0.51	0.22	0.22	0.06
GA-ANN-8	8-9-1	Training	0.89	0.90	0.94	0.56	0.31	0.15	0.11	0.03
		Testing	0.95	0.96	0.97	0.45	0.42	0.13	0.10	0.01
GA-ANN-9	7-8-1	Training	0.77	0.74	0.88	0.89	0.79	0.25	0.16	0.07
		Testing	0.71	0.54	0.78	0.79	1.01	0.45	0.45	0.09

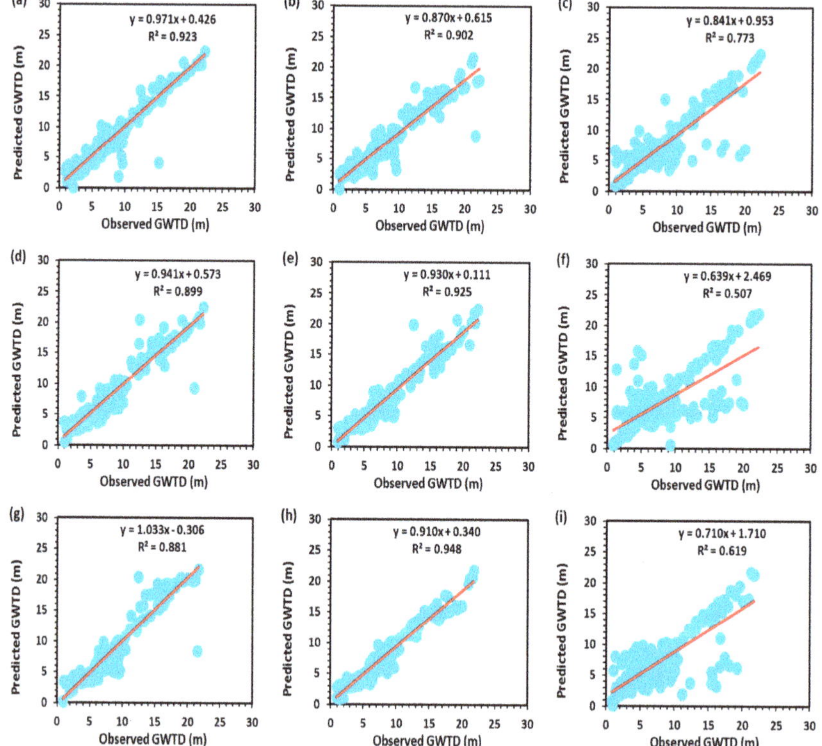

Figure 10. Scatter diagram of observed and predicted GWTD by (**a**) GA-ANN-1, (**b**) GA-ANN-2, (**c**) GA-ANN-3, (**d**) GA-ANN-4, (**e**) GA-ANN-5, (**f**) GA-ANN-6, (**g**) GA-ANN-7, (**h**) GA-ANN-8, and (**i**) GA-ANN-9 models in the testing period for the post-monsoon season.

This study's outcomes follow the studies carried out in other parts of the world to predict the groundwater level with slightly different input parameters [44,62,65,68,69,82,83], and found the performance of GAs implemented with ANN promising for the prediction of groundwater table depth in various regions. Shiri et al. [84] predicted groundwater depth (GWD) fluctuations of two coastal aquifers located in Donghae City, Korea, by employing six heuristic models: boosted regression tree (BRT), random forests (RF), multivariate adaptive regression spline (MARS), ANN, support vector machine (SVM), and gene expression programming (GEP). They found the GEP model with tide and rainfall data provided better estimates than the other models. Some findings also showed the potential capability of genetic algorithm in conjunction with other machine learning techniques in various water resources problems [85–87].

This study's overall findings revealed that the hybrid GA-ANN models performed well in seasonal groundwater table prediction with varying input variables in the study area. These models were more reliable, robust, dynamic, and time-saving than the simple one. This study would help the hydrologists and geologists formulate a smart, intelligent system for effective planning and management of groundwater resources for operating the various drives in the study region. Thus, this study proved the feasibility of the hybrid GA-ANN model in predicting the seasonal GWTD in the area between the Ganga and the Hindon rivers in Uttar Pradesh.

4. Conclusions

With climate change and overexploitation situations, groundwater table fluctuations' accurate predictions are essential for managing groundwater resources. The present study aimed to investigate the comparative potential of the hybrid GA-ANN models against the traditional GA models to predict the seasonal groundwater table depth in the area between the Ganga and the Hindon rivers. The ability of developed models was evaluated by using the statistical indicators (coefficient of determination, coefficient of efficiency, correlation coefficient, mean absolute deviation, root mean square error, coefficient of variation of error residuals, absolute prediction error, and performance index), as well as through visual inspection. The analysis results demonstrate that the GA models recognized the groundwater table depth trend efficiently but failed to predict the groundwater table depth because the maximum coefficient of determination was only 0.47. Simultaneously, the GA-ANN models' performance was found to be superior to the GA models for GWTD prediction in both the seasons, with the highest coefficient of determination values of 0.94 and 0.95, respectively. It was also concluded that the more significant number of input parameters enhanced the predictive rationality of applied GA-ANN models. Thus, the GA-ANN based models may be successfully functional in the field of groundwater to predict the groundwater table fluctuations with reasonably good accuracy.

The efficient models found in this study confirm promising outcomes and proved to be reliable and time-saving technologies for optimal planning and management of groundwater resources in the study area. Our proposed model could be readily transferable or adapted to other areas, specifically those with similar hydrogeological conditions. The accessibility and quantity of data are challenging. In future research, the authors will project to establish a wireless sensor network for near real-time monitoring of groundwater levels and meteorological data in the study area.

Author Contributions: Conceptualization, K.P. and A.M.; methodology, K.P.; software, K.P.; validation, K.P., A.M., S.K. and A.K.; formal analysis, K.P., and A.M.; investigation, K.P., A.M., S.K. and A.K.; data curation, K.P.; writing—original draft preparation, K.P., A.M., and A.K.; writing—review and editing, K.P., A.M., S.K. and A.K.; visualization, K.P., A.M., S.K. and A.K.; supervision, A.M., S.K. and A.K.; project administration, A.K.; funding acquisition, A.K. All authors have read and agreed to the published version of the manuscript.

Funding: This research received no external funding.

Conflicts of Interest: The authors declare no conflict of interest.

References

1. Nunno, F.D.; Granata, F. Groundwater level prediction in Apulia region (Southern Italy) using NARX neural network. *Environ. Res.* **2020**, *190*, 110062. [PubMed]
2. Amarasinghe, U.A.; Smakhtin, V. Global water demand projections: Past, present and future. *Int. Water Manag. Inst. (IWMI) Colombo Sri Lanka* **2014**, *156*, 1–24.
3. Haas, J.C.; Birk, S. Characterizing the spatiotemporal variability of groundwater levels of alluvial aquifers in different settings using drought indices. *Hydrol. Earth Syst. Sci.* **2017**, *21*, 2421–2448. [CrossRef]
4. Yu, H.; Wen, X.; Feng, Q.; Deo, R.C.; Si, J.; Wu, M. Comparative Study of Hybrid-Wavelet Artificial Intelligence Models for Monthly Groundwater Depth Forecasting in Extreme Arid Regions, Northwest China. *Water Resour. Manag.* **2018**, *32*, 301–323. [CrossRef]
5. Goldman, M.; Neubauer, F.M. Groundwater exploration using integrated geophysical techniques. *Surv. Geophys.* **1994**, *15*, 331–361. [CrossRef]
6. Singh, A.; Malik, A.; Kumar, A.; Kisi, O. Rainfall-runoff modeling in hilly watershed using heuristic approaches with gamma test. *Arab. J. Geosci.* **2018**, *11*, 1–12. [CrossRef]
7. Malik, A.; Tikhamarine, Y.; Souag-Gamane, D.; Kisi, O.; Pham, Q.B. Support vector regression optimized by meta-heuristic algorithms for daily streamflow prediction. *Stoch. Environ. Res. Risk Assess.* **2020**. [CrossRef]
8. Tikhamarine, Y.; Souag-Gamane, D.; Ahmed, A.N.; Sammen, S.S.; Kisi, O.; Huang, Y.F.; El-Shafie, A. Rainfall-runoff modelling using improved machine learning methods: Harris hawks optimizer vs. particle swarm optimization. *J. Hydrol.* **2020**, *589*, 125133. [CrossRef]
9. Malik, A.; Kumar, A.; Singh, R.P. Application of Heuristic Approaches for Prediction of Hydrological Drought Using Multi-scalar Streamflow Drought Index. *Water Resour. Manag.* **2019**, *33*, 3985–4006. [CrossRef]
10. Malik, A.; Kumar, A. Meteorological drought prediction using heuristic approaches based on effective drought index: A case study in Uttarakhand. *Arab. J. Geosci.* **2020**, *13*, 1–17. [CrossRef]
11. Malik, A.; Kumar, A.; Salih, S.Q.; Kim, S.; Kim, N.W.; Yaseen, Z.M.; Singh, V.P. Drought index prediction using advanced fuzzy logic model: Regional case study over Kumaon in India. *PLoS ONE* **2020**, *15*, e0233280. [CrossRef] [PubMed]
12. Malik, A.; Kumar, A.; Kisi, O. Monthly pan-evaporation estimation in Indian central Himalayas using different heuristic approaches and climate based models. *Comput. Electron. Agric.* **2017**, *143*, 302–313. [CrossRef]
13. Malik, A.; Kumar, A.; Kisi, O. Daily Pan Evaporation Estimation Using Heuristic Methods with Gamma Test. *J. Irrig. Drain. Eng.* **2018**, *144*, 04018023. [CrossRef]
14. Malik, A.; Rai, P.; Heddam, S.; Kisi, O.; Sharafati, A.; Salih, S.Q.; Al-Ansari, N.; Yaseen, Z.M. Pan Evaporation Estimation in Uttarakhand and Uttar Pradesh States, India: Validity of an Integrative Data Intelligence Model. *Atmosphere* **2020**, *11*, 553. [CrossRef]
15. Malik, A.; Kumar, A.; Kim, S.; Kashani, M.H.; Karimi, V.; Sharafati, A.; Ghorbani, M.A.; Al-Ansari, N.; Salih, S.Q.; Yaseen, Z.M.; et al. Modeling monthly pan evaporation process over the Indian central Himalayas: Application of multiple learning artificial intelligence model. *Eng. Appl. Comput. Fluid Mech.* **2020**, *14*, 323–338. [CrossRef]
16. Malik, A.; Kumar, A.; Ghorbani, M.A.; Kashani, M.H.; Kisi, O.; Kim, S. The viability of co-active fuzzy inference system model for monthly reference evapotranspiration estimation: Case study of Uttarakhand State. *Hydrol. Res.* **2019**, *50*, 1623–1644. [CrossRef]
17. Tikhamarine, Y.; Malik, A.; Kumar, A.; Souag-Gamane, D.; Kisi, O. Estimation of monthly reference evapotranspiration using novel hybrid machine learning approaches. *Hydrol. Sci. J.* **2019**, *64*, 1824–1842. [CrossRef]
18. Tikhamarine, Y.; Malik, A.; Souag-Gamane, D.; Kisi, O. Artificial intelligence models versus empirical equations for modeling monthly reference evapotranspiration. *Environ. Sci. Pollut. Res.* **2020**, *27*, 30001–30019. [CrossRef]
19. Huang, X.; Gao, L.; Crosbie, R.S.; Zhang, N.; Fu, G.; Doble, R. Groundwater Recharge Prediction Using Linear Regression, Multi-Layer Perception Network, and Deep Learning. *Water* **2019**, *11*, 1879. [CrossRef]
20. Chen, L.-H.; Chen, C.-T.; Pan, Y.-G. Groundwater Level Prediction Using SOM-RBFN Multisite Model. *J. Hydrol. Eng.* **2010**, *15*, 624–631. [CrossRef]

21. Chen, L.-H.; Chen, C.-T.; Lin, D.-W. Application of Integrated Back-Propagation Network and Self-Organizing Map for Groundwater Level Forecasting. *J. Water Resour. Plan. Manag.* **2011**, *137*, 352–365. [CrossRef]

22. Panahi, M.; Sadhasivam, N.; Pourghasemi, H.R.; Rezaie, F.; Lee, S. Spatial prediction of groundwater potential mapping based on convolutional neural network (CNN) and support vector regression (SVR). *J. Hydrol.* **2020**, *588*, 125033. [CrossRef]

23. Gong, Y.; Zhang, Y.; Lan, S.; Wang, H. A Comparative Study of Artificial Neural Networks, Support Vector Machines and Adaptive Neuro Fuzzy Inference System for Forecasting Groundwater Levels near Lake Okeechobee, Florida. *Water Resour. Manag.* **2016**, *30*, 375–391. [CrossRef]

24. Arabameri, A.; Lee, S.; Tiefenbacher, J.P.; Ngo, P.T.T. Novel Ensemble of MCDM-Artificial Intelligence Techniques for Groundwater-Potential Mapping in Arid and Semi-Arid Regions (Iran). *Remote Sens.* **2020**, *12*, 490. [CrossRef]

25. Natarajan, N.; Sudheer, C. Groundwater level forecasting using soft computing techniques. *Neural Comput. Appl.* **2020**, *32*, 7691–7708. [CrossRef]

26. Pradhan, S.; Kumar, S.; Kumar, Y.; Sharma, H.C. Assessment of groundwater utilization status and prediction of water table depth using different heuristic models in an Indian interbasin. *Soft Comput.* **2019**, *23*, 10261–10285. [CrossRef]

27. Deb, P.; Kiem, A.S.; Willgoose, G. A linked surface water-groundwater modelling approach to more realistically simulate rainfall-runoff non-stationarity in semi-arid regions. *J. Hydrol.* **2019**, *575*, 273–291. [CrossRef]

28. Deb, P.; Kiem, A.S. Evaluation of rainfall–runoff model performance under non-stationary hydroclimatic conditions. *Hydrol. Sci. J.* **2020**, *65*, 1667–1684. [CrossRef]

29. Tayyab, M.; Zhou, J.; Zeng, X.; Adnan, R. Discharge Forecasting By Applying Artificial Neural Networks At The Jinsha River Basin, China. *Eur. Sci. J. ESJ* **2016**, *12*, 108. [CrossRef]

30. Khan, M.Y.A.; Hasan, F.; Panwar, S.; Chakrapani, G.J. Neural network model for discharge and water-level prediction for Ramganga River catchment of Ganga Basin, India. *Hydrol. Sci. J.* **2016**, *61*, 2084–2095. [CrossRef]

31. Zhou, T.; Wang, F.; Yang, Z. Comparative Analysis of ANN and SVM Models Combined with Wavelet Preprocess for Groundwater Depth Prediction. *Water* **2017**, *9*, 781. [CrossRef]

32. Muhammad, R.; Yuan, X.; Kisi, O.; Yuan, Y. Streamflow Forecasting Using Artificial Neural Network and Support Vector Machine Models. *Am. Sci. Res. J. Eng. Technol. Sci.* **2017**, *29*, 286–294.

33. Gaur, S.; Ch, S.; Graillot, D.; Chahar, B.R.; Kumar, D.N. Application of Artificial Neural Networks and Particle Swarm Optimization for the Management of Groundwater Resources. *Water Resour. Manag.* **2013**, *27*, 927–941. [CrossRef]

34. Alizamir, M.; Sobhanardakani, S. An Artificial Neural Network - Particle Swarm Optimization (ANN- PSO) Approach to Predict Heavy Metals Contamination in Groundwater Resources. *Jundishapur J. Health Sci.* **2018**, *10*, e67544. [CrossRef]

35. Afzaal, H.; Farooque, A.A.; Abbas, F.; Acharya, B.; Esau, T. Groundwater Estimation from Major Physical Hydrology Components Using Artificial Neural Networks and Deep Learning. *Water* **2019**, *12*, 5. [CrossRef]

36. Cho, K.H.; Sthiannopkao, S.; Pachepsky, Y.A.; Kim, K.-W.; Kim, J.H. Prediction of contamination potential of groundwater arsenic in Cambodia, Laos, and Thailand using artificial neural network. *Water Res.* **2011**, *45*, 5535–5544. [CrossRef] [PubMed]

37. Wagh, V.M.; Panaskar, D.B.; Muley, A.A.; Mukate, S.V.; Lolage, Y.P.; Aamalawar, M.L. Prediction of groundwater suitability for irrigation using artificial neural network model: A case study of Nanded tehsil, Maharashtra, India. *Model. Earth Syst. Environ.* **2016**, *2*, 1–10. [CrossRef]

38. Chen, S.; Fang, G.; Huang, X.; Zhang, Y. Water Quality Prediction Model of a Water Diversion Project Based on the Improved Artificial Bee Colony–Backpropagation Neural Network. *Water* **2018**, *10*, 806. [CrossRef]

39. Azimi, S.; Azhdary Moghaddam, M.; Hashemi Monfared, S.A. Prediction of annual drinking water quality reduction based on Groundwater Resource Index using the artificial neural network and fuzzy clustering. *J. Contam. Hydrol.* **2019**, *220*, 6–17. [CrossRef]

40. Das, A.; Maiti, S.; Naidu, S.; Gupta, G. Estimation of spatial variability of aquifer parameters from geophysical methods: A case study of Sindhudurg district, Maharashtra, India. *Stoch. Environ. Res. Risk Assess.* **2017**, *31*, 1709–1726. [CrossRef]

41. Thomas, A.; Eldho, T.I.; Rastogi, A.K.; Majumder, P. A comparative study in aquifer parameter estimation using MFree point collocation method with evolutionary algorithms. *J. Hydroinform.* **2019**, *21*, 455–473. [CrossRef]

42. Delnaz, A.; Rakhshandehroo, G.; Nikoo, M.R. Confined Aquifer's Hydraulic Parameters Estimation by a Generalized Regression Neural Network. *Iran. J. Sci. Technol. Trans. Civ. Eng.* **2020**, *44*, 259–269. [CrossRef]

43. Mohanty, S.; Jha, M.K.; Kumar, A.; Sudheer, K.P. Artificial Neural Network Modeling for Groundwater Level Forecasting in a River Island of Eastern India. *Water Resour. Manag.* **2010**, *24*, 1845–1865. [CrossRef]

44. Jha, M.K.; Sahoo, S. Efficacy of neural network and genetic algorithm techniques in simulating spatio-temporal fluctuations of groundwater. *Hydrol. Process.* **2015**, *29*, 671–691. [CrossRef]

45. Nourani, V.; Ejlali, R.G.; Alami, M.T. Spatiotemporal Groundwater Level Forecasting in Coastal Aquifers by Hybrid Artificial Neural Network-Geostatistics Model: A Case Study. *Environ. Eng. Sci.* **2011**, *28*, 217–228. [CrossRef]

46. Chitsazan, M.; Rahmani, G.; Neyamadpour, A. Forecasting groundwater level by artificial neural networks as an alternative approach to groundwater modeling. *J. Geol. Soc. India* **2015**, *85*, 98–106. [CrossRef]

47. Van Ty, T.; Van Phat, L.; Van Hiep, H. Groundwater Level Prediction Using Artificial Neural Networks: A Case Study in Tra Noc Industrial Zone, Can Tho City, Vietnam. *J. Water Resour. Prot.* **2018**, *10*, 870–883. [CrossRef]

48. Chang, J.; Wang, G.; Mao, T. Simulation and prediction of suprapermafrost groundwater level variation in response to climate change using a neural network model. *J. Hydrol.* **2015**, *529*, 1211–1220. [CrossRef]

49. Banadkooki, F.B.; Ehteram, M.; Ahmed, A.N.; Teo, F.Y.; Fai, C.M.; Afan, H.A.; Sapitang, M.; El-Shafie, A. Enhancement of Groundwater-Level Prediction Using an Integrated Machine Learning Model Optimized by Whale Algorithm. *Nat. Resour. Res.* **2020**, *29*, 3233–3252. [CrossRef]

50. Majumder, P.; Eldho, T.I. Artificial Neural Network and Grey Wolf Optimizer Based Surrogate Simulation-Optimization Model for Groundwater Remediation. *Water Resour. Manag.* **2020**, *34*, 763–783. [CrossRef]

51. Mohanty, S.; Jha, M.K.; Raul, S.K.; Panda, R.K.; Sudheer, K.P. Using Artificial Neural Network Approach for Simultaneous Forecasting of Weekly Groundwater Levels at Multiple Sites. *Water Resour. Manag.* **2015**, *29*, 5521–5532. [CrossRef]

52. Shiri, J.; Kisi, O.; Yoon, H.; Lee, K.-K.; Hossein Nazemi, A. Predicting groundwater level fluctuations with meteorological effect implications—A comparative study among soft computing techniques. *Comput. Geosci.* **2013**, *56*, 32–44. [CrossRef]

53. Kisi, O.; Yaseen, Z.M. The potential of hybrid evolutionary fuzzy intelligence model for suspended sediment concentration prediction. *Catena* **2019**, *174*, 11–23. [CrossRef]

54. Tikhamarine, Y.; Souag-Gamane, D.; Najah Ahmed, A.; Kisi, O.; El-Shafie, A. Improving artificial intelligence models accuracy for monthly streamflow forecasting using grey Wolf optimization (GWO) algorithm. *J. Hydrol.* **2020**, *582*, 124435. [CrossRef]

55. Tikhamarine, Y.; Souag-Gamane, D.; Kisi, O. A new intelligent method for monthly streamflow prediction: Hybrid wavelet support vector regression based on grey wolf optimizer (WSVR–GWO). *Arab. J. Geosci.* **2019**, *12*, 540. [CrossRef]

56. Singh, R.M.; Datta, B. Identification of Groundwater Pollution Sources Using GA-based Linked Simulation Optimization Model. *J. Hydrol. Eng.* **2006**, *11*, 101–109. [CrossRef]

57. Jain, A.; Bhattacharjya, R.K.; Sanaga, S. Optimal Design of Composite Channels Using Genetic Algorithm. *J. Irrig. Drain. Eng.* **2004**, *130*, 286–295. [CrossRef]

58. Şen, Z.; Öztopal, A. Genetic algorithms for the classification and prediction of precipitation occurrence. *Hydrol. Sci. J.* **2001**, *46*, 255–267. [CrossRef]

59. Ni, Q.; Wang, L.; Zheng, B.; Sivakumar, M. Evolutionary Algorithm for Water Storage Forecasting Response to Climate Change with Small Data Sets: The Wolonghu Wetland, China. *Environ. Eng. Sci.* **2012**, *29*, 814–820. [CrossRef]

60. Aytek, A.; Kişi, Ö. A genetic programming approach to suspended sediment modelling. *J. Hydrol.* **2008**, *351*, 288–298. [CrossRef]

61. Jalalkamali, A.; Sedghi, H.; Manshouri, M. Monthly groundwater level prediction using ANN and neuro-fuzzy models: A case study on Kerman plain, Iran. *J. Hydroinform.* **2011**, *13*, 867–876. [CrossRef]

62. Dash, N.B.; Panda, S.N.; Remesan, R.; Sahoo, N. Hybrid neural modeling for groundwater level prediction. *Neural Comput. Appl.* **2010**, *19*, 1251–1263. [CrossRef]

63. Supreetha, B.S.; Prabhakar Nayak, K.; Narayan Shenoy, K. Groundwater level prediction using hybrid artificial neural network with genetic algorithm. *Int. J. Earth Sci. Eng.* **2015**, *8*, 2609–2615.

64. Hosseini, Z.; Nakhaie, M. Estimation of groundwater level using a hybrid genetic algorithm-neural network. *Pollution* **2015**, *1*, 9–21.

65. Jalalkamali, A. Groundwater modeling using hybrid of artificial neural network with genetic algorithm. *Afr. J. Agric. Res.* **2011**, *6*, 5775–5784. [CrossRef]

66. Roshni, T.; Jha, M.K.; Drisya, J. Neural network modeling for groundwater-level forecasting in coastal aquifers. *Neural Comput. Appl.* **2020**, *32*, 12737–12754. [CrossRef]

67. Karamouz, M.; Tabari, M.M.R.; Kerachian, R. Application of Genetic Algorithms and Artificial Neural Networks in Conjunctive Use of Surface and Groundwater Resources. *Water Int.* **2007**, *32*, 163–176. [CrossRef]

68. Wibowo, A.; Arbain, S.H.; Abidin, N.Z. Combined multiple neural networks and genetic algorithm with missing data treatment: Case study of water level forecasting in Dungun River—Malaysia. *IAENG Int. J. Comput. Sci.* **2018**, *45*, 1–9.

69. Li, H.; Lu, Y.; Zheng, C.; Yang, M.; Li, S. Groundwater Level Prediction for the Arid Oasis of Northwest China Based on the Artificial Bee Colony Algorithm and a Back-propagation Neural Network with Double Hidden Layers. *Water* **2019**, *11*, 860. [CrossRef]

70. CGWB. *Status Report on Review of Ground Water Resources Estimation Methodology*; Central Ground Water Board: Faridabad, India, 2009; pp. 1–66.

71. Goldberg, D.E. *Genetic Algorithms in Search, Optimization, and Machine Learning*; Addison Wesley, Reading: Boston, MA, USA, 1989.

72. Malhotra, R.; Singh, N.; Singh, Y. Genetic Algorithms: Concepts, Design for Optimization of Process Controllers. *Comput. Inf. Sci.* **2011**, *4*, 39–54. [CrossRef]

73. Chang, Y.-T.; Lin, J.; Shieh, J.-S.; Abbod, M.F. Optimization the Initial Weights of Artificial Neural Networks via Genetic Algorithm Applied to Hip Bone Fracture Prediction. *Adv. Fuzzy Syst.* **2012**, *2012*, 1–9. [CrossRef]

74. Tahmasebi, P.; Hezarkhani, A. A hybrid neural networks-fuzzy logic-genetic algorithm for grade estimation. *Comput. Geosci.* **2012**, *42*, 18–27. [CrossRef]

75. Mattioli, F.; Caetano, D.; Cardoso, A.; Naves, E.; Lamounier, E. An Experiment on the Use of Genetic Algorithms for Topology Selection in Deep Learning. *J. Electr. Comput. Eng.* **2019**, *2019*, 1–12. [CrossRef]

76. Castillo, P.A.; Merelo, J.J.; Prieto, A.; Rivas, V.; Romero, G. G-Prop: Global optimization of multilayer perceptrons using GAs. *Neurocomputing* **2000**, *35*, 149–163. [CrossRef]

77. Gerken, W.C.; Purvis, L.K.; Butera, R.J. Genetic algorithm for optimization and specification of a neuron model. *Neurocomputing* **2006**, *69*, 1039–1042. [CrossRef]

78. MWRI. *Report of the Ground Water Resource Estimation Committee*; Ministry of Water Resources: New Delhi, India, 2009; p. 133.

79. Chandra, S. Estimation and measurement of recharge to groundwater from rainfall, irrigation and influent seepage. In Proceedings of the International Seminar on Development and Management of Groundwater Resources, Roorkee, India, 5–20 November 1979; pp. 9–17.

80. ARDC. *Report of Ground Water over Exploitation Committee*; Agricultural Refinance and Development Corporation: Mumbai, Indian, 1979; pp. 211–232.

81. Adnan, R.M.; Malik, A.; Kumar, A.; Parmar, K.S.; Kisi, O. Pan evaporation modeling by three different neuro-fuzzy intelligent systems using climatic inputs. *Arab. J. Geosci.* **2019**, *12*, 606. [CrossRef]

82. Coulibaly, P.; Anctil, F.; Aravena, R.; Bobée, B. Artificial neural network modeling of water table depth fluctuations. *Water Resour. Res.* **2001**, *37*, 885–896. [CrossRef]

83. Das, U.K.; Roy, P.; Ghose, D.K. Modeling water table depth using adaptive Neuro-Fuzzy Inference System. *ISH J. Hydraul. Eng.* **2019**, *25*, 291–297. [CrossRef]

84. Shiri, J.; Kisi, O.; Yoon, H.; Kazemi, M.H.; Shiri, N.; Poorrajabali, M.; Karimi, S. Prediction of groundwater level variations in coastal aquifers with tide and rainfall effects using heuristic data driven models. *ISH J. Hydraul. Eng.* **2020**, *26*, 1–11. [CrossRef]

85. Cai, X.; McKinney, D.C.; Lasdon, L.S. Solving nonlinear water management models using a combined genetic algorithm and linear programming approach. *Adv. Water Resour.* **2001**, *24*, 667–676. [CrossRef]

86. Chiu, Y.-C.; Chang, L.-C.; Chang, F.-J. Using a hybrid genetic algorithm–simulated annealing algorithm for fuzzy programming of reservoir operation. *Hydrol. Process.* **2007**, *21*, 3162–3172. [CrossRef]
87. Sharafati, A.; Tafarojnoruz, A.; Shourian, M.; Yaseen, Z.M. Simulation of the depth scouring downstream sluice gate: The validation of newly developed data-intelligent models. *J. Hydro-Environ. Res.* **2020**, *29*, 20–30. [CrossRef]

 *sustainability*

Article

Environmental Flows Assessment in Nepal: The Case of Kaligandaki River

Naresh Suwal [1,*], Alban Kuriqi [2,*], Xianfeng Huang [3], João Delgado [2], Dariusz Młyński [4] and Andrzej Walega [4]

1. Department of Civil Engineering, Khwopa College of Engineering, Bhaktapur 44800, Nepal
2. CERIS, Instituto Superior Técnico, Universidade de Lisboa, Av. Rovisco Pais 1, 1049-001 Lisbon, Portugal; joao.borga.delgado@tecnico.ulisboa.pt
3. College of Water Conservancy and Hydropower Engineering, Hohai University, Gulou District, Nanjing 210098, China; hxfhuang2005@163.com
4. Department of Sanitary Engineering and Water Management, University of Agriculture in Krakow, St. Mickiewicza 24–28, 30-059 Krakow, Poland; dariusz.mlynski@urk.edu.pl (D.M.); andrzej.walega@urk.edu.pl (A.W.)
* Correspondence: suwal.naresh@khwopa.edu.np (N.S.); alban.kuriqi@tecnico.ulisboa.pt (A.K.); Tel.: +977-9841-61-6346 (N.S.)

Received: 10 September 2020; Accepted: 19 October 2020; Published: 22 October 2020

Abstract: Environmental flow assessments (e-flows) are relatively new practices, especially in developing countries such as Nepal. This study presents a comprehensive analysis of the influence of hydrologically based e-flow methods in the natural flow regime. The study used different hydrological-based methods, namely, the Global Environmental Flow Calculator, the Tennant method, the flow duration curve method, the dynamic method, the mean annual flow method, and the annual distribution method to allocate e-flows in the Kaligandaki River. The most common practice for setting e-flows consists of allocating a specific percentage of mean annual flow or portion of flow derived from specific percentiles of the flow duration curve. However, e-flow releases should mimic the river's intra-annual variability to meet the specific ecological function at different river trophic levels and in different periods over a year covering biotas life stages. The suitability of the methods was analyzed using the Indicators of Hydrological Alterations and e-flows components. The annual distribution method and the 30%Q-D (30% of daily discharge) methods showed a low alteration at the five global indexes for each group of Indicators of Hydrological Alterations and e-flows components, which allowed us to conclude that these methods are superior to the other methods. Hence, the study results concluded that 30%Q-D and annual distribution methods are more suitable for the e-flows implementation to meet the riverine ecosystem's annual dynamic demand to maintain the river's health. This case study can be used as a guideline to allocate e-flows in the Kaligandaki River, particularly for small hydropower plants.

Keywords: dynamic flow releases; flow regime; hydrological methods; Indicators of Hydrological Alteration; river health

---

1. Introduction

1.1. Overview of E-Flow Concept

Climate change, rapid globalization, economic boost, and ever-increasing populations demand better management of water resources. To meet humankind's demands (i.e., energy, water), many water conservancy and hydropower projects have been built [1]. Several hydropower plants are under construction or planned to be constructed, which will modify the rivers' natural flow regime either through impoundments such as dams and barrages or through diversion work for agriculture or urban supplies [2]. These projects contribute to tackling the climate change effect (drought, floods) and improve the living standard of citizens (rent standards). However, its adverse effects, such as reducing total flow discharge left to the river, changing the seasonal flow regime [3,4], altering the magnitude and frequency of floods, and modifying the groundwater table are a severe threat to the aquatic ecosystem existence [5,6]. Most rivers with dams and diversion projects are at the ecological tipping point, which means act now or face the projects' worst effects soon. Hence, along with humankind's benefits, it is required to protect the rivers, natural lakes, and groundwater of the watershed by mitigating the riverine ecosystem [7].

The escalating hydrological alterations of the rivers flow regime, and its resulting severe impacts on the riverine ecosystem's health is recognized globally [8–10]. With the advent of growing public consciousness in the river health and its hydrological alteration causing adverse impacts, river scientists developed the science of environmental flows (e-flows) assessments, which aid in determining the quality and quantity of water required for the protection of the riverine ecosystem and its inhabitants. The US clean water act 1977, 1992 European Commission (EC), Habitat Directive, and 2000 EC Water Framework Directives are examples that stressed mandatory river restoration to improve rivers' ecological status. The Water Sustainability Act (2016) mandated that streamflow should not reduce below the environmental flow needed due to groundwater pumping in British Columbia and Canada. Environmental flow has been mandated in British Columbia, Canada, and California, USA [11]. An "Instream Flow Requirement (IFR)," "environmental flows," "Environmental Flows Requirement (EFR)," or "environmental water demand (EWD)" are used as interchangeable terminology for the e-flows [2]. E-flows represent the flow regime provided within a river downstream to maintain a certain acceptable river health level and are widely used as environmental protection measures in many water conservancies projects [12,13]. E-flows is defined as *the quantity, timing, and quality of freshwater flows and levels necessary to sustain aquatic ecosystems which, in turn, support human cultures, economies, sustainable livelihoods, and well-being,* according to the Brisbane Declaration [14]. The process of defining the e-flows is known as "environmental flow assessment (EFA)" [15]. E-flows represent the water needed for the river ecosystem. However, there are other demands as well, such as residential, agriculture, and farming. Hence, all respective sectors' demand should be considered to understand the relationship between water availability and water stress on the river ecosystem. For instance, Xu et al. [16] integrated the e-flow requirement in water-stress impact analysis to inform energy system deployment. This study did not consider any water needs except for hydropower production. All of the basin's existing facilities' water needs should be integrated to conduct water-stress analysis, giving us an idea about the basin's energy system deployment. Groundwater plays a vital role in river hydrology and ecology at various scales (spatial and temporal), evaluating the spatial and temporal pumping effects on streamflow depletion while allocating environmental flows [17,18]. Gleeson and Richter [18] suggested that "high levels of ecological protection will be provided if groundwater pumping decreases monthly natural baseflow by less than 10% through time". However, the discussed e-flows methodologies do not consider the impact of groundwater pumping. The study recommends integrating the impact of groundwater pumping and the impact of water infrastructure on environmental flows.

Environmental flow methodologies stem from a need to conserve mostly rivers and wetlands sustainably with the appropriate ecological balance in the water system close to the natural flow regime. This concern led to the development of different approaches in the environmental flows' methodologies worldwide. Tharme [8] reviewed more than 200 methodologies spread across different geographical spectra. Though all of these methodologies, in principle, have the same goal to achieve a suitable environmental flow regime of water bodies, they differ in their working principle and assumptions made during derivation. These vast methodologies can be classified into four major categories: (1) hydrological, (2) hydraulic rating, (3) habitat simulation, and (4) holistic methodologies. The first two categories, hydrological and hydraulic rating, are based on the assumption that water bodies' habitats/ecosystem functions (e.g., rivers) degrade with reducing water availability in the river. Whereas, habitat simulation assumes an "optimal" flow in the river, sustaining the river ecology in balance [9,19]. Holistic methods are more comprehensive methods with environmental flows designed to mimic the natural hydrograph [20]. Table 1 shows a short overview of all types of methods. The different factors, namely, types of the river (e.g., perennial, seasonal, high base flow, flashy); ecological importance of the river; stakeholders involved; cost; and difficulty of obtaining data decide the preference of the EFA methods [21]. For brevity, the concepts and details of these methods are described in the literature [19,20,22–25].

Table 1. Selection criteria for different environmental flow calculation methodologies [26].

Method Category	Resolution Level	Ecosystem	Time	Cost
Hydrologic	Very Low/Low	River	Short	Less
Hydraulic rating	Low	River	Short/Long	Less/Medium
Habit simulation	Medium/High	River	Medium/Long	Medium/High
Holistic	High	Wetland, floodplains,	Long	High

Hydrological methods use historical flow data records (e.g., daily, monthly, seasonal, yearly flow) to recommend an e-flows setting for maintaining the desired level of river health [24]. Rather than focusing on the optimized environment for single species, this method encompasses the conservation of rivers' overall ecological integrity. The hydrological methods assume that there exists a relationship between flow parameters and biological attributes [27]. Based on this, different parameters are developed to optimize the flow regime. However, hydrological methods have some limitations. They can be used only in gauged catchments, are sensitive to hydrological data, and assume that all the aquatic organisms need the same quantity of water for survival. Due to its easiness of application, rapid assessment, low cost, and fewer field visits with most of the data readily arrived or can be simulated; hydrological methods are widely used to calculate environmental flow [28]. About 30% of all methods are hydrological-based [29]. Most of the preliminary and planning studies' hydrological methods are used [30]. Suwal et al. [31] used the desktop hydrological method "Global Environmental Flow Calculator" (GEFC) to calculate the e-flow for different classes. Pastor et al. [32] recommended the suitability of the Variable Monthly Flow (VMF) and Tessman methods, especially for the variable flow regimes river.

1.2. Environmental Flows Practices in Nepal

The world is focusing and giving prime importance to the ecological impact assessment of water-related projects like hydropower and developing a robust e-flows setting methodology to ensure rivers' ecological integrity [27,33,34]. The concept of e-flows is burgeoning on a global scale. However, in Nepal, the e-flows concept is still in its infancy as investors and the government are not serious about the rivers' e-flows requirements while constructing water conservancy projects on the rivers in Nepal. Many hydropower projects (storage, run of the river, and peaking run of the river) are at the construction phase. Most of the projects are located on most of Nepal's major rivers', namely in Kaligandaki, Karnali, and Mahakali. Developed countries have used different advanced and robust methodological approaches for e-flows regulation. However, developing nations like Nepal usually use hydrological methods. Though hydropower development goes back a century, environmental flow concern was not significant until the Water Resource Act, 1992 in Nepal. This act gave the basic idea about the "environmental study" but did not encompass the broader environmental flows aspect. The Environmental Protection Act (EPA), 1997, was the turning point for e-flows in Nepal as it gave basic guidelines and highlighted the need for Initial Environmental Examination (IEE) and Environmental Impact Assessment (EIA) based on the installed capacity of projects. The EPA made the Environmental Protection Rules (EPR, 1997), making it mandatory to conduct an EIA for projects above 50 MW, and an IEE study for projects below 50 MW [35]. The introduction of the Hydropower Development Policy, 2001, became the paradigm shift in the e-flows setting in Nepal as it exclusively defined the minimum flow requirements for hydropower projects built onwards. Ecological, downstream, or environmental flows are the terms used in Nepal for the e-flows. The policy stated that "Provision should be made to release such quantum of water which is higher of either at least 10% of the minimum monthly average discharge of the river/stream or the minimum required quantum as identified in the environmental impact assessment study report" [35]. The working Policy for Construction and Operation of Physical Infrastructure within Protected Area (2009) further defined different provisions of e-flows that is: If the headworks is within the conservation areas, at least 50% of the monthly flow is considered as e-flows and if the headworks are not within conservation areas, but the downstream flow through the conservation areas, at least 10% of the monthly flow is considered as e-flows [35]. With the support of the International Water Management Institute (IWMI) and the World-Wide Fund for Nature (WWF), e-flows assessment in Nepal is undergoing in many rivers [36,37]. About 50% of all large rivers are affected by dam construction. Studies have identified that cold-water fish in Nepal are threatened due to block connectivity by dam blockages of hydropower projects [38]. Snow trout and gold mahseer are critically endangered fish species, and dark mahseer and *Gangetic ailia* are the vulnerable fish species according to the International Union for Conservation of Nature (IUCN).

Furthermore, the irrigation policy (2014) [39] gave the working direction and decided to use only the residual water for irrigation purposes after maintaining the minimum required water flow in the river and creeks. During the last two decades, many hydropower projects had been constructed. More projects are in the construction phase and the pipeline [40]. The EIA reports show the requirement of e-flows in the projects; however, due to a lack of monitoring resources, none of Nepal's projects has been following the policy regarding the e-flows implementation, which is a severe threat to the downstream ecosystem of the projects. For instance, the Modi River did not get environmental flows as prescribed in EIA and IEE reports except for the wet season [41,42].

The case study on the environmental flow assessment discussed in this paper is located in Nepal's Himalayan region. Table 1 shows that hydrological methods are best suited for Nepal on the primary investigation of e-flows. The main objectives of the present study are: (i) study different existing hydrological e-flow methods (EFMs) and allocate e-flows using six different e-flow methods; (ii) comparison of 6 EFMs; (iii) compute flow alteration using the Indicators of Hydrological Alterations (IHA) indicators and e-flows components (EFC) and application of global indexes; and (iv) to suggest better e-flows assessment methods (hydrological). The Kaligandaki River is considered as a case study for this investigation. The study is organized as follows: Section 2 presents details about the study area and applied methodologies. Section 3 shows the main results of the study. The discussion and conclusion are developed in Sections 4 and 5, respectively.

2. Materials and Method

2.1. Study Area

Kaligandaki is one of Nepal's major rivers and the Narayani River (Figure 1). Narayani River joins the Ganges River in India as a left-bank tributary, eventually draining at the Bay of Bengal.

This paper's study station is Kotagaon Shringe hydrometric station located at 27°45′00″N latitude and 84°20′50″E longitude at an elevation of 198 m [6]. The station lies downstream of the powerhouses of the Kaligandaki-A hydropower station. The daily flow values recorded at the Kotagaon station were obtained from the Department of Hydrology and Meteorology (DHM), Nepal (http://dhm.gov.np/). The mean daily flows data from the year 1964–2015 were used in the study. The details of the Kaligandaki River are shown in Table 2. These flow values were processed in MS-Excel.

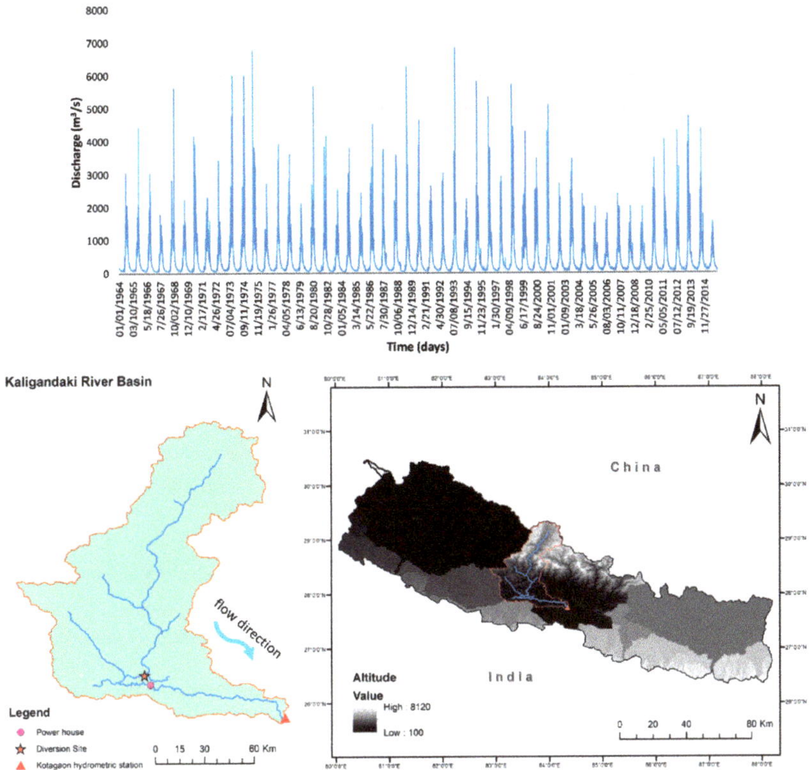

Figure 1. Hydropower and hydrometric station locations within the Kaligandaki River Basin, Nepal. The upper figure shows the mean daily time series of flow discharge from the year 1964–2015.

Table 2. Flow data and watershed description of Kaligandaki River, Nepal.

Name	Details
Elevation	190 m to 8168 m
Total catchment area	11.851 km^2
Location	82°52.8′ E to 84°26.3′ E, 27°43.2′ N to 29°19.8′ N
Mean annual precipitation	1396 mm
Flow data Series	1 January 1964–31 December 2015
Min flow (m^3/s)	46
Mean flow (m^3/s)	449.7
Max flow (m^3/s)	6840
Min average monthly flow (m^3/s)	90
10% of min average monthly flow (m^3/s)	9

2.2. Methodology

In this study, six different hydrologic-based EFA methods were used to evaluate e-flows in the Kaligandaki River and later compared its effectiveness and influence on the natural flow regime using IHA indicators. The leading cause of using hydrological methods is the lack of ecological information of the basin, which is a must in other advanced methods, such as holistic habitat simulation methods. The figurative workflow of the study is shown in Figure 2. The key features of the used methods are discussed in the following sections.

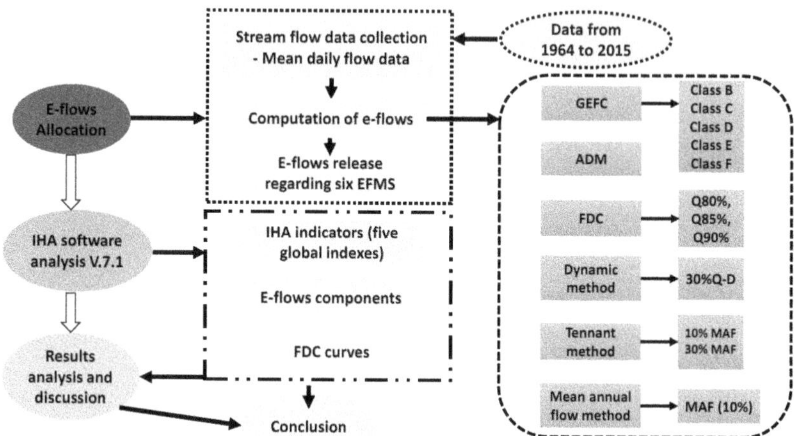

Figure 2. The flow chart shows the main steps and e-flow methods applied in this study.

2.2.1. Annual Distribution Method

The Annual Distribution Method (ADM) method is based on the mean ratio index, a ratio between annual average runoff and minimum annual average runoff. The ADM assumes that the minimum monthly average runoff of rivers could meet the essential water needs to maintain essential ecological, environmental function, give a favourable environment for the survival of the aquatic organism and riverine ecosystem will not suffer severe irreversible damage; however, it cannot reflect the hydrological characteristics of the river. Nevertheless, for many years, the average monthly discharge process can better reflect the overall historical river discharge process, such as timing, duration, frequency, and flow rate change. The mean ratio index is determined and quantified in the same period. According to the long series data of the natural average monthly flow of hydrological station, the average annual discharge $\overline{Q}$ and the minimum annual discharge $\overline{Q}_{\min}$ are calculated respectively [43]. The calculation formulas are given below:

$$\overline{Q} = \frac{1}{12}\sum_{i=1}^{12} \overline{q}_i \tag{1}$$

$$\overline{q}_i = \frac{1}{n}\sum_{j=1}^{12} q_{ij} \tag{2}$$

$$\overline{Q}_{\min} = \frac{1}{12}\sum_{i=1}^{12} q_{\min(i)} \tag{3}$$

$$q_{\min(i)} = \min(q_{ij}), j = 1, 2, \ldots, n \tag{4}$$

where $\overline{q}_i$ is the average monthly discharge of ith month during series of n years (m^3/s); $q_{\min(i)}$ is the minimum monthly discharge of ith month for n years (m^3/s); q_{ij} is the average discharge of ith month jth year; n is the number of years data available.

Calculate the mean ratio index (η) using an annual average discharge ($\overline{Q}$) and minimum annual average discharge ($\overline{Q}_{\min}$) at the same time.

$$\eta = \frac{\overline{Q}_{\min}}{\overline{Q}} \tag{5}$$

$$Q_i = \overline{q}_i \times \eta \tag{6}$$

Calculate the basic ecological flow for each month (Q_i) using Equation (6), where i is the month, i.e., $i = 1, 2\text{–}12$. The ADM method calculates monthly environmental flow; however, we needed daily flow for a year to compute IHA flow alteration. The study used Equation (7) to convert monthly environmental flow into daily flow [44].

$$Q^i_{env,j} = \frac{Q^i_{n,j}}{Q^i_n} \times Q^i_{env} \tag{7}$$

where $Q^i_{env,j}$ is the minimum environmental flow on the jth day of the ith month, $Q^i_{n,j}$ is the natural flow on the jth day of the ith month, Q^i_n is the average value of the natural flow of the ith month, and Q^i_{env} is the minimum environmental flow of the ith month.

2.2.2. Global Environmental Flow Calculator

The study used the "Global Environmental Flow Calculator" (GEFC) software [13] to calculate the environmental flow of different Environmental Management Classes (EMC). The GEFC software implements the "FDC Shifting" method. The details about the software are described in Smakhtin and Eriyagama [13]. The method describes 'Environmental Management Classes' (EMC). Table 3 shows the details about EMC and its corresponding ecological description with a management perspective. It classified EMC into six classes giving six similar environmental flow levels to each class. The software gives the percentage of mean annual flow (MAF) in each EMC. The study considers only EMC from B to F.

Table 3. Description of Environmental Management Classes (EMC) used in the Global Environmental Flow Calculator [2,13].

EMC	Most likely Ecological Condition	Management Perspective
A (Natural)	Same as natural rivers with insignificant modification of instream and riparian habitat	Protected rivers and basins. Reserves and national parks. No new water projects (dams, diversions, etc.) allowed.
B (Slightly modified)	Largely intact biodiversity and habitats despite anthropogenic activities (dam, diversion, basin modifications)	Water supply schemes or irrigation development present and/or allowed.
C (Moderately modified)	The biota's habitats and movement have been impacted, but essential ecosystem functions are still unmodified; some sensitive species are vanished and/or reduced in extent; alien species survived.	Multiple disturbances (for instance, dams, diversions, habitat modification, and reduced water quality) related to the need for socio-economic development
D (Largely modified)	Substantial changes in natural habitat, biota, and essential ecosystem functions have occurred; a lower than expected species richness; the much-lowered presence of intolerant species; alien species prevail.	Significant and precise visible disturbances (such as dams, diversions, transfers, habitat modification, and water quality degradation) associated with basin and water resources development
E (Seriously modified)	Habitat diversity and availability have declined; species richness is strikingly lower than expected; only tolerant species remain; indigenous species can no longer breed; alien species have invaded the ecosystem.	High human population density and extensive water resources exploitation. This class is not suitable as a management goal. The management team should move to a higher class to restore the flow pattern of the river.
F (Critically modified)	Modifications have reached a tipping point; the ecosystem has been completely modified with an almost complete loss of natural habitat and biota; in the worst case, the underlying ecosystem functions have been destroyed, and changes are irreversible.	This status is not acceptable from the management perspective. Management interventions are necessary to restore flow patterns and river habitats (if still possible/feasible) to 'move' a river to a higher management category.

2.2.3. Flow Duration Curve Analysis

The Flow Duration Curve Analysis (FDCA) method is another popular method used in environmental flow calculations. A typical flow duration curve represents the proportion of flow exceeded for a particular time in the river section. Based on this exceedance curve, the minimum threshold is defined to preserve the ecological integrity of rivers. Generally, indices related to flow duration curves are developed. The maximum abstraction level of water from the river can be subsequently calculated. This method is useful in setting environmental flows downstream of hydropower. The equation used to compute the exceedance probability, which also is referred to as the flow-duration percentile, is given as:

$$P = 100 \times (m/(n+1)) \tag{8}$$

where P is the exceedance probability, m is the ranking, from highest to lowest, of all daily mean flows for the specified period of record, and n is the total number of daily means flows in the recorded period.

In this study, $Q_{80\%}$, $Q_{85\%}$, and $Q_{90\%}$ are used as low flow indices.

2.2.4. Tennant Method

The Tennant method [45,46] suggested that specific percentages of the average annual flow (AAF) are necessary to maintain a river ecosystem's biological integrity. It assumed that the aquatic organisms' water requirement depends on different life cycles, such as the reproductive stage, global growth stage, etc. These stages have different water requirements. Hence, the method divided the whole year into a spawning period (April–September) and the general growth period (October–March). Table 4 shows the percentage standards of an aquatic organism's water requirement in different life cycle stages.

Table 4. Flow recommendations as per the Tennant method, based on Tennant [45].

Aquatic-Habitat Condition for Small Stream	Recommended Base Flow (% of MAF)	
	General Period (October–March)	Fish Spawning Period (April–September)
Flushing or maximum	200% of the average flow	
Optimum range	60–100	60–100
Outstanding	40	60
Excellent	30	50
Good	20	40
Fair or degrading	10	30
Poor or minimum	10	10
Severe degradation	<10% of average flow to zero flow	

2.2.5. Dynamic Methods

30% of Mean Daily Flow (30%Q-D).

This is based on the concept of minimum daily flow. It releases 30% of mean daily flow, considering a long series of interannual mean daily flow data, allowing for dynamic e-flows releases [47–49].

2.2.6. Mean Annual Flow

At least 10 or 25% of the Mean Annual Flow (MAF) must be released to the downstream depending upon the degree of environmental protection in the river reach. For satisfactory results, at least five years of continuous daily flow data are needed. For more representative results, a long time series of interannual mean daily flow data is required for this method [47].

2.2.7. Indicators of Hydrological Alteration (IHA) and Global Indexes

Richter et al. [50] used 32 "ecologically relevant" hydrological indexes to develop a set of Indicators of Hydrologic Alteration (IHA). The IHA indexes represent five essential parameters of the natural flow regime: magnitude, frequency, duration, timing, and rate of change [51]. The IHA indexes captured most of the variation and information described by 171 indexes, which makes IHA indexes the best choice to represent alteration on rivers [52]. The study used the Richter et al. [53] approach to assess the flow regime and e-flows component (EFC) alteration using relative mean deviation between the natural flow regime (NFR) and e-flows release from the 6 EFMs.

Kuriqi et al. [47] proposed a global index for a separate group of IHA indicators to simplify the analysis. Initially, the alteration of each indicator was computed using Equation (9), which is the relative mean difference between natural flow regime (NFR) and altered flow regimes (AFR) divided by *nfr*. Here, *afr* is the regime that will be obtained after releasing the e-flows instead of natural flows. After that, each group's new global index was computed using Equation (10), each group's average. The details about indexes are shown in Table 5.

$$HI_{i,j} = \left| \frac{HI_{i,j}(nfr) - HI_{i,j}(afr)}{HI_{i,j}(nfr)} \right| \tag{9}$$

$$I_j = \frac{\sum HI_{i,j}}{N} \tag{10}$$

where $HI_{i,j}$ is the relative mean difference of i hydrological indicator of j group. I_j: global alteration index of j group, $HI_{i,j}(nfr)$: hydrological indicator related to *nfr*, $HI_{i,j}(afr)$: hydrological indicator related to the altered flow regime, and N; the total number of IHA indexes for each group. Here, $i = 1,2$–12 depend upon the group and $j = 1,2,3,4,5$. The five indices are classified, as shown in Table 6.

Table 5. List of the Indicators of Hydrologic Alteration (IHA) parameters, their ecological significance, and regime characteristics and global indexes for each IHA group.

Global Index for Each Group	IHA Parameters	Regime Characteristic (Specific Alteration)	Ecological Significance
Mean Monthly Flow Alteration Index (I_{mm})	**Group 1:** Mean value of each calendar month	Magnitude (increased variation)	Guaranteed favourable habitat conditions and flow regime (quantity, quality, and temperature) for aquatic and terrestrial organisms. Availability of food and cover for fur-bearing mammals.
Magnitude and Duration of Extreme Flow Alteration Index (I_{MDE})	**Group 2:** Annual minima, 1, 3, 7, 30, 90 day means Annual maxima, 1,3,7,30,90 day means Number of zero-flow days Baseflow index: 7 day minimum flow/mean flow for the year	Magnitude and Duration (prolonged low flows; altered inundation duration; prolonged inundation)	Structuring of aquatic ecosystems by abiotic and biotic factors. The shaping of river channel morphology and physical habitat conditions.
Timing of Extreme Flow Alteration Index (I_{TE})	**Group 3:** Julian date of each annual 1 day maximum Julian date of each annual 1 day minimum	Timing (oss of seasonal flow peaks)	Disrupt cues for fish: (spawning, egg hatching, migration) [54]. Evolution of the life history and behaviour mechanism of the aquatic organisms [48].

Table 5. *Cont.*

Global Index for Each Group	IHA Parameters	Regime Characteristic (Specific Alteration)	Ecological Significance
Frequency and Duration Alteration Index (I_{FD})	**Group 4:** No. of high pulses each year No. of low pulses each year Mean duration of high pulses within each year (days) Mean duration of low pulses within each year (days)	Frequency and Duration (flow stabilization)	Availability of floodplain habitats for aquatic organisms. Influences bedload transport, channel sediment textures, and duration of substrate disturbance (high pulses). Nutrient and organic matter exchanges between river and floodplain.
Rate and Frequency Alteration Index (I_{RF})	**Group 5:** Means of all positive differences between consecutive daily values Means of all negative differences between consecutive daily values Reversals	Rate of change and Frequency (rapid changes in river stage; accelerated flood recession)	Wash out and stranding of aquatic species [55]. Failure of seedling establishment [56].

Note: All the ecological significance of the IHA parameters is not listed in the table. A few are listed for more ecological significance; the authors recommend literature related to IHA parameters [51,53,57,58].

Table 6. Range of alteration for global indexes. Where '0' means no alteration while '1' implies the highest alteration.

Range	0.00–0.25	0.25–0.50	0.50–0.75	0.75–1.00
Alteration	Low	Mild	Moderate	High

2.2.8. Environmental Flow Components

The five essential components of flow, namely: low flows, extreme low flows, high flow pulses, small floods, and large floods, have been identified ecologically essential and have been incorporated in IHA software as "environmental flow components" [51]. Each of the flow components has a respective role in the life of an organism. For instance, extreme low flows reduce water connectivity, restricting organisms' movement; high flow pulses help aquatic mobile organisms move upstream and downstream of rivers. Different species, different geographic conditions, and different rivers (perennial, ephemeral) could be life-threatening, causing the death of organisms or may provide favorable conditions for aquatic organisms for their life stages [51]. The study used the e-flows component (EFC) to assess the alteration using each parameter's relative mean difference.

2.2.9. Limitations of the Methodology

The hydrological methods are the most simple, straightforward, and data-friendly methods that have been used extensively for the preliminary study of the e-flows allocation [48]. So, when there are several available solutions for the same problem, there is no guarantee of the best solution from both fixed and scientific perspectives. Single flow indices and other hydrological methods (such as Tennant, GEFC, FDC, ADM) have been applied globally, having many advantages. These methods simplify river basin planning work, needless time, and money. They require a low-level of knowledge related to the eco-hydrology of the basins [29,57]. The study's analysis was limited to hydrological-based e-flows methods, which required only time-series data of mean daily flows [47]. Other sophisticated methods are also available such as habitat simulation and holistic approach, which are more directly involved in ecological concepts. They were not used in the analysis due to a lack of data in the present case study. However, the study can be continued further using those methods if the required data are available.

3. Results

3.1. E-Fows Allocation

Different hydrological methods are used to compute the potential e-flows in the river Table 7 presents the estimated e-flows from different methods. Daily flow data collected over 51 years (between 1964 and 2015) were used to calculate e-flows. The GEFC was used to calculate e-flows for different management classes.

Table 7. E-flow allocation of Kaligandaki River regarding all applied e-flows methods (EFMs).

Method	Classes	(%of MAF)	E-flows (m^3/s)
GEFC	Class B	47.8	214.46
	Class C	32.8	147.16
	Class D	23.7	106.33
	Class E	18.6	83.45
	Class F	15.7	70.44
Tennant	Oct–Mar	10	44.87
	Apr–Sept	30	134.6
FDC	Q80% FDCA,		49.04
	Q85% FDCA,		45.65
	Q90% FDCA,		43.05
Mean annual flow	10%MAF		44.97
Dynamic methods	30%Q-D		30% of daily flow

						Annual Distribution Method						
Month	Jan	Feb	March	April	May	June	July	August	Sep	Oct	Nov	Dec
E-flow (m^3/s)	60.57	50.05	45.07	49.75	73.54	204.58	646.33	785.59	584.67	274.83	127.61	83.71

This case study only considered Class B classes to Class F, as Class A flows could not be considered as e-flows for any project. The lowest class, 'F', shows 15.7% of MAF, that is, 70.44 m^3/s, which indicates that below this amount would characterize the river as a dead environment. An average annual e-flows allocation of 147.16 m^3/s (32.8% of MAF) is expected to maintain the essential ecosystem functions. Figures 3 and 4 show the interannual discharge and monthly mean flow of all EFMs.

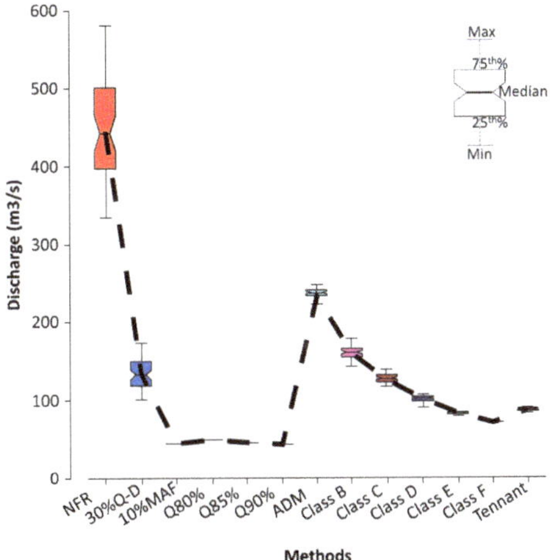

Figure 3. Interannual discharge regarding all applied EFMs. The dashed line connects the mean values of e-flows discharge for each EFMs.

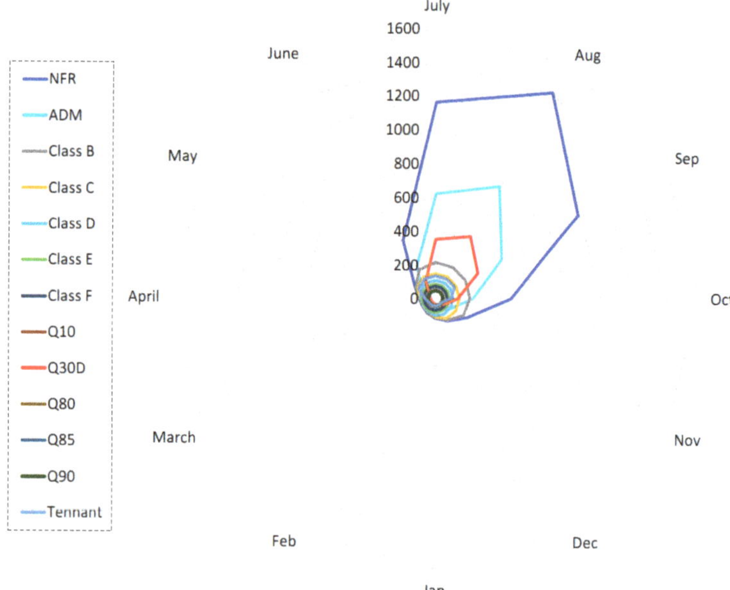

Figure 4. Mean monthly e-flows regarding all EFMs. The year is divided into two parts; the first part refers to high flow seasons (July–December). The second part refers to the low flow seasons (January–June). The reading of the left-vertical axis corresponds to the first part. The reading of the right-vertical axis corresponds to the second part of the year.

The study considers the Tennant method's fair and degrading condition criteria, which allocates 10% of the MAF (equivalent to 44.87 m^3/s) for the flow from October to March and 30% of the MAF (equivalent to 134.6 m^3/s) for the flow from April to September. The method is simple and easy to implement, which lets it become more user friendly. It must be noted that 10% of the 6 months' flow is the lowest among all the methods studied. Hence, it may create critical conditions for the river ecosystem during a 10% period.

The mean annual flow method that is 10% MAF allocated 44.97 m^3/s to maintain the health of river reach, which is the second-lowest allocation showing its vulnerability of implementation. The dynamic methods of 30%Q-D methods follow the natural hydrograph pattern showing its applicability in the EFs allocation.

The FDC curve could be determined using whole multi-year data, or the FDC curve could be determined for each year separately. Młyński et al. [58] showed significant differences between the Q_p ('p' percentage exceedance discharge) for the multi-year curve and Q_p for mean or median annual curves. The study used multi-year curves to determine the e-flows of the Kaligandaki River. FDC or percentile methods such as Q_{80}, Q_{85}, and Q_{90} was considered for the study. This method allocated e-flows as the mean daily discharge that is equaled or exceeded by 80% (Q_{80}), 85% (Q_{85}), and 90% (Q_{90}). The e-flows suggested by these methods are Q_{80}= 49.04 m^3/s, Q_{85} = 45.65 m^3/s, and Q_{90} = 43.05 m^3/s. The EF estimations from the FDCA methods considered here are less than Class F (equivalent to 70.44 m^3/s), which critically modified the river ecosystem. Hence, it can be concluded that FDCA methods are not suitable for the EF estimations.

The ADM allocated the lowest e-flows at 45.07 m^3/s for March and the highest for August at 785.59 m^3/s, which is shown in Table 5. The method considered the intra-annual variation of the flow regimes, which can meet the actual need to sustain the ecological function of the river instead of considering the specific percentages of average annual runoff or specific guarantee rate of frequency curve of natural runoff as e-flows.

3.2. Interannual and Seasonal E-Flows Characterization

Figures 5 and 6 show that the constructed FDC's shape for different EF methods differs from NFR for mean annual flow and seasonal flow (autumn, spring, summer, and winter). The FDC curve of ADM and 30%Q-D methods show a slope, which means this method tries to maintain the rivers' variability, but not by other methods. From the figure, we can see all methods on the FDC curve are below the mean NFR.

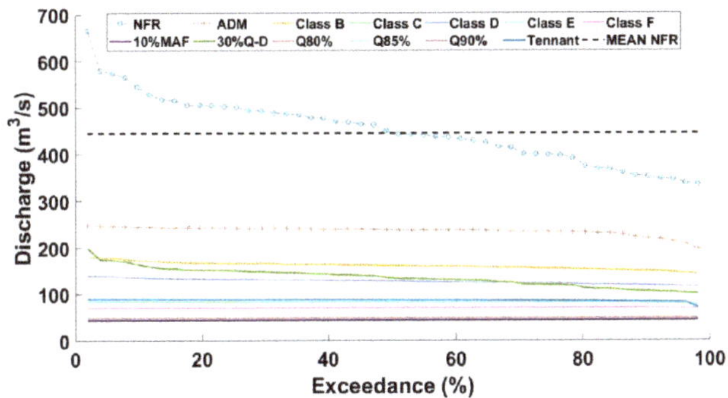

Figure 5. Flow duration curve of interannual mean daily e-flows from 1964 to 2015, regarding of all applied EFMs.

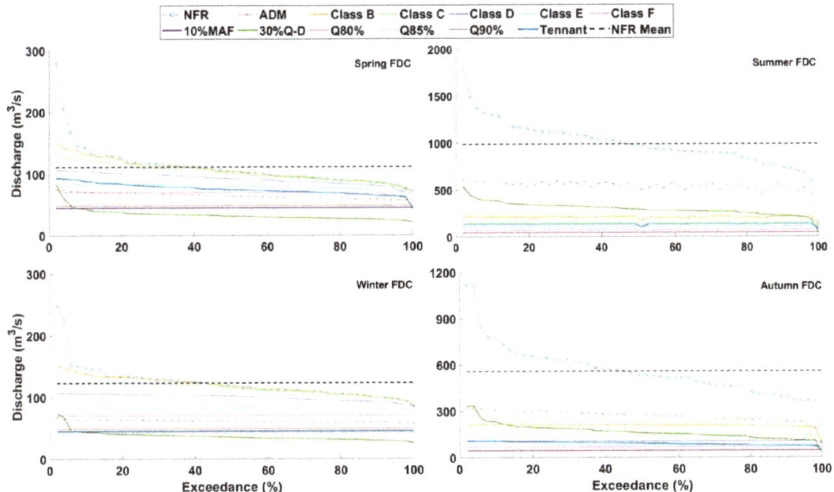

Figure 6. Flow duration curve of seasonal mean daily e-flows regarding of all applied EFMs. Summer (June, July, August), autumn (September, October, November), winter (December, January, February), and spring (March, April, May).

The FDC curve of ADM and 30%Q-D show a similar pattern as NFR but with lower values. Simultaneously, all methods are low flow fixed values methods, represented in a straight line rather than normal FDC curves. The annual mean FDC curves show that all methods value is less than low flows of the NFR, which gives us an idea about the insufficiency of the flow in the river. The seasonal FDC of all methods shows that autumn and summer FDC is the same as the mean annual FDC;

however, FDC of spring and winter shows that Class B and Class C e-flows are higher than mean seasonal flows, which demonstrates suitability for the low flow seasons. In contrast, all other methods are below the mean seasonal flows. The ADM and 30%Q-D methods show a similar pattern as NFR FDC; however, they give very low e-flows during spring and winter seasons. Most of the e-flows methods give a straight line FDC, which means it does not consider the river's flow variability.

The methods which mimic the shape of NFR will be the better methods to sustain the health of the river. In this way, we can choose the different methods for a different season or even different months to take each method's strength and to give one robust method for e-flows calculation in different periods.

3.3. Flow Regime Alteration

3.3.1. IHA Alteration

Figure 7 shows the e-flows regime alteration degree using five indices developed by Kuriqi et al. [47] for all e-flow methods used in the study. The detailed information about the e-flows regime alteration regarding each e-flow method in Appendix A (Table A1). Each method has its strengths and weaknesses, allowing a certain percentage of flow into the rivers. Figure 7 shows the flow alteration due to different EFMs, which may have a high impact on the hydro-ecosystem and related hydro-ecological process.

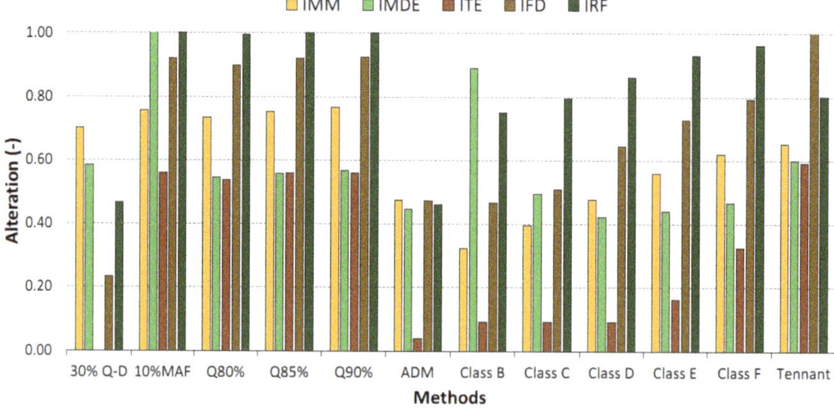

Figure 7. Global indices derived from indicators of hydrological alteration regarding all applied EFMs. Namely, Mean Monthly Flow Alteration Index (IMM), Magnitude and Duration of Extreme Flow Alteration Index (IMDE), Timing of Extreme Flow Alteration Index (ITE), Frequency and Duration Alteration Index (IFD), and Rate and Frequency Alteration Index (IRF). The five indexes are classified into four categories, low (0–0.25), mild (0.25–0.5), moderate (0.5–0.75), and high alteration (0.75–1).

For instance, 10% MAF, Q85%, and Q90% show a high Mean Monthly Flow Alteration Index (I_{MM}). Whereas, 30%Q-D, Q80%, Class E, Class F, and Tennant show moderate I_{MM}. While ADM, Class B, Class C, and Class D show mild I_{MM}. For the second global index, Magnitude and Duration of Extreme Flow Alteration Index (I_{MDE}), only Class B shows great alteration. ADM and Class D show mild alteration while all remaining methods show moderate alteration. Looking at the Timing of Extreme Flow Alteration Index (I_{TE}), 10% MAF, Q80%, Q85%, Q90%, and Tennant show moderate alteration. Only Class B shows mild alteration, but all remaining EFMs methods show low alteration. Frequency and Duration Alteration Index (I_{FD}) shows that environmental management classes Class C, Class D, and Class E show moderate. In contrast, 30%Q-D, ADM, and Class B show low alteration, but other EFMs, for instance, 10%MAF, Q80%, Q85%, Q90%, Class F, and Tennant show high alteration. The Rate and Frequency Alteration Index (I_{RF}), all except ADM and 30%Q-D, show high alteration while ADM and 30% Q-D show low alteration showing these methods effectiveness in e-flows determination.

3.3.2. E-flows Components (EFC)

The study calculated relative changes in the mean of e-flows component (EFC) regarding six EFMs against NFR using the values obtained from IHA software. The results are shown in Table 6. The results show that most of the EFC low flow changes 100% negatively except for 30% Q-D, ADM, and Tennant methods. For 30% Q-D, all EFC Low Flows changes −70%, while in ADM, it varies between −31% during June and 65% during August. The Tennant method varies between −6% during April and −100% for the winter season, extending up to March month. Table 6 shows the relative mean change of the EFC parameters against NFR. Most of the methods, namely 10% MAF, Q80%, Q85%, Q90%, Class B to Class F, show all EFC parameters changes of −100%. While 30%Q-D method shows duration, timing, and frequency of extreme low, high flow, small flood, and enormous flood vary by 0%, and remaining parameters show −70% relative changes against NFR. The ADM method shows a variation of relative mean change within a range between −2% of high flow frequency and +260% of significant flood frequency.

In comparison, the Tennant method shows a variation between −39% extremely low peak and 1393% extremely low duration. The variability of flow is high when the positive and relative change of C_v is high. In contrast, flow variability is low when Cv's negative and relative change is of low value.

4. Discussion

The results show that MAF, Tennant (Oct–Mar), and FDC methods allocate e-flows less than the GEFC class F (critically modified) method; hence these methods are not recommended for the e-flows assessment for the present case-study. Further, Class B (Slightly modified) and Class C (moderately modified) allocated a considerable amount of e-flows; however, they gave a fixed value which is a certain percentage of MAF. Nevertheless, river flow is dynamic with interannual variability. Hence, ADM and dynamic method (30%Q-D) are recommended for e-flows allocation. These methods consider the changing characteristics of natural runoff. Hence, to ensure the sustainability of the river health, we must choose the methods that maintain the river's flow dynamics rather than an absolute fixed percentage of MAF of the river.

NFR of many rivers worldwide had been altered by anthropogenic activities such as diversions and impoundment work [5,6,44]. For the rivers with limited data, time, and funding, the hydrological methods are the best method to allocate e-flows. The applied five alteration indexes showed that there is a considerable difference in alteration regarding each EFM. The degree of alteration was varied from EFM to EFM. The EFM, such as ADM and 30%Q-D, appeared to be less altered. This may be due to the consideration of the flow variability of the river [48]. Here, the I_{TE}, I_{FD}, and I_{RF} which have an essential role in sustaining the different ecological processes [54,59], were preserved near NFR, which means they are less altered only for two EFMs, that is, ADM and 30%Q-D. Overall, the results obtained by taking an average of five global indexes showed that the 10% MAF, $Q_{80\%, 85\%, 90\%}$, and Tennant methods gave high alteration, class D, E, and F showed moderate alterations, indicating their unsuitability in the e-flows allocation. The remaining methods showed mild alterations, indicating the suitability of e-flows allocation. The ADM and 30%Q-D methods showed the lowest and second-lowest alteration among all considered methods with values of 0.380 and 0.406, respectively. This is because the ADM and dynamic methods were the only methods that considered the concept of a dynamic pattern of the river like the NFR.

All six EFMs showed a high alteration of an e-flows component (EFC). Among those methods, except ADM, 30%Q-D, and the Tennant method, all others showed a −100% alteration, confirming the methods' applicability. ADM gave a lower EFC alteration; it showed a monthly low flow alteration between −31% to −65%, resulting from less water in the river than the naturally available water in the riverine ecosystem. The low alteration of monthly flows means it maintains the temperature, flow velocity, and connectivity needed for most of the aquatic habitat than other methods [51]. Extreme low flows, high flow pulses, small floods, and large floods low alterations further give strong supports to the suitability of the method, which is shown in Table 8, because these components play a crucial role in maintaining the health of the rivers [51].

The FDC is a plot of the observed historical river flow data collected with different temporal resolutions daily, weekly, monthly, seasonally, and annually at the gauge station and when that flow is equaled or exceeded [60–62]. The FDC is an informative method that shows the characteristics of flow. The slope of the flow-duration curve is a quantitative measure of the flow regime variability [62]. The NFR FDC generally showed low flows exceeded most of the time while high flows are exceeded infrequently. However, the paper's studied e-flows methods showed low flow exceedance 100% of the time, except for the ADM and 30%Q-D. The FDC curve plotted for annual mean and mean seasonal flows showed a drastic reduction in river flows, especially the high flood. However, the ADM and 30%Q-D methods try to mimic the FDC curve of NFR, which might help sustain the riverine ecosystems' health. Other methods that release fixed minimum flow rather than dynamic release gives a straight line FDC, removing high floods, small floods and it is lower than NFR over the year. The FDC is useful to evaluate the relationship between magnitude and frequency of the river flow; however, it does not maintain temporal sequences of flows and so is unsuccessful in meeting the criteria of the timing or duration of the e-flows [63]. Nevertheless, as recommend by Kuriqi et al. [9,49], to guarantee suitable habitat conditions during low flow periods, AMD and 30%Q-D should be combined with other methods by setting a minimum of e-flows to be released downstream of the water intake. The results of this case study are aligned with the recommendations of the investigations above.

Table 8. Relative change (%) of the EFC regarding six EFMs against the natural flow regime (NFR), represented by the mean values. The sign (+) symbolizes an increase and (−) a decrease.

E-Flows Components (EFC)	Dynamic E-Flows	Minimum Annual	FDC Curve			ADM	Global Environmental Flow Calculator					Tennant
EFC Low Flows	30%Q-D	10%MAF	Q80%	Q85%	Q90%		Class B	Class C	Class D	Class E	Class F	
July—Low Flow	−70	−100	−100	−100	−100	−46	−100	−100	−100	−100	−100	−70
August—Low Flow	−70	−100	−100	−100	−100	−65	−100	−100	−100	−100	−100	−61
September—Low Flow	−70	−100	−100	−100	−100	−35	−100	−100	−100	−100	−100	−72
October—Low Flow	−70	−100	−100	−100	−100	−40	−100	−100	−100	−100	−100	−64
November—Low Flow	−70	−100	−100	−100	−100	−49	−100	−100	−100	−100	−100	−41
December—Low Flow	−70	−100	−100	−100	−100	−49	−100	−100	−100	−100	−100	−100
January—Low Flow	−70	−100	−100	−100	−100	−48	−100	−100	−100	−100	−100	−100
February—Low Flow	−70	−100	−100	−100	−100	−49	−100	−100	−100	−100	−100	−100
March—Low Flow	−70	−100	−100	−100	−100	−50	−100	−100	−100	−100	−100	−100
April—Low Flow	−70	−100	−100	−100	−100	−48	−100	−100	−100	−100	−100	−6
May—Low Flow	−70	−100	−100	−100	−100	−34	−100	−100	−100	−100	−100	−15
June—Low Flow	−70	−100	−100	−100	−100	−31	−100	−100	−100	−100	−100	−52
EFC Parameters												
Extreme low peak	−70	−100	−100	−100	−100	−41	−100	−100	−100	−100	−100	−39
Extreme low duration	0	−100	−100	−100	−100	−55	−100	−100	−100	−100	−100	1393
Extreme low timing	0	−100	−100	−100	−100	−3	−100	−100	−100	−100	−100	234
Extreme low freq.	0	−100	−100	−100	−100	82%	−100	−100	−100	−100	−100	−74
High flow peak	−70	−100	−100	−100	−100	−62	−100	−100	−100	−100	−100	−100
High flow duration	0	−100	−100	−100	−100	−30	−100	−100	−100	−100	−100	−100
High flow timing	0	−100	−100	−100	−100	26	−100	−100	−100	−100	−100	−100
High flow frequency	0	−100	−100	−100	−100	−2	−100	−100	−100	−100	−100	−100
High flow rise rate	−70	−100	−100	−100	−100	−74	−100	−100	−100	−100	−100	−100
High flow fall rate	−70	−100	−100	−100	−100	−81	−100	−100	−100	−100	−100	−100
Small Flood peak	−70	−100	−100	−100	−100	−68	−100	−100	−100	−100	−100	−100
Small Flood duration	0	−100	−100	−100	−100	11	−100	−100	−100	−100	−100	−100
Small Flood timing	0	−100	−100	−100	−100	6	−100	−100	−100	−100	−100	−100
Small Flood frequency	0	−100	−100	−100	−100	−64	−100	−100	−100	−100	−100	−100
Small Flood rise rate	−70	−100	−100	−100	−100	−96	−100	−100	−100	−100	−100	−100
Small Flood fall rate	−70	−100	−100	−100	−100	−85	−100	−100	−100	−100	−100	−100
Large flood peak	−70	−100	−100	−100	−100	−74	−100	−100	−100	−100	−100	−100
Large flood duration	0	−100	−100	−100	−100	−18	−100	−100	−100	−100	−100	−100
Large flood timing	0	−100	−100	−100	−100	−2	−100	−100	−100	−100	−100	−100
Large flood frequency	0	−100	−100	−100	−100	260	−100	−100	−100	−100	−100	−100
Large flood rise rate	−70	−100	−100	−100	−100	−88	−100	−100	−100	−100	−100	−100
Large flood fall rate	−70	−100	−100	−100	−100	−71	−100	−100	−100	−100	−100	−100

5. Conclusions

In developing countries like Nepal, where ecological information is insufficient and has inadequate baseline data regarding e-flows assessment, the hydrological methods can estimate e-flows requirements for planning and study phases. This study aimed to discuss e-flows calculation methodologies (hydrological) and to discuss the present status of the e-flows in Nepal. We compared six hydrological-based EFMs to allocate e-flows, to evaluate flow alteration, to estimate relative change (%) of the EFC against NFR, and to characterize the interannual and seasonal e-flows of the Kaligandaki River.

The results of the study showed that the global indexes such as Frequency and Duration Alteration Index (I_{FD}) and the Rate and Frequency Alteration Index (I_{RF}) showed a high alteration for all methods, except for the ADM and dynamic method (30%Q-D), which in turn showed a low alteration. The remaining three indexes, namely the Mean Monthly Flow Alteration Index (I_{MM}), the Magnitude and Duration of Extreme Flow Alteration Index (I_{MDE}), and the Timing of Extreme Flow Alteration Index (I_{TE}) showed moderate and mild alteration for all hydrological-based EFMs investigated in this case study. In the overall analysis, it can be seen that the flow alteration of five indexes and the e-flows component (EFC) is lower for ADM and dynamic methods compared to other hydrological methods considered in this study. Furthermore, the FDC of annual mean flows and annual seasonal mean flows showed a dramatic decrease in river flows, especially the high flows in most e-flow methods except for ADM and 30%Q-D methods. This concludes the practicability of the ADM and dynamic methods; it reflects the interannual variability of the river to meet the specific ecological function of the different sections of the river in different periods. Nevertheless, we suggest that the application of those methods should be made under biota requirements at a given river.

The ADM method used in the study was specially designed for large and medium-size perennial rivers, not for temporary or seasonal rivers. The runoff process of these rivers may be intervened more often, which can create a large error in the calculation procedure. Furthermore, many researchers highlighted that the interannual and intra-annual variability of flow in the river must be maintained to sustain the ecological biodiversity. Hence, within hydrological methods, a method that considers the dynamic nature of the flow regime of the river is recommended for the e-flows allocation.

Author Contributions: Conceptualization, methodology, software, N.S. and A.K.; formal analysis, validation, data curation, writing—original draft preparation, N.S. and A.K.; writing—review and editing, visualization, N.S., A.K., X.H., J.D., D.M., and A.W.; funding acquisition and supervision, A.K. and A.W. All authors have read and agreed to the published version of the manuscript.

Funding: This research received no external funding.

Acknowledgments: The first author would like to thank Kamal Prasad Pandey for his invaluable information on Nepal's hydrology and data acquisition. The first author is highly indebted to Khwopa College of Engineering for rendering a warm environment for the research work. Alban Kuriqi was supported by a Ph.D. scholarship granted by Fundação para a Ciência e a Tecnologia, I.P. (FCT), Portugal, under the Ph.D. Program FLUVIO–River Restoration and Management, grant number PD/BD/114558/2016. Thanks to the anonymous reviewers for their comments and suggestions to further improve the manuscript.

Conflicts of Interest: The authors declare no conflict of interest.

Appendix A

Table A1. IHA indicators are showing flow alteration regarding all applied EFMs against NFR.

IHA Parameters	Mean											
	30% Q-D	10% MAF	Q80%	Q85%	Q90%	ADM	Class B	Class C	Class D	Class E	Class F	Tennant
Group #1												
July	0.70	0.96	0.96	0.96	0.96	0.46	0.82	0.87	0.91	0.93	0.94	0.89
August	0.70	0.97	0.97	0.97	0.97	0.46	0.85	0.90	0.92	0.94	0.95	0.91
September	0.70	0.95	0.95	0.95	0.96	0.54	0.79	0.85	0.89	0.92	0.93	0.87
October	0.70	0.90	0.89	0.90	0.91	0.50	0.54	0.68	0.77	0.82	0.85	0.79
November	0.70	0.80	0.78	0.80	0.81	0.50	0.12	0.35	0.53	0.63	0.69	0.79
December	0.70	0.70	0.68	0.70	0.72	0.50	0.03	0.10	0.30	0.45	0.54	0.70
January	0.70	0.62	0.58	0.61	0.63	0.50	0.05	0.06	0.14	0.29	0.40	0.62
February	0.70	0.55	0.51	0.54	0.57	0.52	0.05	0.06	0.08	0.18	0.29	0.55
March	0.70	0.50	0.45	0.49	0.52	0.49	0.04	0.05	0.06	0.14	0.23	0.50
April	0.70	0.54	0.50	0.54	0.56	0.46	0.04	0.06	0.10	0.20	0.30	0.35
May	0.70	0.69	0.66	0.68	0.70	0.34	0.05	0.14	0.29	0.42	0.51	0.19
June	0.70	0.89	0.88	0.88	0.89	0.43	0.51	0.64	0.73	0.79	0.82	0.67
Group #2												
1-day minimum	0.70	1.10	0.09	0.07	0.06	0.02	0.69	0.00	0.00	0.02	0.06	0.32
3-day minimum	0.70	0.39	0.34	0.38	0.42	0.42	1.00	1.00	0.00	0.03	0.10	0.39
7-day minimum	0.70	0.41	0.35	0.40	0.43	0.42	1.00	0.00	0.00	0.04	0.11	0.41
30-day minimum	0.70	0.46	0.41	0.45	0.48	0.46	1.00	0.02	0.03	0.09	0.18	0.46
90-day minimum	0.70	0.51	0.47	0.50	0.53	0.47	1.00	0.03	0.05	0.14	0.24	0.51
1-day maximum	0.70	0.99	0.99	0.99	0.99	0.64	1.00	0.96	0.97	0.98	0.98	0.97
3-day maximum	0.70	0.99	0.98	0.98	0.99	0.59	0.93	0.95	0.96	0.97	0.98	0.96
7-day maximum	0.70	0.98	0.98	0.98	0.98	0.53	0.91	0.94	0.95	0.96	0.97	0.94
30-day maximum	0.70	0.97	0.97	0.97	0.97	0.51	0.87	0.91	0.94	0.95	0.96	0.92
90-day maximum	0.70	0.96	0.96	0.96	0.96	0.49	0.83	0.88	0.91	0.93	0.94	0.89
Number of zero days	0.00	0.00	0.00	0.00	0.00	0.00	0.00	0.00	0.00	0.00	0.00	0.00
Base flow index	0	4.580357	0.0008	0.000801	0	0.8071	1.455677	0.255647	0.253194	0.184867	0.080163	0.452621
Group #3												
Date of minimum	0.000	0.94	0.90	0.94	0.94	0.05	0.00	0.000	0.000	0.148	0.475	1.000
Date of maximum	0.00	0.18	0.18	0.18	0.18	0.03	0.18	0.18	0.185	0.178	0.178	0.183

Table A1. *Cont.*

IHA Parameters	Mean											
	30% Q-D	10% MAF	Q80%	Q85%	Q90%	ADM	Class B	Class C	Class D	Class E	Class F	Tennant
Group #4												
Low pulse count	0	1	0.988495	1	1	0.709692	0	C.067454	0.272659	0.580616	0.736016	1
Low pulse duration	0.00	1.00	0.91	1.00	1.00	0.41	0.00	0.09	0.56	0.61	0.75	1.00
High pulse count	0.00	1.00	1.00	1.00	1.00	0.29	1.00	1.00	1.00	1.00	1.00	0.44
High pulse duration	0.00	1.00	1.00	1.00	1.00	0.46	1.00	1.00	1.00	1.00	1.00	4.56
Low Pulse Threshold	0.70	0.57	0.53	0.56	0.59	0.45	0.01	0.05	0.14	0.25	0.34	1.00
High Pulse Threshold	0.70	0.96	0.95	0.96	0.96	0.52	0.79	0.85	0.89	0.92	0.93	0.88
Group #5												
Rise rate	0.70	1.00	0.99	1.00	1.00	0.67	0.91	0.93	0.96	0.97	0.98	0.86
Fall rate	0.70	1.00	0.99	1.00	1.00	0.63	0.90	0.92	0.94	0.95	0.97	0.68
Number of reversals	0.00	1.00	1.00	1.00	1.00	0.08	0.44	0.53	0.69	0.86	0.94	0.87

References

1. Couto, T.B.; Olden, J.D. Global Proliferation of Small Hydropower Plants—Science and Policy. *Front. Ecol. Environ.* **2018**, *16*, 91–100. [CrossRef]
2. Karimi, S.S.; Yasi, M.; Eslamian, S. Use of Hydrological Methods for Assessment of Environmental Flow in a River Reach. *Int. J. Environ. Sci. Technol.* **2012**, *9*, 549–558. [CrossRef]
3. Kuriqi, A.; Ali, R.; Pham, Q.B.; Gambini, J.M.; Gupta, V.; Malik, A.; Linh, N.T.T.; Joshi, Y.; Anh, D.T.; Nam, V.T.; et al. Seasonality Shift and Streamflow Flow Variability Trends in Central India. *Acta Geophys.* **2020**, *68*, 1461–1475. [CrossRef]
4. Ali, R.; Kuriqi, A.; Abubaker, S.; Kisi, O. Long-Term Trends and Seasonality Detection of the Observed Flow in Yangtze River Using Mann-Kendall and Sen's Innovative Trend Method. *Water* **2019**, *11*, 1855. [CrossRef]
5. Huang, X.; Suwal, N.; Fan, J.; Pandey, K.P.; Jia, Y. Hydrological Alteration Assessment by Histogram Comparison Approach: A Case Study of Erdu River Basin, China. *J. Coast. Res.* **2019**, *93*, 139–145. [CrossRef]
6. Gao, Y.; Pandey, K.P.; Huang, X.; Suwal, N.; Bhattarai, K.P. Estimation of Hydrologic Alteration in Kaligandaki River Using Representative Hydrologic Indices. *Water* **2019**, *11*, 688.
7. Suwal, N.; Huang, X.; Pandey, K.P.; Bhattarai, K.P. Assessment of Hydrological Alteration and Selection of Representative Hydrological Indicators in Erdu River. In Proceedings of the ICWRER 2019, Nanjing, China, 14–18 June 2019.
8. Tharme, R.E. A Global Perspective on Environmental Flow Assessment: Emerging Trends in the Development and Application of Environmental Flow Methodologies for Rivers. *River Res. Appl.* **2003**, *19*, 397–441. [CrossRef]
9. Kuriqi, A.; Pinheiro, A.N.; Sordo-Ward, A.; Garrote, L. Water-Energy-Ecosystem Nexus: Balancing Competing Interests at a Run-of-River Hydropower Plant Coupling a hydrologic–ecohydraulic Approach. *Energy Convers. Manag.* **2020**, *223*, 113267. [CrossRef]
10. Ali, R.; Kuriqi, A.; Abubaker, S.; Kisi, O. Hydrologic Alteration at the Upper and Middle Part of the Yangtze River, China: Towards Sustainable Water Resource Management Under Increasing Water Exploitation. *Sustainability* **2019**, *11*, 5176. [CrossRef]
11. Li, Q.; Gleeson, T.; Zipper, S.C.; Kerr, B. Too Many Streams and Not Enough Time or Money? New Analytical Depletion Functions for Rapid and Accurate Streamflow Depletion Estimates. OSF Preprints 2020. Available online: https://osf.io/gfhym (accessed on 21 October 2020).
12. Dyson, M.; Bergkamp, G.; Scanlon, J. *Flow: The Essentials of Environmental Flows*; IUCN: Gland, Switzerland; Cambridge, UK, 2003; pp. 20–87.
13. Smakhtin, V.; Eriyagama, N. Developing a Software Package for Global Desktop Assessment of Environmental Flows. *Environ. Model. Softw.* **2008**, *23*, 1396–1406. [CrossRef]
14. Arthington, A.H.; Bhaduri, A.; Bunn, S.E.; Jackson, S.E.; Tharme, R.E.; Tickner, D.; Young, B.; Acreman, M.; Baker, N.; Capon, S.; et al. The Brisbane Declaration and Global Action Agenda on Environmental Flows. *Front. Environ. Sci.* **2018**, *6*, 6. [CrossRef]
15. Pittock, J.; Lankford, B.A. Environmental Water Requirements: Demand Management in an Era of Water Scarcity. *J. Integr. Environ. Sci.* **2010**, *7*, 75–93. [CrossRef]
16. Xu, H.; Lee, U.; Coleman, A.M.; Wigmosta, M.S.; Sun, N.; Hawkins, T.R.; Wang, M.Q. Balancing Water Sustainability and Productivity Objectives in Microalgae Cultivation: Siting Open Ponds by Considering Seasonal Water-Stress Impact Using AWARE-US. *Environ. Sci. Technol.* **2020**, *54*, 2091–2102. [CrossRef] [PubMed]
17. De Graaf, I.E.M.; Gleeson, T.; Van Beek, L.P.H.R.; Sutanudjaja, E.H.; Bierkens, M.F.P. Environmental Flow Limits to Global Groundwater Pumping. *Nat. Cell Biol.* **2019**, *574*, 90–94. [CrossRef]
18. Gleeson, T.; Richter, B. How Much Groundwater Can We Pump and Protect Environmental Flows through Time? Presumptive Standards for Conjunctive Management of Aquifers and Rivers. *River Res. Appl.* **2017**, *34*, 83–92. [CrossRef]
19. Jowett, I.G. Instream Flow Methods: A Comparison of Approaches. *Regul. Rivers Res. Manag.* **1997**, *13*, 115–127. [CrossRef]
20. Williams, J.G.; Moyle, P.B.; Webb, J.A.; Kondolf, G.M. *Environmental Flow Assessment: Methods and Applications*; John Wiley & Sons: Hoboken, NJ, USA, 2019.
21. Lumbroso, D.M.; Sakamoto, D.; Johnstone, W.M.; Tagg, A.F.; Lence, B.J. Development of a Life Safety Model to Estimate the Risk Posed to People by Dam Failures and Floods. *Dams Reserv.* **2011**, *21*, 31–43. [CrossRef]

22. Acreman, M.C.; Dunbar, M.J. Defining Environmental River Flow Requirements—A Review. *Hydrol. Earth Syst. Sci.* **2004**, *8*, 861–876. [CrossRef]

23. Shokoohi, A.; Hong, Y. Using Hydrologic and Hydraulically Derived Geometric Parameters of Perennial Rivers to Determine Minimum Water Requirements of Ecological Habitats (case Study: Mazandaran Sea Basin-Iran). *Hydrol. Process.* **2011**, *25*, 3490–3498. [CrossRef]

24. Fuladipanah, M.; Jorabloo, M. Hydrological Method to Evaluate Environmental Flow (case Study: Gharasou River, Ardabil). *Int. J. Environ. Ecol. Eng.* **2015**, *9*, 62–65.

25. Dubey, A.; Singh, O.; Shekhar, S.; Pohshna, C. Assessment of Environmental Flow Requirement Using Environmental Management Classes-Flow Duration Curve for Narmada River. *Int. J. Curr. Microbiol. Appl. Sci.* **2019**, *8*, 891–897. [CrossRef]

26. Pandey, K.P. Study on Hydrologic Alteration and Alteration Parameter Reduction Methods. Master's Dissertation, Hohai University, Nanjing, China, 2019.

27. Smakhtin, V.U.; Shilpakar, R.L.; Hughes, D.A. Hydrology-Based Assessment of Environmental Flows: An Example from Nepal. *Hydrol. Sci. J.* **2006**, *51*, 207–222. [CrossRef]

28. Suwal, N. Research on Optimal Operation of Cascade Hydropower Stations Considering Ecological Flows. Master's Dissertation, Hohai University, Nanjing, China, 2019.

29. Młyński, D.; Operacz, A.; Walega, A. Sensitivity of Methods for Calculating Environmental Flows Based on Hydrological Characteristics of Watercourses Regarding the Hydropower Potential of Rivers. *J. Clean. Prod.* **2020**, *250*, 119527. [CrossRef]

30. Operacz, A.; Wałęga, A.; Cupak, A.; Tomaszewska, B. The Comparison of Environmental Flow Assessment—The Barrier for Investment in Poland or River Protection? *J. Clean. Prod.* **2018**, *193*, 575–592. [CrossRef]

31. Suwal, N.; Huang, X.; Kuriqi, A.; Chen, Y.; Pandey, K.P.; Bhattarai, K.P. Optimisation of Cascade Reservoir Operation Considering Environmental Flows for Different Environmental Management Classes. *Renew. Energy* **2020**, *158*, 453–464. [CrossRef]

32. Pastor, A.V.; Ludwig, F.; Biemans, H.; Hoff, H.; Kabat, P. Accounting for Environmental Flow Requirements in Global Water Assessments. *Hydrol. Earth Syst. Sci.* **2014**, *18*, 5041–5059. [CrossRef]

33. Smakhtin, V.; Anputhas, M. *An Assessment of Environmental Flow Requirements of Indian River Basins*; IWMI: Colombo, Sri Lanka, 2006; Volume 107.

34. Poff, N.L.; Tharme, R.E.; Arthington, A.H. Evolution of Environmental Flows Assessment Science, Principles, and Methodologies. In *Water for the Environment*; Academia Press: Cambridge, MA, USA, 2017; pp. 203–236.

35. Gaudel, P. Environmental Assessment of Hydropower Development in Nepal: Current Practices and Emerging Challenges. *Vidyut.* Febraury 2015. Available online: https://www.researchgate.net/publication/316080737_Environmental_Assessment_of_Hydropower_Development_in_Nepal_Current_Practices_and_Emerging_Challenges (accessed on 21 October 2020).

36. Doody, T.; Cuddy, S.; Bhatta. *Connecting Flow and Ecology in Nepal: Current State of Knowledge for the Koshi Basin*; Sustainable Development Investment Portfolio (SDIP) Project; CSIRO: Canberra, Australia, 2016.

37. Oglethorpe, J.; Regmi, S.; Bartlett, R.; Dongol, B.S.; Wikramanayake, E.; Freeman, S.J.O. The Value of a River Basin Approach in Climate Adaptation. In Proceedings of the International Conference on Climate Change Innovation and Resilience for Sustainable Livelihoods, Kathmandu, Nepal, 12–14 January 2015.

38. Gubhaju, S.R. Impact of Damming on the Aquatic Fauna in Nepalese Rivers. In *Cold Water Fisheries in Thetrans-Himalayan Countries*; Petr, T., Swar, S.B., Eds.; Food and Agriculture Organization of the United Nations: Rome, Italy, 2002; pp. 129–145.

39. Panta, S.K.; Resurrección, B.P. Gender and Caste Relations Amidst a Changing Political Situation in Nepal: Insights from a Farmer-Managed Irrigation System. *Gender Technol. Dev.* **2014**, *18*, 219–247. [CrossRef]

40. International Hydropower, A. *Hydropower Status Report: Sector Trends and Insights*; IHA: London, UK, 2018.

41. Jalsrot Vikas Sanstha; GWP Nepal. *Assessment of the Environmental Flow in the Gandaki River Basin: A Case of Modi Khola*; Jalsrot Vikas Sanstha: Kathmandu, Nepal; GWP Nepal: Hetauda, Nepal, 2016.

42. Rijal, N.; Shrestha, H.K.; Bruins, B. Environmental Flow Assessment of Hewa Khola A and Lower Hewa Khola Hydropower Projects in Nepal. *Hydro Nepal J. Water Energy Environ.* **2018**, *23*, 71–78. [CrossRef]

43. Zha-Rong, P.A.N.; Xiao-Hong, R.U.A.N.; Jing, X.U. A New Calculation Method of Instream Basic Ecological Water Demand. *J. Hydraul. Eng.* **2013**, *44*, 119–126.

44. Zhang, H.; Chang, J.; Gao, C.; Wu, H.; Wang, Y.; Lei, K.; Long, R.; Zhang, L. Cascade Hydropower Plants Operation Considering Comprehensive Ecological Water Demands. *Energy Convers. Manag.* **2019**, *180*, 119–133. [CrossRef]

45. Tennant, D.L. Instream Flow Regimens for Fish, Wildlife, Recreation and Related Environmental Resources. *Fish* **1976**, *1*, 6–10. [CrossRef]

46. Wałęga, A.; Mlynski, D.; Kokoszka, R.; Miernik, W. Possibilities of Applying Hydrological Methods for Determining Environmental Flows in Select Catchments of the Upper Dunajec Basin. *Pol. J. Environ. Stud.* **2015**, *24*, 2663–2676. [CrossRef]

47. Kuriqi, A.; Pinheiro, A.N.; Sordo-Ward, A.; Garrote, L. Influence of Hydrologically Based Environmental Flow Methods on Flow Alteration and Energy Production in a Run-of-River Hydropower Plant. *J. Clean. Prod.* **2019**, *232*, 1028–1042. [CrossRef]

48. Kuriqi, A.; Pinheiro, A.N.; Sordo-Ward, A.; Garrote, L. Flow Regime Aspects in Determining Environmental Flows and Maximizing Energy Production at Run-of-River Hydropower Plants. *Appl. Energy* **2019**, *256*, 113980. [CrossRef]

49. Bejarano, M.; Sordo-Ward, A.; Gabriel-Martin, I.; Garrote, L. Tradeoff Between Economic and Environmental Costs and Benefits of Hydropower Production at Run-of-River-Diversion Schemes under Different Environmental Flows Scenarios. *J. Hydrol.* **2019**, *572*, 790–804. [CrossRef]

50. Richter, B.D.; Baumgartner, J.V.; Powell, J.; Braun, D.P. A Method for Assessing Hydrologic Alteration Within Ecosystems. *Conserv. Biol.* **1996**, *10*, 1163–1174. [CrossRef]

51. Mathews, R.; Richter, B.D. Application of the Indicators of Hydrologic Alteration Software in Environmental Flow Setting1. *JAWRA J. Am. Water Resour. Assoc.* **2007**, *43*, 1400–1413. [CrossRef]

52. Olden, J.D.; Poff, N.L. Redundancy and the Choice of Hydrologic Indices for Characterizing Streamflow Regimes. *River Res. Appl.* **2003**, *19*, 101–121. [CrossRef]

53. Fausch, K.D.; Bestgen, K.R. Ecology of Fishes Indigenous to the Central and Southwestern Great Plains. In *Ecological Studies*; Springer: New York, NY, USA, 1997; pp. 131–166.

54. Rood, S.B.; Mahoney, J.M. River Damming and Riparian Cottonwoods along the Marias River, Montana. *Rivers* **1995**, *5*, 195–207.

55. Richter, B.; Baumgartner, J.; Wigington, R.; Braun, D. How Much Water Does a River Need? *Freshw. Biol.* **1997**, *37*, 231–249. [CrossRef]

56. Richter, B.D.; Baumgartner, J.V.; Braun, D.P.; Powell, J. A Spatial Assessment of Hydrologic Alteration within a River Network. *Regul. Rivers Res. Manag.* **1998**, *14*, 329–340. [CrossRef]

57. Książek, L.; Woś, A.; Florek, J.; Wyrębek, M.; Młyński, D.; Wałęga, A. Combined Use of the Hydraulic and Hydrological Methods to Calculate the Environmental Flow: Wisloka River, Poland: Case Study. *Environ. Monit. Assess.* **2019**, *191*, 254. [CrossRef] [PubMed]

58. Młyński, D.; Wałęga, A.; Ozga-Zielinski, B.; Ciupak, M.; Petroselli, A. New Approach for Determining the Quantiles of Maximum Annual Flows in Ungauged Catchments Using the EBA4SUB Model. *J. Hydrol.* **2020**, *589*, 125198. [CrossRef]

59. Cushman, R.M. Review of Ecological Effects of Rapidly Varying Flows Downstream from Hydroelectric Facilities. *North Am. J. Fish. Manag.* **1985**, *5*, 330–339. [CrossRef]

60. Verma, R.K.; Murthy, S.; Verma, S.; Mishra, S.K. Design Flow Duration Curves for Environmental Flows Estimation in Damodar River Basin, India. *Appl. Water Sci.* **2016**, *7*, 1283–1293. [CrossRef]

61. Vogel, R.M.; Fennessey, N.M. Flow-Duration Curves. I: New Interpretation and Confidence Intervals. *J. Water Resour. Plan. Manag.* **1994**, *120*, 485–504. [CrossRef]

62. Searcy, J.K. *Flow-Duration Curves*; manual of hydrology. Part 2. US Geol. Survey Water Supply Paper 1542-A., Low flow techniques; United States Government Printing Office: Washington, DC, USA, 1959.

63. Jain, S.K.; Kumar, P. Environmental Flows in India: Towards Sustainable Water Management. *Hydrol. Sci. J.* **2014**, *59*, 751–769. [CrossRef]

Publisher's Note: MDPI stays neutral with regard to jurisdictional claims in published maps and institutional affiliations.

sustainability

Article

New Insights on Flood Mapping Procedure: Two Case Studies in Poland

Andrea Petroselli [1], Jacek Florek [2], Dariusz Młyński [3], Leszek Książek [2] and Andrzej Wałęga [3,*]

1   Department of Economy, Engineering, Society and Business (DEIM), Tuscia University, Via San Camillo De Lellissnc, 01100 Viterbo, Italy; petro@unitus.it
2   Department of Hydraulics Engineering and Geotechnics, University of Agriculture in Krakow, St. Mickiewicza 24–28, 30-059 Krakow, Poland; rmflorek@cyf-kr.edu.pl (J.F.); leszek.ksiazek@urk.edu.pl (L.K.)
3   Department of Sanitary Engineering and Water Management, University of Agriculture in Krakow, St. Mickiewicza 24–28, 30-059 Krakow, Poland; dariusz.mlynski@urk.edu.pl
*   Correspondence: andrzej.walega@urk.edu.pl; Tel.: +48-12-662-4029

Received: 27 September 2020; Accepted: 13 October 2020; Published: 14 October 2020

Abstract: The use of the Mike11 one-dimensional (1D) hydraulic model, together with official hydrology, represents a standard approach of the National Water Management Authority (NWMA) in Poland for flood mapping procedures. A different approach, based on the hydrological Event-Based Approach for Small and Ungauged Basins (EBA4SUB) model and the Flood-2 Dimensional (FLO-2D) hydraulic model has here been investigated as an alternative procedure. For the analysis, two mountainous rivers in Poland were selected: Kamienica Nawojowska is characterized by a narrow valley, while Skawinka has a broad valley. It was found that the flood zones can enormously differ locally, with larger zones generated by the Mike11/NWMA model in some cases and by the EBA4SUB/FLO-2D model in other situations. The benefits of using the two-dimensional (2D) model are consistent in areas without drainage and where the connection to the main channel is insufficient. The use of 1D modeling is preferred for the possibility of mapping the entire river network in a short computational time.

Keywords: flood hazard zone; EBA4SUB model; FLO-2D; Mike11

---

1. Introduction

Floods are a natural phenomenon that occurred in the past and that will be repeated in the future. For a natural ecosystem, periodic flooding from the river plays a positive role. In situations of intensive urban development, when an increase in urbanization of areas located in the proximity of rivers is observed, floods can cause considerable damages. Adequate flood protection can only be implemented by combining technical measures with proper urban development planning [1,2]. According to the provisions of the Floods Directive [3], flood risk management plans, and their derived flood-prone areas, are crucial elements of flood hazard reduction. In Poland, actions have been carried out for many years in order to delineate such flood-prone areas. The National Security System (ISOK) has the necessary task to provide an effective system for protecting the country from extraordinary hazards, which is particularly essential due to the growing number of such events and the increasing scale of their economic and social impact. Following this task, flood hazard management plans were developed for the majority of catchments in Poland, mainly the ones having discharge gauging stations, allowing local authorities to make decisions about the identification of flood hazard zones.

When determining the flood hazard zones, hydraulic models are used to forecast the water flow surface during the flood wave [4]. When using one-dimensional (1D) models to determine the extension of flood hazard zones, the digital elevation model (DEM) should be subtracted from the water surface model (WSM), which is based on water surface profiles in river cross-sections. The accuracy of determining flood hazard zones is affected by many factors, which include accuracy of data acquisition, type of hydraulic model, DEM accuracy, identification of land cover, the accuracy of mapping the configuration of the riverbed [4], and the accuracy and quality of hydrological data [5].

In the case of hydraulic modeling, 1D or two-dimensional (2D) models, like FLO-2D, can be used [6]. Petroselli et al. [7] and Vojtek et al. [8] showed the main problems with the use of both hydrologic and hydraulic models in the generation of flood mapping. The main problems are the role of DEM resolution, the choice of hydrologic modeling, and the fact that 1D hydraulic models do not provide information about the direction of the flow field or the course of flowing around obstacles (e.g., buildings), which is a predominant characteristic in urban areas. As specified in the local spatial development plans, flood hazard mapping constitutes an essential instrument in the decision-making process performed by local authorities to tackle areas that are particularly threatened with flooding [9]. Numerical models allowing forecasting the water surface level during the passing of the flood wave are used in the determination of the flood hazard zone. In practice, the model is a compromise between obtaining a solution and obtaining a sufficient number of parameters describing the object and the accuracy of the result. The stages of developing a 1D model include implementation of the river network, cross-sections, and hydraulic structures; identification of land cover; setting of initial conditions; and finally, the calibration and verification of the model [10]. In order to define the boundaries of flood zones, the water surface model, which is based on the water surface level in cross-sections, should be combined with the digital elevation model (DEM). The accuracy of determining the water surface level consists of many factors, including the quality of hydrological data, the type of numerical model, the accuracy of the DEM [11], the identification of land cover, or the accuracy of mapping the bottom of the water wetted channel [12].

In this work, hydrological and hydraulic calculations that allow the generation of flood hazard zones are key issues. Hydrological rainfall–runoff models are very popular in the assessment of the design discharge and hydrograph to be propagated by the hydraulic models. Design hydrographs may be derived for instance by synthetic means such as SCS Unit Hydrograph (UH) or Snyder UH [13–15]. Geomorphological instantaneous unit hydrograph (GIUH) is also commonly used to assess the design hydrograph [16]. After having defined the design hydrograph, in order to calculate the water level for different return periods, hydraulic 1D models are commonly used, like HEC-RAS or Mike11 (Hydro River). Polish experiences from the use of such rainfall–runoff models showed some problems. The main problems concern the following issues: lack of recommended rainfall hyetographs [17], high sensitivity of synthetic UH to the distribution of rainfall gauge stations, and errors in rainfall data [18]. The review of the critical problems in the application of the hydrological model in the Polish situation shows the need to develop a method that is more objective in practical applications. The Event-Based Approach for Small and Ungauged Basins (EBA4SUB) rainfall–runoff model, introduced by Piscopia et al. [19] and Petroselli and Grimaldi [20], is an example of a model that may be used for ungauged catchments. This model is based on geographic information systems and the optimization of the topographic information in the DEM, and it uses the same input data necessary to apply the well-known rational formula. Młyński et al. [21] provided an initial study in the application of this model under Polish conditions. In the present study, the EBA4SUB/FLO-2D framework was used to create flood zones for the following reasons: the framework provides an accurate estimation of the design hydrograph, minimizing at the same time the subjectivity of the practitioner related to the choice of the input parameters. Indeed, the EBA4SUB model reduces the user subjectivity in modeling both the infiltration process (thanks to the automatic estimation of the Curve Number (CN) parameter) and both the propagation process (thanks to the automatic estimation of concentration time). In brief, the EBA4SUB model proposes a framework that provides similar results when two analysts apply it at different times for the same watershed and input data. Generally, using the EBA4SUB model the two following benefits can be achieved. First, in excess rainfall estimation, thanks to the Curve Number for Green-Ampt (CN4GA) internal routine and its automatic calibration, we benefit from the accuracy of a physically based infiltration scheme (the Green-Ampt equation) mixed with the simplicity of an empirical approach (the CN method). Second, for excess rainfall-direct runoff transformation, we determine the instantaneous unit hydrograph (IUH) shape using detailed geomorphological information pixel by pixel, avoiding the use of synthetic methods [20].

Bearing in mind the limitations associated with the hydrological rainfall–runoff models used in Poland for determining flood zones, this study aimed to analyze the possibilities of using the EBA4SUB model to determine the extent of flood zones. The EBA4SUB model was here coupled with FLO-2D, and the results have been compared with the official procedure used in Poland by the National Water Management Authority (i.e., the use of Mike11/NWMA). The novelty of this work is the use of a combined approach based on the hydrological EBA4SUB model with a 2D hydraulic model, FLO-2D, to generate flood zones in two mountainous catchments in Poland. It should be emphasized that there has been no analysis so far regarding the possibility of using this approach to determine the design hydrographs in this region. The study was carried out for selected catchments of the Upper Vistula river basin. Therefore, the conducted study will allow us to determine whether this model may be an alternative to commonly used methodological approaches. Moreover, comparing the official procedure with the proposed approach based on the combination of EBA4SUB and FLO-2D will allow us to understand if an update of the existing flood-prone areas maps is possible, minimizing the user subjectivity.

2. Materials and Methods

2.1. Study Area

The study regarded two catchments of the Upper Vistula basin: Skawinka and Kamienica Nawojowska. The Upper Vistula basin accounts for about 25% of the entire Vistula basin and about 15% of Poland's area and covers the south-eastern region of the country. It covers part of the Carpathians, Subcarpathian valleys, and Małopolska Uplands. The Upper Vistula basin area shows large variations in elevation [21]. Physiographic parameters of the studied catchments (Skawinka and Kamienica Nawojowska, respectively) are: catchment area (A) 316.0 and 238.0 km^2, main river length (L) 34.0 and 33.1 km, average main river slope (I) 10.3 and 17.3‰, average catchment slope (Ψ) 18.6 and 31.0‰, soil imperviousness index (N) 34.0 and 33.0%, and runoff coefficient (Φ) 0.62 and 0.82. The Skawinka catchment area includes agricultural areas (75.6%), forests (20.8%), and artificial surfaces (3.6%). The Kamienica Nawojowska catchment is characterized by forests and seminatural areas (58.6%), agricultural areas (36.4%), and artificial surfaces (5.0%). The studied catchments have a different hydrographic system and the analyzed zones include estuary river sections with a length of about 9 km (Figure 1). The layout of the river network indicates differences between Kamienica Nawojowska and Skawinka. The tributaries of Kamienica are concentrated in the central part of the basin and the supply is mainly on the left bank. Skawinka has tributaries distributed more evenly along the entire length.

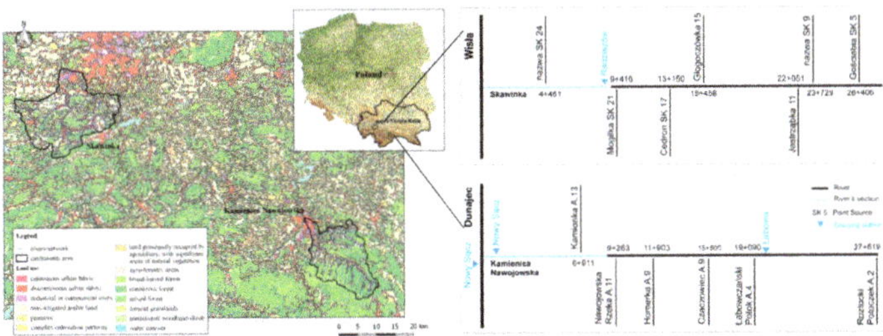

Figure 1. River network layout for Kamienica Nawojowska and Skawinka. The modelled sections included in both methods (Mike11/National Water Management Authority (NWMA) and FLO-2D) modeling are marked blue.

2.2. Materials

The physiographic characteristics for the analyzed catchments were determined on the basis of the Hydrographic Division Map for Poland, the DEM with a resolution of 100 m, the Corine Land Cover database, and the Soil Map for Poland. The source of DEM, aerial photos, and images is the Polish Central Office of Geodesy and Cartography [GUGiK], which provides open access services. In particular, the open access services used in the present study are: http://www.gugik.gov.pl/pzgik/zamow-dane/ortofotomapa, http://www.gugik.gov.pl/pzgik/dane-bez-oplat/dane-dotyczace-numerycznego-modelu-terenu-o-interwale-siatki-co-najmniej-100-m-nmt_100, and http://www.gugik.gov.pl/pzgik/zamow-dane/numeryczny-model-terenu. The source of river channel data was the Mike11 model, which was developed in 2015 in connection with the implementation of the flood hazard maps and flood risk maps. The Polish National Water Management Authority implemented the project. The daily rainfall and flow data for the calculation of the 100-years return period of annual maximum daily precipitation and annual maximum flow were obtained from the Institute of Meteorology and Water Management in Warsaw. The obtained data covered the period from 1981 to 2014. Starting from the available data, hydrologic calculations were first performed employing the EBA4SUB rainfall–runoff model. Then, hydraulic calculations were performed employing one 1D model (Mike11) and one 2D model (FLO-2D), as explained in the following sections. Flood-prone areas were determined using both the hydraulic models assuming a return period of 100 years.

2.3. Methods

2.3.1. The Proposed Approach

The data needed to apply the EBA4SUB hydrologic model include the selection of a gross rainfall design hyetograph, the DEM of the area, and data on the catchment land use and soil properties [20]. The gross rainfall cumulative value was here assessed on the based on the maximum daily precipitation for a 100-year return period, calculated by log-normal distribution on observed data. It was here assumed that rainfall causes the highest runoff from catchment with a duration equal to concentration time (Tc), so the calculated maximum daily precipitation was rescaled to a rainfall duration equal to Tc. In this approach we assume that the critical rainfall duration is equal to the catchment concentration time, following the theoretic assumption that this choice causes the maximum peak discharge at the outlet compared to shorter or longer rainfall durations [22]. It is noteworthy, anyway, that this hypothesis is debated in the literature. In many practical applications, rainfall durations 2–3 times larger than the concentration time are used in order to maximize the peak discharge [23].

The gross rainfall design hyetographs in this study were determined distributing, within *Tc*, the gross rainfall cumulative value using the Chicago, the beta distribution, and the Deutscher Verband für Wasserwirtschaft und Kulturbau (DVWK) methods, respectively. For beta distribution, the following parameters were assumed: 5 and 2; 2 and 2; 2 and 3. The choice of these parameters resulted from the verification of the impact of changing the shape of the hyetograph of precipitation on the peak culmination and, in consequence, on the extent of the flood zones. The excess rainfall hyetograph was determined according to the Curve Number for Green-Ampt (CN4GA) procedure proposed by Grimaldi et al. [24]. CN4GA is based on the Curve Number (CN) method and the Green-Ampt (GA) equation. The CN method first estimates the excess rainfall cumulative value with the following formula:

$$P_n = \begin{cases} \frac{(P - 0.2S)^2}{P + 0.8S} & \text{when } P > 0.2\,S \\ 0 & \text{when } P < 0.2\,S \end{cases} \tag{1}$$

where: P_n—excess rainfall cumulative value [mm], P—gross rainfall cumulative value [mm], S—maximum potential catchment retention [mm] that is determined based on the CN value of the investigated area.

The Green-Ampt equation is then used to determine the excess rainfall hyetograph [25]:

$$q_0(t) = \begin{cases} i(t) \text{ for } t < t_{pon} \\ K_s\left(1 + \frac{\Delta\theta\Delta H}{I(t)}\right) \text{ for } t > t_{pon} \end{cases} \tag{2}$$

where: $q_{0(t)}$—infiltration rate, t_{pon}—ponding time, $I(t)$—cumulative infiltration, K_s—saturated hydraulic conductivity, $\Delta\theta$—change in soil-water content between the initial value and the field saturated soil-water content, ΔH—the difference between the pressure head at the soil surface and the matrix pressure head at the moving wetting front.

The CN4GA automatically assures that the calculated excess rainfall event has the same cumulative value and the same initial abstraction value derived with the CN method. However, it presents a physically based time distribution.

The CN parameter is the only parameter governing the relationship among the gross rainfall cumulative value and the excess rainfall cumulative value. Its value is determined based on soil type, land cover, and land management condition, according to the official tables provided in the the United States Department of Agriculture (USDA) [26]. The The Antecedent Moisture Condition (AMC) II (average condition for soil moisture) was here assumed. The soil types were established based on the Soil Map for Poland. The land cover and land management were established from the Corine Land Cover base.

Runoff hydrograph is then determined using the DEM that was adequately preprocessed, using the width function based instantaneous unit hydrograph (*WFIUH*). In particular, the DEM preprocessing analysis was performed as follows. First, pits and flat areas were removed. Then, flow paths were defined using an optimized flow direction algorithm [27]. Third, the river network was extracted using the drop analysis and the *WFIUH* was calculated using the following equation [28]:

$$WFIUH(t) = \frac{L_c(x)}{V_c(x)} + \frac{L_h(x)}{V_h(x)} \tag{3}$$

where: L_c, L_h—length of the path for the channel and hill slope cell of the DEM, V_c, V_h—surface flow velocity for the channel and hillslope cell.

Regarding V_c and V_h, the hillslope surface flow velocity is linked according to empirical formulas to the local slope and land cover data. In contrast to the channel surface flow, velocity is calibrated so that the projection on the time axis of the WFIUH centre of mass is equal to the basin lag time(T_L). The lag time is expressed as the 60% of T_c, that is calculated using the Giandotti formula [29]. After having defined the WFIUH, the design hydrograph $Q(t)$ can be calculated with the following equation:

$$Q(t) = A\int_0^t WFIUH(t - \tau)P_n(\tau)d\tau \tag{4}$$

where: A—catchment area [km^2], t—precipitation duration [h], $P_n(\tau)$—excess rainfall determined with the CN4GA method [mm].

In order to characterize the design hydrograph shape, the values of wave slenderness coefficients were determined from the relationship [30]:

$$\alpha = \frac{t_o}{t_s} \tag{5}$$

where: t_o—wave fall time [h], t_s—wave rise time [h].

The quality of the simulations obtained using the EBA4SUB model was assessed with the mean absolute percentage error (MAPE), described by the relationship [30]:

$$MAPE = \frac{Q_{s,max} - Q_{m,max}}{Q_{s,max}} \cdot 100 \ [\%] \tag{6}$$

where: $Q_{m,max}$—maximum flow with a specified return period, calculated using EBA4SUB [m$^3 \cdot$s^{-1}], $Q_{s,max}$—maximum flow with the same return period, calculated using the log-normal distribution [m$^3 \cdot$s^{-1}].

The 2D FLO-2D model was used to determine flood-prone areas. FLO-2D employs the dynamic wave momentum equation solved on a numerical grid of square cells, whose resolution depends on the adopted hydrograph peak discharge [31]. In the present study, aerial photos and images allowed the estimation of hydraulic parameters such as floodplain roughness coefficients. In particular, Manning's roughness coefficients of 0.04 and 0.045 were assigned for the Kamienica Nawojowska main channel and the Skawinka main channel, respectively, with the value of 0.08 assigned for the floodplain areas of both case studies. The channel cross sections were approximated, starting from the existing Mike11 data, to a trapezoidal shape (45° side slope) with a maximum width of 45 m and a maximum depth of 2 m for Kamienica Nawojowska, and to a triangular shape with a maximum width of 20 m and a maximum depth of 4 m for Skawinka. The selected hydraulic domains are represented by the last 9 km of the Skawinka main channel (the total hydraulic modeling area is equal to 27 km^2) and by the last 9 km for the Kamienica Nawojowska main channel (the total hydraulic modeling area is equal to 18 km^2). The duration of computer simulations can take a few hours for each one. In the FLO-2D model, the flood-prone areas are expressed in a gridded way, and they are the result of the maximum flow depth occurring in the specific pixel in the whole simulation. In both investigated areas, the key parameters affecting the extension of the zones and their shape are the design hydrograph peak discharge and its total volume. The spatial resolution for the EBA4SUB data was 100 m, for FLO-2D input data 8m, for FLO-2D output data 50 m.

2.3.2. The Standard Approach

The process of generating flood hazard maps is usually carried out in Poland using several stages (Mike11/NWMA procedure). (1) Identification of input data regarding the catchment, river system, floodplains, and hydraulics structures. (2) Performing geodetic measurements in the channel and floodplain terraces. (3) Hydrological and hydraulic calculations (Mike11). (4) Calibration and verification of the model. (5) Determination of flood-prone areas [32].

Transformation of the excess rainfall into direct runoff is performed using the Soil Conservation Service Unit Hydrograph (SCS-UH). The excess rainfall hyetograph is determined according the Curve Number (SCS-CN) method described above. The shape of rainfall hyetograph is assumed based on beta distribution [21]. In regards to the SCS-UH method, in order to determine the Synthetic Unit Hydrograph (SUH) shape from the nondimensional q/q_p versus the t/tp hydrograph, the peak discharge q_p and the time to peak t_p are computed, as:

$$q_p = \frac{2.08 \cdot A}{t_p} \tag{7}$$

$$t_p = \frac{t_r}{2} + T_l \tag{8}$$

where: A—the catchment area, t_r—the duration of rainfall T_l—the lag time from centroid of rainfall to peak discharge.

The T_l can be calculated from watershed characteristics using main stream length L, watershed slope s, and curve number (CN):

$$T_l = \frac{(3.28 \cdot L \cdot 1000)^{0.8} \left(\frac{1000}{CN} - 9\right)^{0.7}}{1900 \cdot s^{0.5}} \tag{9}$$

In this approach, the 1D Mike11 model was used for calculations of unsteady flow in watercourses based on continuity equation and the Saint-Venant's equation of momentum conservation [33]:

$$\frac{\partial Q}{\partial x} + \frac{\partial A}{\partial t} = q \tag{10}$$

$$\frac{\partial Q}{\partial t} + \frac{\partial \left(\alpha_c Q^2 A^{-1}\right)}{\partial x} + gA\frac{\partial h}{\partial x} + \frac{gQ|Q|}{C^2 AR} = 0 \tag{11}$$

where: Q—discharge, A—cross-section of the bed, q—lateral tributary, h—ordinate of the water surface, x—longitudinal coordinate measured alongside the river bed, R—hydraulic radius, α_c—Coriolis coefficient, t—time, g—gravitational acceleration.

The last equation considers incompressibility and homogeneity of water, small bottom declines, flux parallel to the bottom, and steady motion. For a rapid flow, a reduced equation is used:

$$\frac{\partial Q}{\partial t} + gA\frac{\partial h}{\partial x} + \frac{gQ|Q|}{C^2 AR} = 0 \tag{12}$$

The exactness of water surface reproduction in the Mike11 software results from the exactness of iterative discharge calculation (10^{-4}) and the area (10^{-3}) and subsequent recalculation for the water surface elevation [34]. The duration of computer simulations last a few seconds (from 4 to 17). The Mike11 model was used for computation of water surface levels at cross-sections in main channel and flood zones. River models accounted for a river of the length of significance in terms of flood protection, i.e., for Kamienica Nawojowska 27.6 km and for Skawinka 35.0 km. Boundary conditions covered all required tributaries concentrated and distributed as provided for in the guidelines [35]. In Mike11 flood-prone areas are determined from the intersection of the numerical water surface model (WSM) with the DEM. WSM was generated using ordinates in cross-sections. Moreover, areas without a hydraulic connection in the riverbed were omitted, as were areas where the water depth is less than the accuracy of DEM. Regarding the input data, for Mike 11 flood zones generation procedure the resolution is less than 1 m.

For both the proposed approach and the standard approach, the source of DEM, aerial photos and images was the Polish Central Office of Geodesy and Cartography [36], which provides open access services. The source of river channel data was a Mike11 model, which was developed in 2015 in connection with the implementation of the flood hazard maps and flood risk maps. The Polish National Water Management Authority implemented the project. Regarding the land cover data and FLO-2D, references are [37,38]. In both the approaches, the final modelled grids were resampled at 50 m × 50 m resolution to provide a direct comparison.

Regarding the calibration of both the approaches, in the proposed approach EBA4SUB is characterized by two advantages. First, the excess rainfall calculation is automatically performed matching the cumulate excess rainfall values computed by applying Equation (2) and Equation (1). It is noteworthy that this approach combines the accuracy of a physically based infiltration scheme (the Green-Ampt equation) with the simplicity of an empirical approach (the SCS-CN method) employing only one parameter (the CN). Second, the WFIUH is calculated based on the basin concentration time that is automatically estimated by the model. Regarding the hydraulic modeling, the proposed approach calibration was not performed (note: for calibration we mean the fitting of parameters of a model to achieve results of simulation more similar to observation). In case of the standard approach, the calibration process was performed during project "Flood protection programme in the Upper Vistula (2014–2015)". The results of calibration of Mike11/NWMA procedure follow the National Water

Management Authority rules, where the used metrics are the correlation coefficient (that were higher than 0.70), special correlation coefficient ($\geq$0.70), total square error ($\leq$10%) and errors of culmination level ($\leq$0.15 m), culmination flow ($\leq$10%), culmination dislocation ($\leq 1\frac{1}{2}$ h), and flood wave volume ($\leq$10%). In case of calibration, the model error must qualify it to the range of excellent–good, and for verification to the range of excellent–quite good.

3. Results

3.1. Initial Analysis

The initial analysis of hydrological data was conducted for annual maximum flows time series for Kamienica Nawojowska and Skawinka. The analysis included the determination of some descriptive statistics: positional measures: min, average, and max; measures of dispersion: standard deviation (S) and coefficient of variation (C_s), and measures of shape of the studied variate distribution: coefficient of skewness (Ske) and kurtosis (K). The results are presented in Table 1.

Table 1. Values of descriptive statistics for the annual maximum flows time series for analyzed catchments.

Catchment	Min	Average	Max	S	Cs	Ske	K
Kamienica	26.2	147.3	405.0	110.8	0.8	1.0	−0.2
Skawinka	13.0	92.4	346.0	82.8	0.9	1.6	2.2

Based on the results summarized in Table 1, it was found that the differences between minimum and maximum observed annual maximum flow was 94% for Kamienica Nawojowska and 96% for Skawinka. The dynamics of changes in maximum annual flows remained at a high level, which was evidenced by the coefficient of variation. The coefficients of variations for analyzed catchments are strongly related to the high variability of maximum precipitation, which has an enormous impact on flooding. Flood size in this region depends on multiple factors such as geology, soils, geomorphological evolution, landscape structure, or land use. These factors turn precipitation transformation into the runoff. Analyzing the values of the skewness coefficients, for Kamienica Nawojowska and Skawinka, it was shown that it is greater than 0. Therefore, right-sided asymmetry of the empirical distributions of random variables was found. This is because in the analyzed time–series most observations are smaller than their average value. Therefore, the average values are greater than the medians of time series. In turn, the kurtosis values indicated the platykurtic (negative value) and leptokurtic (positive value) empirical distributions for Kamienica Nawojowska and Skawinka respectively.

3.2. EBA4SUB Design Hydrographs

The EBA4SUB model was used to determine the design hydrographs to be propagated at the beginning of the investigated river sections. Hydrographs were determined for the following gross rainfall distributions: beta, DVWK, and Chicago. The characteristics of the modelled design hydrographs are presented in Table 2. The hydrographs shapes are shown in Figure 2a,b.

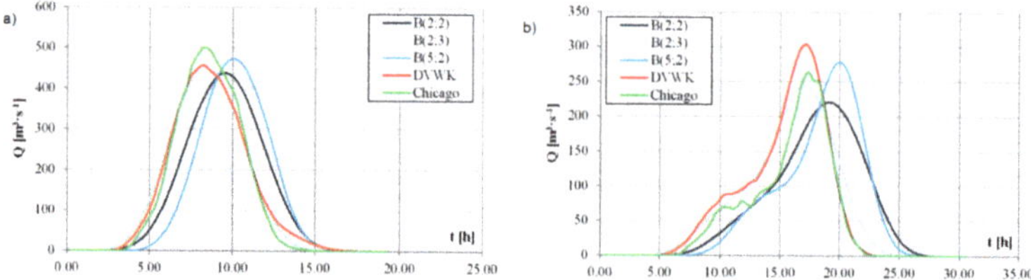

Figure 2. Design hydrographs determined with the Event-Based Approach for Small and Ungauged Basins (EBA4SUB) model, return period 100 years: (**a**) The Kamienica Nawojowska River, (**b**) The Skawinka River.

Table 2. Design hydrographs characteristics determined with the EBA4SUB model.

Parameters	Standard *	Advanced Approach				
		B(2;2)	B(2;3)	B(5;2)	DVWK	Chicago
Kamienica Nawojowska						
Q_{max} [m^3·s^{-1}]	576.39	438.8	450.6	472.7	456.4	499.8
V [mln·m^3]	15.08			8.751		
t [h]	34.0			19.25		
α [–]		1.026	1.265	0.878	0.925	1.333
MAPE [%]		23.0	21.0	17.0	20.0	12.0
Skawinka						
Q_{max} [m^3·s^{-1}]	470.38	220.8	244.9	277.9	303.2	263.5
V [mln·m^3]	26.41			6.815		
t [h]	89.4			30.5		
α [–]		0.605	0.794	0.525	0.768	0.794
MAPE [%]		45.0	39.0	31.0	25.0	35.0

* alues from the Mike11 model (NWMA).

Based on the results, it was found that the highest peak discharge value for the Kamienica Nawojowska river was obtained when a Chicago hyetograph is assumed. The lowest peak discharge value was obtained for the beta hyetograph with parameters 2 and 2. The difference between the highest and lowest value in the peak discharge value is 12%. The values of the wave slenderness coefficient for the obtained hydrographs were close to 1.000. This means that the volumes of the rising and falling parts are similar to each other. The MAPE error values indicate that the peak discharge determined by the EBA4SUB model is close to the value from the statistical method for the hydrograph where precipitation distribution was assumed according to the Chicago method. In the case of the Skawinka River, the highest peak discharge value from the EBA4SUB model was obtained for the hydrograph where the hyetograph was assumed according to the DVWK method. The lowest peak discharge value, similarly to the Kamienica Nawojowska river, was obtained for the beta distribution with parameters 2 and 2. The difference between the highest and the smallest size of the peak discharge is 27%. The slenderness coefficient values are below 1.000. It means that the volumes of the rising part are larger than the corresponding of the falling part. The MAPE error values indicate that the peak discharge determined by the EBA4SUB model is close to the value from the statistical method if precipitation distribution is assumed according to the DVWK method. This statement is close to early work of the authors [22], where it was showed that peak discharges from the EBA4SUB model is closer to quantiles of peak discharges achieved from statistical method than Qp from other rainfall–runoff models, like Snyder or SCS-UH. A second work [21] showed that Qp from the EBA4SUB model where gross rainfall hyetograph was determined using beta distribution was closer to Qp from Pearson III type distribution. A third work [39] shows that the EBA4SUB model can be used as part of new method to generate design peak discharges in mountainous catchments in Poland. The presented results confirm that it is necessary to calibrate and verify hydrological models to eliminate errors that affect the quality of the obtained results [37].

3.3. Hydraulic Modeling and Flood-Prone Areas Determination

Flood-prone areas modeled employing the previously described methodologies are shown in Figure 3.

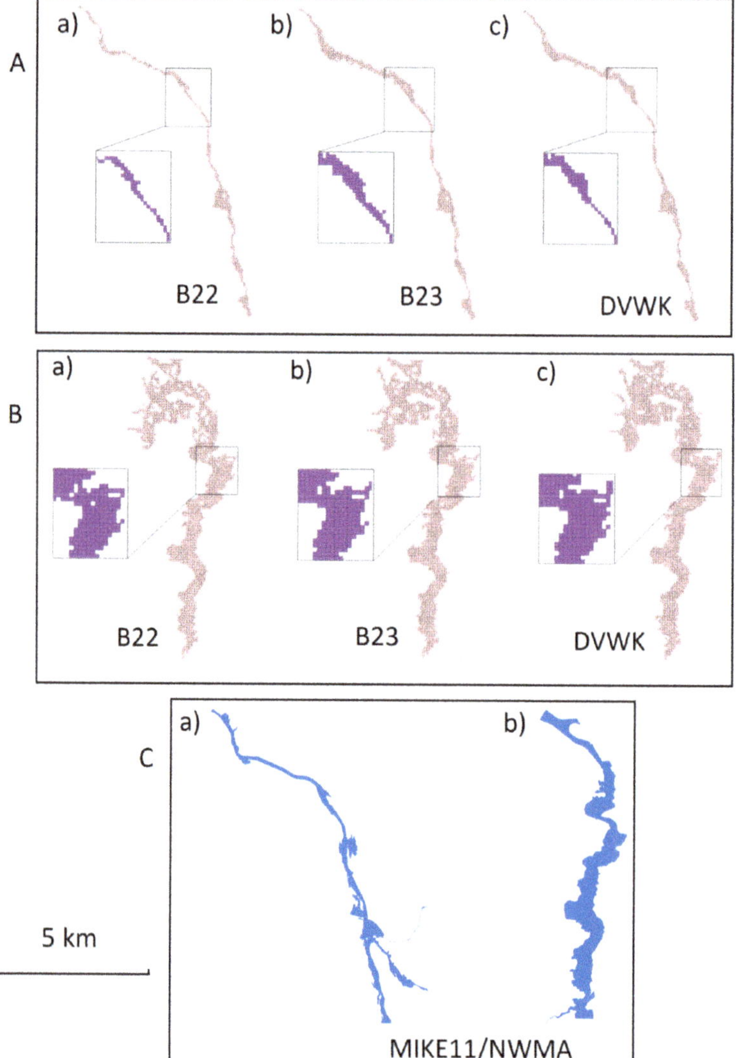

Figure 3. (**A**) Flood-prone areas; the Kamienica Nawojowska River—FLO-2D: (a) B(2;2), (b) B(2;3), (c) DVWK (we report only the cases with the major differences). (**B**) Flood-prone areas; the Skawinka River—model FLO-2D: (a) B(2;2), (b) B(2;3), (c) DVWK (we report only the cases with the major differences). (**C**) Flood-prone areas; Mike11/NWMA: (a) Kamienica Nawojowska, (b) Skawinka.

Table 3 summarizes the total areas of flood hazard map covering the investigated river sections with a length of about 9 km. In the case of Skawinka, the flood-prone area obtained with Mike11/NWMA procedure is the largest, with a total area A = 2.380 km^2 and the smallest (FLO-2D with B(2;2)) has the value A = 1.973 km^2. FLO-2D with B (2;2) flood-prone area is 17.1% smaller than the corresponding obtained with Mike11/NWMA. The largest zone for Kamienica Nawojowska was also generated based on Mike11/NWMA, A = 1.156 km^2, and the smallest based on the FLO-2D application (again with B(2;2)) presents A = 0.666 km^2, with a difference of 42.4%. The smallest differences with Mike 11/NWMA were found in the case of Skawinka with the DVWK model and with the B(2;3) model in the case of Kamienica Nawojowska. As for the differences between the EBA4SUB models, the differences are up to 8.2% for Skawinka and 26.8% for Kamienica Nawojowska.

Table 3. Total area of flood-prone areas.

Model	Case	Kamienica Nawojowska [m^2]	Skawinka [m^2]
EBA4SUB/FLO-2D	B(2;2)	666,025	1,972,525
	B(2;3)	910,125	2,027,650
	B(5;2)	733,525	2,094,875
	Chicago	732,475	2,065,725
	DVWK	872,525	2,148,300
Mike11/NWMA	SCS-CN	1,156,400	2,380,275

A detailed analysis was carried out by identifying areas and focusing on those locations where the zones generated by the FLO-2D model are the same, or underestimated (Figure 4a) or overestimated (Figure 4b), compared to the Mike11/NWMA application. On average, the difference in the areas of FLO-2D-B(2;2) and Mike11/NWMA zones is 17.1%, with local differences that may not occur or be much larger in specific locations.

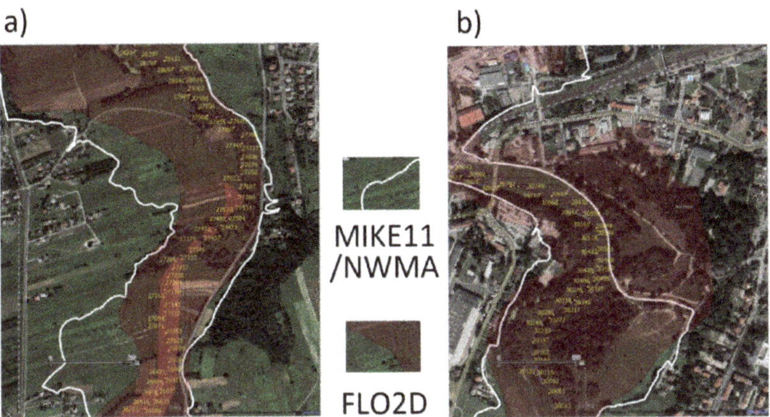

Figure 4. Comparison of flood hazard zones of Skawinka between the Mike11/NWMA procedure (white line) and FLO-2D B22, (**a**) underestimation, (**b**) overestimation.

4. Discussion

The final result of the procedures, i.e., the map of flood-prone areas, consists of the work of many people, including specialists in spatial analysis, topography, hydrology and, finally, hydraulic modeling. The selection of the hydroinformatics tools affects not only the accuracy of results but, most of all, the computational time of the calculations needed for the map generation. In the case of 1D models, the short calculation time is counterbalanced by the need to obtain the most accurate information from various databases, like a detailed survey of the river cross-section. In the case of 2D models, the increased computation time is counterbalanced by a simplified model preparation and by a limited postprocessing operational time.

In the selected case studies, the flood-prone areas obtained routing on the topography with FLO-2D and the different design hydrographs obtained with the EBA4SUB model slightly differ from each other (Figure 3). In the case of the Mike11/NWMA application, the modelled zones (Figure 3) significantly differ from the corresponding obtained with the EBA4SUB/FLO-2D application. The nature of these differences is not the same in the case of Skawinka and Kamienica Nawojowska rivers. This circumstance is due to the different structure of the river valley and floodplains. Differences between individual zones are more evident at the local scale. They can be found in both rivers due to the use of different calculation methods.

In order to present the diversity of the flood-prone area over the entire section and, thus, the differences in the use of particular methods, the entire river section was divided into sectors (Kamienica Nawojowska into 23 sectors, Skawinka into 22 sectors, see Figure 5), and the areas of flood-prone zones in the single sectors were calculated. This allowed providing the analysis of zone extension variability along the length of the longitudinal profile of both streams (Figure 6). The flood-prone area map generated based on the EBA4SUB/FLO-2D modeling approach takes on a larger or smaller area compared to the Mike11/NWMA approach, for the specific sector. The size of the zones in the different sectors can be influenced by both the hydrological and hydraulic model. If the primary source of differences between the zones is due to the hydraulic model, in which a more realistic flood propagation can be expected in the 2D model, two separate trends should result. The first is the difference in flow capacity of the section. The second is created when some parts of the Mike 11 floodplain are always removed in the same way, i.e., in the case that sufficient depth is reached to ensure contact of the zone with the rest of the flood risk area. The observed tendency presenting something different, i.e., variability, indicates the inclusion of other factors found both in the Mike11/NWMA and FLO-2D methods.

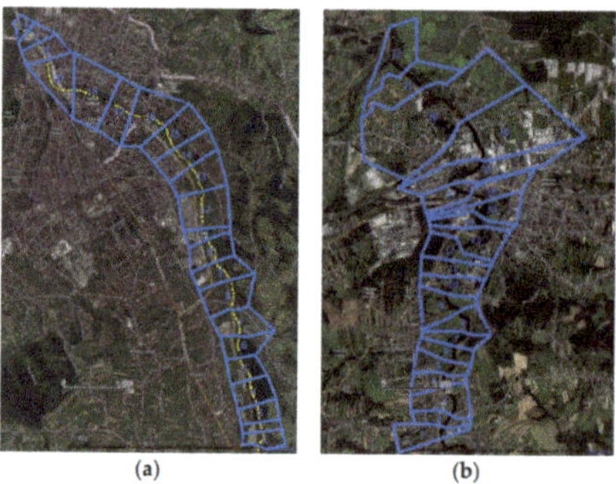

Figure 5. Sectors: (**a**) Kamienica Nawojowska, (**b**) Skawinka.

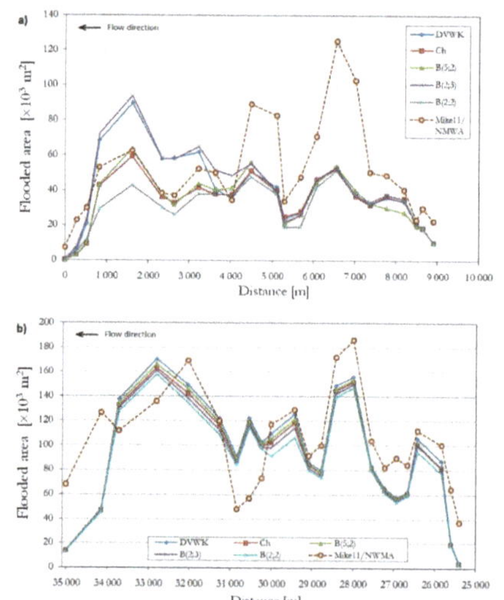

Figure 6. Variability of flood-prone areas: (**a**) Kamienica Nawojowska, (**b**) Skawinka.

In the case of Kamienica Nawojowska, the dimensions of the Mike11/NWMA zones are consistent with the zones obtained from the FLO-2D-B(5;2) application, where the zones are small, almost exclusively narrow, and the section relatively compact and under the influence of pungent and throw pressure manifested by floodplain terraces. In the upper part of the investigated section, where the zones are created above hydrotechnical structures (bridges and water steps), the differences between the zones obtained using the B(2;2), B(2,3) and B(5;2) hydrographs are small and the Mike11/NWMA zone is substantially larger. This is due to the application of insufficient accuracy of the model grid in FLO-2D, which under these conditions generalizes the actual layout of the terrain. This approach is necessary due to the practical usability of the software in modeling of significant fragments of the riverbed and allowing avoiding increasing computation time compared to the 1D model over 1000 times [40]. At the bottom of Kamienica Nawojowska, the regulated riverbed in the FLO-2D model has been generalized, making water access to floodplains possible. The different size of zones due to the characteristics of the flood wave is also visible here: the FLO-2D-B(2;3) application presents the largest zone.

In the case of Skawinka, we are dealing with a watercourse that creates larger zones and the relative differences between the distributions are not reflected in the simulations with FLO-2D, which present similar values. In the outlet section, the differences considering the Mike11/NWMA zones result from merging with the receiver zone following the procedure and in other places with insufficient representation of the terrain layout by generalization of embankments Their absence allows for the creation of zones that the standard procedure does not produce. Regarding the spatial distribution of the flood zone areas B5.2, B2.3, and B2.2, the differences are due to the larger flow capacity in the B2.3 zone. Another aspect for the Mike11/NWMA approach zone is to consider water depth in the grid in relation to the accuracy of DEM. In this case, since the DEM is characterized by an accuracy inside the range up to 0.2 m, if we remove B2.3 grid nodes having the depth less than 0.2 m, the zones would be similar to the Mike11/NWMA approach.

In the traditional approach of generating flood-prone areas, i.e., intersecting the WSM obtained from a 1D model with DEM on floodplain terraces, areas that require expert interpretation emerge [39]. This applies in particular to areas without drainage and those where the connection in the main channel is too small to fill with water during the peak of the wave. The FLO-2D model has a functionality that allows the analysis of details of the generated floodplains [41]. This is particularly important when verifying the physical connection of part of the zone in the main channel with the zone in the floodplain. The advantage of the FLO-2D model is the possibility of performing the procedure in an automated manner and without a great subjective assessment [42]. However, it depends mainly on the accuracy of the data. Moreover, the 2D model allows the user to check the results of hydraulic calculations, e.g., flow depth and local speed, at each pixel of the investigate area. It is noteworthy that this possibility can be time consuming due to the high number of pixels contained in the gridded structure of the model.

Final considerations can be done concerning the calibration issue. Commonly known hydrological approaches used to design flood-prone areas need calibration of parameters in both hydrological and hydraulic models. Calibration is a critical component of model development because parameters generally cannot be determined directly from measurement but are instead inferred indirectly by calibrating the model to observed responses [43,44]. In the last years in Poland, a new approach was applied taking into account hydrological, hydraulic, social, environmental, economic, and implementation aspects. In order to choose a solution, a multicriteria analysis was used, which is based on 13 indicators that use the results of mathematical modeling of flood waves for the formulated variants compared to the current state. This method was applied for the main tributaries of the Vistula River and its results were used in the Flood Risk Management [45–47]. The EBA4SUB and FLO-2D model can alternatively be used to determine flood-prone areas in ungauged catchments. The main advantage of the approach is the simplicity of its use. Concerning EBA4SUB, the simplicity means that it can be efficiently and successfully used by practitioners to determine the size of credible flows for estimating flood-prone areas or for designing hydrotechnical constructions. This approach also resolves the problem of having a high uncertainty in hydrological and hydraulic calculations, because parameters of both models have physical representation in real catchments characteristics and can be easily inferred from topography and land use, and in doing so minimizing the subjectivity of the user during the calibration procedure. With the introduction of Water Framework Directive of UE (WFD) in Poland over ten years ago, all water bodies are subject now to the same procedure. Standardized, simplified with use of 1D modeling (Mike 11 was commonly used) rivers and some streams were modelled and hazard zones were generated (and later risk zones). The authors of this study participated in many projects where flood zones were designed in Polish conditions [44–48]. This had two significant consequences. First is that in a short period of time (about 3 years) all rivers were to some degree measured and analyzed. Second is that the attention of researchers was focused elsewhere with the job done thanks to the country offices financial involvement. Now with the perspective of obligatory repeating every several years, the procedure is repelling many attempts of scientific involvement. Years pass and new capabilities in the field of remote survey and modeling are accessible. If we look at the Mike11/NWMA procedure itself, the growing labor costs in Poland, and on the opposite side the better 2D modeling capabilities, we cannot expect this to be the official procedure in the future for much longer. We believe any attempt to use alternative procedures needs to go under the careful eye of the science world. Moreover, practitioners around the world need modern and quick tools to assess flood risk because they are responsible for protecting the infrastructure and maintaining its functionality.

5. Conclusions

The differences in the flood-prone area maps created using different hydrological and hydraulic models have been investigated in the present manuscript. The obtained results showed that the proposed modeling approach, characterized by the combined use of the hydrological EBA4SUB model and the hydraulic FLO-2D model, represent an accurate alternative, although much more demanding in terms of computational time, to the standard approach used in Poland. Finding a solution capable of the same results without the 2D modeling effort is the main issue of the research today. While FLO-2D requires a high computational effort, the same way as any 2D modeling tool, it also offers fast access to detailed results. Pixel by pixel verification provides evidence of flow capacity, with FLO-2D modeled flood zones that were here generally narrower in respect to the official procedure. One question arises: is the simplifying of the source data in FLO-2D leading to such differences or is the 1D tool generally inaccurate?

Is FLO-2D discredited? In our opinion: no. Extensions of flood-prone areas appear in places where the official procedure results would not be created without manual intervention (for instance the cutting of parts of the flood zones). An essential aspect of any modeling tool is time effectiveness, which also applies to the FLO-2D model. The procedure using FLO-2D can be utilized as a valid alternative to the official approach. Moreover, in this work a second focus concerned the use of a recently developed rainfall–runoff model, named EBA4SUB, for estimating the design hydrograph to be propagated with the hydraulic model. The obtained results are in line with the official hydrological approach, so the use of the combined approach EBA4SUB–FLO-2D in our opinion can pave the way for a fast update of the flood-prone areas at a national level. Further research is needed, in particular regarding the application of the proposed procedure to a large number of catchments, eventually assessing the quality of the results comparing the modeled flood area maps with occurred ones.

Author Contributions: Conceptualization, A.P., D.M., J.F., L.K. and A.W.; methodology, A.P., D.M., J.F., L.K. and A.W.; software, A.P. and J.F.; validation, A.P., D.M. and J.F.; formal analysis, A.P., D.M. and J.F.; investigation, A.P., D.M., J.F., L.K. and A.W.; resources, D.M., J.F. and L.K.; data curation, A.P., D.M. and J.F.; writing—original draft preparation, A.P., D.M., J.F., L.K. and A.W.; writing—review and editing, A.P., D.M., J.F., L.K. and A.W.; visualization, D.M. and J.F.; supervision, A.P. and A.W.; project administration, D.M., J.F. and A.W.; funding acquisition, D.M., J.F., L.K. and A.W. All authors have read and agreed to the published version of the manuscript.

Funding: This research received no external funding.

Conflicts of Interest: The authors declare no conflict of interest.

References

1. Pellicani, R.; Parisi, A.; Iemmolo, G.; Apollonio, C. Economic Risk Evaluation in Urban Flooding and Instability-Prone Areas: The Case Study of San Giovanni Rotondo (Southern Italy). *Geosciences* **2018**, *8*, 112. [CrossRef]
2. Apollonio, C.; Bruno, M.F.; Iemmolo, G.; Molfetta, M.G.; Pellicani, R. Flood Risk Evaluation in Ungauged Coastal Areas: The Case Study of Ippocampo (Southern Italy). *Water* **2020**, *12*, 1466. [CrossRef]
3. Directive 2007/60/EC on the Assessment and Management of Flood Hazards. Available online: https://eur-lex.europa.eu/legal-content/EN/TXT/PDF/?uri=CELEX:32007L0060&from=EN (accessed on 13 October 2020).
4. Sojka, M.; Wróżyński, R. Impact of digital terrain model uncertainty on flood inundation mapping. *Rocz. Ochr. Śr.* **2013**, *15*, 564–574.
5. Młyński, D.; Wałęga, A.; Książek, L.; Florek, J.; Petroselli, A. Possibility of using selected rainfall-runoff models for determining the design hydrograph in mountainous catchments: A case study in Poland. *Water* **2020**, *12*, 1450. [CrossRef]
6. Mark, O.; Weesakul, S.; Apirumanekul, C.; Aroonnet, S.B.; Djordjevic, S. Potential and limitations of 1D modelling of urban flooding. *J. Hydrol.* **2014**, *299*, 284–299. [CrossRef]
7. Petroselli, A.; Vojtek, M.; Vojteková, J. Flood mapping in small ungauged basins: A comparison of different approaches for two case studies in Slovakia. *Hydrol. Res.* **2018**, *50*, 379–392. [CrossRef]
8. Vojtek, M.; Petroselli, A.; Vojteková, J.; Ashgarynia, S. Flood inundation mapping in small and ungauged basins: Sensitivity analysis using the EBA4SUB and HEC-RAS modeling approach. *Hydrol. Res.* **2019**, *50*, 1002–1019. [CrossRef]
9. Pijanowski, J.M. *System Approach to Planning and Arranging Rural Areas in Poland*; University of Agriculture in Krakow: Krakow, Poland, 2013. (In Polish)
10. Czajkowska, A.; Osowska, J. The use of ArcGIS Desktop and Mike 11 for determining flood hazard zones. In *Geochemia i Geologia Środowiska Terenów Uprzemysłowionych*; Pozzi, M., Ed.; PA NOVA: Gliwice, Poland, 2014; pp. 220–235. (In Polish)
11. Hejmanowska, B. *Data Quality Effect on Risk of Decision Processes Supported by GIS Analyses*; AGH: Krakow, Poland, 2005. (In Polish)
12. Al-Khafaji, M.S.; Al-Sweiti, F.H. Integrated impact of digital elevation model and land cover resolutions on simulated runoff by SWAT Model. *Hydrol. Earth Syst. Sci. Discuss.* **2017**, *653*, 1–26.
13. Gądek, W.; Bodziony, M. The hydrological model and formula for determining the hypothetical flood wave volume in non-gauged basins. *Meteorol. Hydrol. Water Manag.* **2015**, *3*, 3–9. [CrossRef]
14. Egiazarova, D.; Kordzakhia, M.; Wałęga, A.; Drożdżal, E.; Milczarek, M.; Radecka, A. Application of Polish experience in the implementation of the flood directive in Georgia—Hydrological calculations. *Acta Sci. Pol. Form. Circumiectus* **2017**, *16*, 89–110. [CrossRef]
15. Gądek, W.J.; Baziak, B.; Tokarczyk, T. Nonparametric design hydrograph in the gauged cross sections of the Vistula and Odra basin. *Meteorol. Hydrol. Water Manag.* **2017**, *5*, 53–61. [CrossRef]
16. Rodriguez-Iturbe, I.; Valdez, J.B. The geomorphologic structure of hydrology response. *Water Resour. Res.* **1979**, *15*, 1409–1420. [CrossRef]
17. Wałęga, A.; Drożdżal, E.; Piórecki, M.; Radoń, R. Some problems of hydrology modelling of outflow from ungauged catchments with aspects of flood maps design. *Acta Sci. Pol. Form. Circumiectus* **2012**, *11*, 57–68. (In Polish)
18. Wałęga, A.; Książek, L. Influence of rainfall data on the uncertainty of flood simulation. *Soil Water Res.* **2016**, *11*, 277–284. [CrossRef]
19. Piscopia, R.; Petroselli, A.; Grimaldi, S. A software package for the prediction of design flood hydrograph in small and ungauged basins. *J. Agric. Eng.* **2015**, *432*, 74–84.
20. Petroselli, A.; Grimaldi, S. Design hydrograph estimation in small and fully ungauged basins: A preliminary assessment of the EBA4SUB framework. *J. Flood Risk Manag.* **2018**, *8*, 1–14. [CrossRef]
21. Młyński, D.; Petroselli, A.; Wałęga, A. Flood frequency analysis by an event-based rainfall-runoff model in selected catchments of Southern Poland. *Soil Water Res.* **2018**, *13*, 170–176.

22. Šraj, M.; Dirnbek, L.; Brilly, M. The influence of effective rainfall on modeled runoff hydrograph. *J. Hydrol. Hydromech.* **2010**, *58*, 3–14. [CrossRef]

23. Sikorska, A.E.; Viviroli, D.; Seibert, J. Effective precipitation duration for runoff peaks based on catchment modelling. *J. Hydrol.* **2017**, *556*, 510–522. [CrossRef]

24. Grimaldi, S.; Petroselli, A.; Romano, N. Curve-Number/Green-Ampt mixed procedure for streamflow predictions in ungauged basins: Parameter sensitivity analysis. *Hydrol. Process.* **2013**, *27*, 1265–1275. [CrossRef]

25. Green, W.H.; Ampt, G.A. Studies on soil physics. *J. Agric. Sci.* **1911**, *4*, 1–24. [CrossRef]

26. USDA. Estimation of direct runoff from storm rainfall. In *National Engineering Handbook*; Chapter 10, Part 630; United States Department of Agriculture (USDA) Soil Conservation Service: Washington, WA, USA, 2004; pp. 1–22.

27. Nardi, F.; Grimaldi, S.; Santini, M.; Petroselli, A.; Ubertini, L. Hydrogeomorphic properties of simulated drainage patterns using DEMs: The flat area issue. *Hydrol. Sci. J.* **2008**, *53*, 1176–1193. [CrossRef]

28. Grimaldi, S.; Petroselli, A.; Nardi, F. A parsimonious geomorphological unit hydrograph for rainfall-runoff modelling in small ungauged basins. *Hydrol. Sci. J.* **2012**, *57*, 73–83. [CrossRef]

29. Giandotti, M. Previsione delle piene e delle magre dei corsi d'acqua (Estimation of floods and droughts of rivers). *Ist. Poligr. Dello Stato* **1934**, *8*, 107–117.

30. Kim, S.; Kim, H. A new metric of absolute percentage error for intermittent demand forecast. *Int. J. Forecast* **2016**, *32*, 669–679. [CrossRef]

31. O'Brien, J.S.; Julien, P.Y.; Fullerton, W.T. Two dimensional water flood and mud flow simulation. *J. Hydraul. Eng.* **1993**, *119*, 244–261. [CrossRef]

32. Książek, L.; Wałęga, A.; Bartnik, W.; Krzanowski, S. Calibration and verification of computational model of The Wislok River by means of flood wave. *Infrastruct. Ecol. Rural Areas* **2010**, *8*, 15–28. (In Polish)

33. Michalik, A.; Książek, L. Dynamics of Water Flow on Degraded Sectors of Polish Mountain Stream Channels. *Pol. J. Environ. Stud.* **2009**, *18*, 665–672.

34. Chow, V.T.; Maidment, D.K.; Mays, L.W. *Applied of Hydrology*; McGRAW Hill Book Company: New York, NY, USA, 1988.

35. Stodolak, R.; Baran, J.; Knap, E. The influence of rain temopration on the results of rainfall-runoff model. *Ecol. Eng.* **2018**, *19*, 87–93. [CrossRef]

36. GUGiK. Polish Central Office of Geodesy and Cartography, State Surveying and Cartographic Resources. 2020. Available online: www.gugik.gov.pl/pzgik (accessed on 19 March 2020).

37. European Commission. *CORINE (Coordination of Information on Environment) Database, a Key Database for European Integrated Environmental Assessment*; Programme of the European Commission; European Environmental Agency (EEA): Copenhagen, Denmark, 2000.

38. FLO-2D (2012) FLO-2D Reference Manual. Available online: https://www.flo-2d.com/download/ (accessed on 5 February 2019).

39. Młyński, D.; Wałęga, A.; Ozga-Zieliński, B.; Ciupak, M.; Petroselli, A. New approach for determining the quantiles of maximum annual flows in ungauged catchments using the EBA4SUB model. *J. Hydrol.* **2020**, *589*, 125198. [CrossRef]

40. Jowett, J.G.; Duncan, M.J. Effectiveness of 1D and 2D hydraulic models for instream habitat analysis in a braided river. *Ecol. Eng.* **2012**, *48*, 92–100. [CrossRef]

41. Gibson, S.A.; Pasternack, G.B. Selecting between one-dimensional and two-dimensional hydrodynamic models for ecohydraulic analysis. *River Res. Appl.* **2016**, *32*, 1365–1381. [CrossRef]

42. Dimitriadis, P.; Tegos, A.; Oikonomou, A.; Pagana, V.; Koukouvinos, A.; Mamassis, N.; Koutsoyiannis, D.; Efstratiadis, A. Comparative evaluation of 1D and quasi-2D hydraulic models based on benchmark and real-world applications for uncertainty assessment in flood mapping. *J. Hydrol.* **2016**, *534*, 478–492. [CrossRef]

43. *Development and Calibration of a One-Dimensional Hydraulic Model and Designation of Flood Hazard Zones in the Wisłoka Catchment Area*; Regional Water Management Authorityin Krakow: Krakow, Poland, 2010.

44. *Development of a One-Dimensional Hydraulic Model of the CzarnaStaszowska Catchment Area; Calibration and Verification, Determination of the Water Surface Level for Discharges with a Probability of Exceedance p=50%, p=20%, p=10%, p=5%, p=2%, p=1%, p=0.5% and p=0.2%*; Regional Water Management Authority in Krakow: Krakow, Poland, 2013.

45. *Development and Verification of Hydraulic Models of Dry Dams and Water Reservoirs Based on Water Management Manuals of Holding Reservoirs*; 2013–2014, Project "Analysis of the investment programme in the catchment area of the Raba River"; Regional Water Management Authority in Krakow: Krakow, Poland, 2014.

46. *Analysis of Condition for the Transformation of Flood Wave in the Catchment Areas of the Soła, the Skawa and the Dunajec*; 2014–2015, Flood Protection Programme in the Upper Vistula; Regional Water Management Authority in Krakow: Krakow, Poland, 2015.

47. Gabryś, Z.; Grela, J.; Laskosz, E.; Piszczek, M.; Wybraniec, K.; Bartnik, W.; Książek, L. Approach to the development of investment programme of flood protection on the Dunajec River including environmental protection aspects. *Acta Hydrol. Slovaca* **2015**, *16*, 142–151.

48. *Risk management in Nature 2000 sites under condition of flooding on the example of "Małopolski Przełom Wisły" (km 254+000-307+000) Tarnobrzeg*; Słupia, project under Norwegian and EEA Funds; National Fund for Environmental Protection and Water Management; University of Agriculture in Krakow: Krakow, Poland, 2017.

Publisher's Note: MDPI stays neutral with regard to jurisdictional claims in published maps and institutional affiliations.

Article

Stream Flow Changes and the Sustainability of Cruise Tourism on the Lijiang River, China

Yuefeng Yao [1],* and Azim Mallik [2],*

[1]   Guangxi Key Laboratory of Plant Conservation and Restoration Ecology in Karst Terrain, Guangxi Institute of Botany, Guangxi Zhuang Autonomous Region and Chinese Academy of Sciences, Guilin 541006, China
[2]   Department of Biology, Lakehead University, Thunder Bay, ON P7B 5E1, Canada
*   Correspondence: yf.yao@gxib.cn (Y.Y.); amallik@lakeheadu.ca (A.M.)

Received: 24 August 2020; Accepted: 15 September 2020; Published: 18 September 2020

Abstract: Water resources play a critical role in the sustainable development of river-based tourism. Reduced stream flow on the Lijiang River, south China, may negatively impact the development of cruise tourism. We explored the effects of stream flow changes on cruise tourism by determining (1) cruise tourism development indicators, (2) stream flow regime characteristics and their impacts on cruise tourism development indicators, and (3) climate variability and socio-economic factors effecting stream flow. Cruise tourism on the river has experienced rapid growth in recent decades. Stream flow regimes displayed no significant changes between 1960 and 2016, although dry season stream flow was significantly lower than in other seasons. We found that stream flow changes did not have a significant impact on the development of cruise tourism. As precipitation has not changed significantly, policies, including regulated stream flow from hydroelectric reservoirs, are assumed to mitigate reduced stream flow. However, increased irrigation and economic development, combined with future climate change, may increase challenges to cruise tourism. Future reservoir operations should prepare for climate change-related increases in temperature and insignificant changes in precipitation, and adopt adaptive measures, such as rationing water use in various sectors, to mitigate water shortages for supporting sustainable tourism development.

Keywords: river-based tourism; reservoir regulation; water availability; climate variability; land use change

---

1. Introduction

Water is a foundational natural and economic resource that plays a critical role in sustainable development [1]. However, with economic expansion and population growth, the demand for water across multiple sectors has increased the stress on regional water availability, particularly in China [2–4]. While the demand for water continues to grow, water availability is declining in many regions, threatening regional sustainable development, particularly in river-based areas where tourism and economic development are reliant on adequate water availability. Furthermore, the expansion of river-based tourism can create additional water stress in areas where water availability is already limited in meeting the increasing demands of economic growth [5]. Therefore, the relationship between water resources and tourism has received particular attention from UNWTO (United Nations World Tourism Organization), UNEP (United Nations Environment Program), OECD (Organization for Economic Co-operation and Development) and related organizations and researchers [6,7].

Since the 1960s, the rapid growth in river-based tourism, and especially cruise tourism ("a socio-economic system generated by the interaction between human, organizational and geographical entities, aimed at producing maritime-transportation-enable leisure experiences" [8]), has led to increasing concern regarding the environmental impacts of this tourism sector [9,10]. The boom in cruise tourism has brought about substantial local and regional economic benefits, including increased employment, infrastructure development and the expansion of urbanization [11]. The consequence of these developments can have serious negative impacts on water resources and the natural environment [12,13]. Most research into the relationship between cruise tourism and water resources focuses on water pollution and other environmental issues [12,14], while the effects of stream flow changes on cruise tourism remains underexplored. The Lijiang River, one of 13 state-level key protected rivers in China, is located in the Guilin region, in the northeast Guangxi Zhuang Autonomous Region. River cruise tourism from Guilin to Yangshuo takes place along an 83 km meandering waterway that runs through the largest and the most spectacular karst tourist attraction of the world (Figure 1). River cruise tourism along the Lijiang River officially commenced in 1973 with five cruise boats with a total capacity of 440 passengers. In 1982, the Lijiang River scenic zone was listed as one of 44 national scenic spots in China. The Lijiang River scenic zone received a 4A national ranking in 2001 and a 5A ranking in 2007. In 2012, there were 234 three- and four-star level cruise boats on the river with a passenger capacity of 18,802 [15]. According to the government report, the carrying capacity for Lijiang River cruise tourism is 19,000 passengers per day [16]. As a corridor, the Lijiang River connects downtown Guilin and Yangshuo County with more than 18 scenic spots in the middle and lower reaches of the river with 3A or higher rankings. As such, it represents the core area of socio-economic growth in the Guilin region [17]. Because of this, cruise tourism from Guilin to Yangshuo plays a critical role in the Guilin–Lijiang River–Yangshuo tourism destination system and in regional economic growth [18]. Tourism development has been booming based on the distribution of tourism resources and accelerating the economic growth in the Guilin region [17,18]. Downtown Guilin is the core zone of a cluster of tourism resources and tourist services. The Merryland Resort in Xing'an County and Longji Terraced Rice Fields in Longsheng County represent the northern core areas of tourism in the Guilin region. Yangshuo County is the southern core areas of tourism, and it contains a number of historic and cultural heritage sites, and karst landscape scenic spots.

However, in an era of rapid tourism and economic growth, the Lijiang River now experiences water shortages, especially during the dry season [19,20]. What was once a year-round 83 km-long waterway has been shortened to 10 km during the dry season, due to a reduction in stream flow. When water discharge is less than 30 m^3/s, cruise boats are unable to navigate the Lijiang River, and a discharge of less than 8 m^3/s imperils the river's riparian ecosystems [21–23]. Given the almost 50 years history of cruise tourism as an economic activity on the Lijiang River, a reconsideration of current approaches to water resource management to support sustainable cruise tourism is required. Few studies have explicitly analyzed the effects of stream flow changes on cruise tourism on the Lijiang River. This has resulted in a major gap in the understanding needed for sustainable tourism development in the Guilin region.

We used the Lijiang River as a demonstration site to determine the effect of stream flow changes on cruise tourism development between 1979 and 2016, based on selected stream flow parameters and cruise tourism indicators. We also explored the effect of climate variability (precipitation, in this case) and socio-economic factors (reservoir operation, land use change, and river engineering) that might affect stream flow and, in turn, cruise tourism. The theoretical findings of this study form the basis of water resource management recommendations that can facilitate the development of sustainable cruise tourism on the Lijiang River.

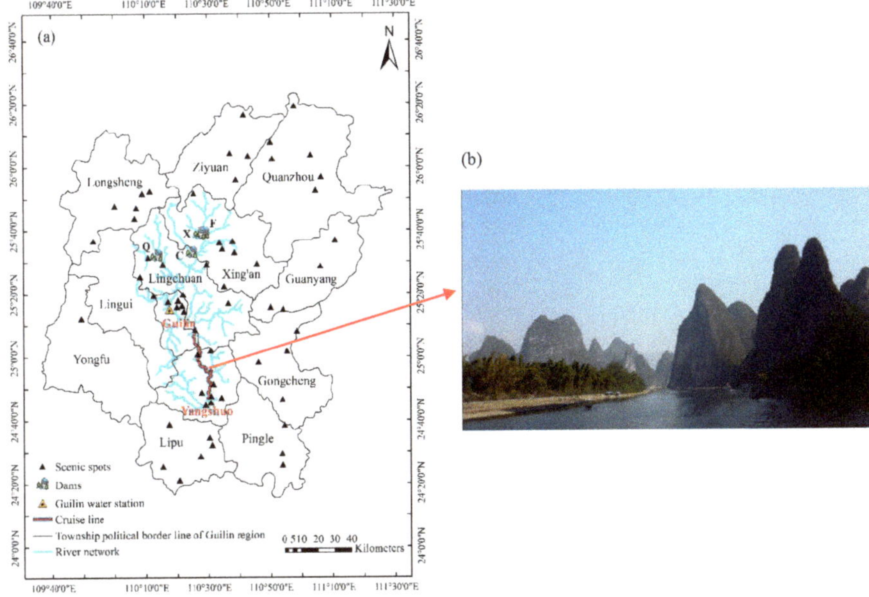

Figure 1. (**a**) The Lijiang River scenic zone, located in the Guilin region, Northeast Guangxi Zhuang Autonomous Region, China. Q, Qingshitan reservoir; C, Chuanjiang reservoir; X, Xiaorongjiang reservoir; F, Fuzikou reservoir; (**b**) Yellow Cloth Shoal, one of the essential scenic spots of the Lijiang River (photographed by Yuefeng Yao).

2. Materials and Methods

2.1. Study Area

The Lijiang River originates at Mao'er Shan Nature Reserve and flows 164 km to Yangshuo County. The cruise line from Guilin to Yangshuo, known as the Gold Waterway of the Guilin region, covers an 83 km stretch of river on a four- to five-hour journey through spectacular karst landscapes (Figure 1). The annual temperature in the Lijiang River area ranges from 17 °C to 20 °C, and annual precipitation ranges from 1400 mm to 2000 mm. Precipitation during the rainy season, from March to August, accounts for approximately 80% of total annual precipitation, while dry season precipitation, from September to February, accounts for the remaining 20%.

The Lijiang River, as one of China's most important scenic and historic sites, is the largest and the most spectacular karst tourist attraction of the world. Both the state and the local government have created a series of programs to preserve and develop the area. To meet the water demands of cruise tourism during the dry season, several additional hydroelectric dams have been constructed in the upper reaches of the Lijiang River [24]. In order to enhance water availability, the local government has implemented the national Grain for Green policy by converting cropland to forest or grassland since the early 2000s [17]. In 2011, the regional government issued the Lijiang River Eco-environmental Protection law, which focuses on the protection and conservation of riparian vegetation, water resources and landscapes [25]. Furthermore, based on the Lijiang River scenic zone, an outline plan (2012–2020) for the development of Guilin into an international tourist attraction was approved by the National Development and Reform Commission in November 2012 [26]. In the past few years, the local government has created bonus policies to promote tourism development and plan to introduce relevant policies to foster high-quality tourism development in the future [27].

2.2. Methods

We used linear regression to analyze recent trends in cruise tourism development and stream flow change, and applied the Pearson correlation analysis and the Granger causality test to evaluate the effects of stream flow changes on cruise tourism, based on selected stream flow parameters and cruise tourism indicators. We then explored the effects of climate variability (in this case, precipitation) and socio-economic factors (in this case, reservoir operation, land use change, and river engineering) on stream flow and, therefore, cruise tourism development. Figure 2 is a simplified flow chart of this study.

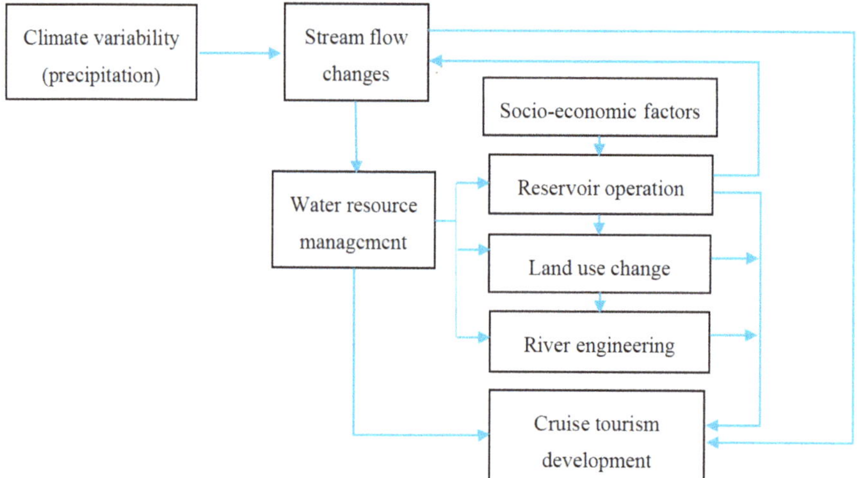

Figure 2. Simplified flow diagram illustrating the framework of this study.

2.2.1. Data Sources

To determine the development of regional cruise tourism we obtained economic data from the Guilin Lijiang Zhi [28] in 2004 and the Guilin Economic and Social Statistical Yearbook [29] from 1990 to 2016. These data included cruise passenger numbers and ticket prices per passenger on cruise boats from Guilin to Yangshuo, total number of tourists and tourism income for the Guilin region, and regional GDP for the Guilin region from 1979 to 2016. Since cruise tourism on the Lijiang River is dependent on the regulation of stream flow based on the water regimes at the Guilin water station during the dry season, we collected daily water discharge data from the Annual Hydrological Report of the Guilin water station from 1960 to 2016. Historical precipitation data from 1960 to 2016 were obtained from the China Meteorological Data Service Center (https://data.cma.cn). A dataset of future averages of annual precipitation, precipitation anomalies, air temperature, and air temperature anomalies for four representation concentration pathways (RCPs) for the 20 year period between 2020 and 2039 was downloaded from the GIS Program and Climate Changes Scenarios of the National Center for Atmospheric Research (https://gisclimatechange.ucar.edu/).

2.2.2. Handling Missing Data

Missing data for total number of tourists and tourism income in the Guilin region from 1979 to 1998 were interpolated using regression analysis based on tourist and tourism income data from 1999 to 2016. To compensate for missing cruise passenger number data for 2010, 2013, and 2014, we interpolated from average data from the preceding and succeeding years. Further, we replaced missing cruise ticket prices from 1979 to 1983 with prices from 1984.

2.2.3. Cruise Tourism Development Indicators and Stream Flow Parameters

The number of cruise passengers, the cruise passenger index (ratio of the number of cruise passengers to the total number of tourists of the Guilin region), cruise ticket revenues (number of cruise passengers multiplied by the ticket price per passenger, from Guilin to Yangshuo), and the cruise ticket revenue index (ratio of cruise ticket revenue to total tourism income for the Guilin region) were used to represent the cruise tourism development trend, and to analyze the effects of stream flow changes on the development of cruise tourism on the Lijiang River. Based on the Indicators of Hydrologic Alteration method [30], we employed twenty-five hydrological parameters to explore changes in stream flow and the impact of those changes on cruise tourism development. These were (1) annual stream flow, (2–13) annual monthly stream flow (from January to December), (14–18) minimum flow (1-, 3-, 7-, 30- and 90-day minimum flow), (19–23) maximum flow (1-, 3-, 7-, 30- and 90-day maximum flow), (24) base flow index, and (25) annual dry season flow (total monthly stream flow from September to February).

2.2.4. Linear Regression

Linear regression analysis was conducted to detect time series data trends, and to interpolate missing data based on recorded data. The trend slope was calculated using the least squares method [31]:

$$Slope = \frac{n \times \sum_{i=1}^{n}(i \times y_i) - \sum_{i=1}^{n} i \times \sum_{i=1}^{n} y_i}{n \times \sum_{i=1}^{n} i^2 - \left(\sum_{i=1}^{n} i\right)^2} \tag{1}$$

where y is a variable such as stream flow parameters, precipitation, and cruise tourism indicators; i is the time/year within the study period; n is the number of years within the study period. A positive slope value means an increasing trend, while a negative value indicates a decreasing trend, and zero signifies no change.

An F test with a confidence level of 95% was used to detect the significant trend of the time series data.

2.2.5. Detecting Abrupt Change of the Time Series Data

An abrupt change of stream flow or cruise tourism time series was detected using the Mann–Kendall test, and the Pettitt test was also applied to verify the abrupt change with the R program [32]. For a time series of stream flow or cruise tourism, the Mann–Kendall test statistic was calculated as follows:

$$S = \sum_{k=1}^{n-1} \sum_{j=k+1}^{n} sgn\left(X_j - X_k\right) \tag{2}$$

where $sgn\left(X_j - X_k\right) = \begin{cases} 1 & X_k < X_j \\ 0 & X_k = X_j \\ -1 & X_k > X_j \end{cases}$.

The null hypothesis (H_0) is that the stream flow or cruise tourism time series is identically distributed, while the alternative hypothesis (H_A) is that the stream flow or cruise tourism time series displays a monotonic trend.

The Pettitt test is defined as:

$$K_T = \max\left|U_{t,T}\right| \tag{3}$$

where

$$U_{t,T} = \sum_{i=1}^{t} \sum_{j=t+1}^{T} sgn\left(X_i - X_j\right) \tag{4}$$

Details of the Mann–Kendall test and Pettitt test can be obtained from the relevant sources [33–35].

2.2.6. Correlation Analysis and Granger Causality Test

Pearson correlation analysis [36] was employed to calculate the correlation coefficient between the stream flow parameter (x) and cruise tourism indicator (y):

$$\rho(x,\, y) = \frac{E(xy)}{\sigma_x \sigma_y} \tag{5}$$

where $E(xy)$ is the cross-correlation between x and y, and σ_x and σ_y are the variances of x and y, respectively. If $\rho(x,\, y)^2$ is closer to 1, the stronger the correlation between x and y. If $\rho(x,\, y)^2$ is equal to 0, x and y are independent.

Lasso analysis shrinks some of the variable coefficients, sets others to zero, and selects the best variables for enhancing the accuracy of predictions [37,38]. We employed Lasso analysis to select the related variables from the 25 stream flow parameters to determine which stream flow parameter has the most significant impact on cruise tourism development.

Granger [29] initially proposed his causality test as a statistical method in economics to explore whether one variable or time series can cause others (known as Granger-cause) (Equation (4)). The test has been applied widely in the fields of economics, biology, and environmental sciences to describe causality and feedback [39–41]. We assumed that stream flow Granger-caused the development of cruise tourism on the Lijiang River, and used the Granger causality test to verify this hypothesis.

$$X_t = \sum_{j=1}^{m} a_j X_{t-j} + \sum_{j=1}^{m} b_j Y_{t-j} + \varepsilon'_t$$

$$Y_t = \sum_{j=1}^{m} c_j X_{t-j} + \sum_{j=1}^{m} d_j Y_{t-j} + \varepsilon''_t \tag{6}$$

where m is the maximum number of lagged observation; a_j, b_j, c_j, and d_j are the coefficients; ε' and ε'' are residual for X_t and Y_t time series, respectively.

2.2.7. Future Climate Change Scenario Analysis

We analyzed temperature and precipitation changes for the next 20 years (2020–2039) under RCP2.6, RCP4.5, RCP6.0, and RCP8.5 scenarios using CCSM4.0 (Community Climate System Model). The CCSM is a geographic information system (GIS) Global Climate Model (GCM), and is one of the coupled climate models (CCM) used to simulate a series of emission scenario experiments in the fifth phase of the Coupled Model Intercomparison Project (CMIP5) [42]. It uses present day (1986–2005) climate datasets as the baseline to analyze future climate anomalies under four RCPs. RCP2.6 is a low forcing level which predicts that radiative forcing will rise to about 3 W/m^2 before 2100 and will then decrease. RCP4.5 and RCP6.0 are two stabilization levels in which radiative forcing will stabilize at approximately 4.5 and 6.0 W/m^2, respectively, after 2100. The radiative forcing level of RCP8.5 will reach more than 8.5 W/m^2 by 2100 [43]. The difference in temperature/precipitation over the Lijiang River basin between the next 20 years and the 1986 to 2006 average was interpolated using inverse distance weighting (IDW).

2.2.8. Effect of Reservoir Operation on Stream Flow

The Qingshitan reservoir has regulated downstream water supply during the dry season since 1987. Therefore, we compared monthly stream flow between 1960 and 1986 and between 1987 and 2016 to estimate the effect of reservoir regulation on monthly stream flow.

The deviation degree [44] was calculated to determine the effects of reservoir operation on stream flow:

$$D = \frac{x_t - x_0}{x_0} \times 100\% \tag{7}$$

where x_0 and x_t are the mean values of stream flow during the pre-regulation (1960 to 1986) and regulation period (1987 to 2016), respectively.

A positive D means an increase in stream flow during the regulation period compared to the pre-regulation period, while a negative D indicates a decrease.

3. Results

3.1. Trend Analysis of Cruise Tourism Development from Guilin to Yangshuo

The number of cruise passengers showed a fluctuated increase from 1979 to 2016, with an abrupt increase in 2000 ($p = 4.79 \times 10^{-5}$) (Figure 3). Prior to 1980, there were only 0.16 million cruise passengers annually, followed by a sharp increase from 0.19 million in 1980 to 1.44 million in 1987. Between 1987 and 2000, the number fluctuated, with a peak value of 1.74 million in 1992. However, following 2000, the number of cruise passengers increased to 2.18 million in 2006 with a dip in 2003. There was a slight decrease in 2007 and a sharp decrease in 2008, followed by relative stability until 2012. This may have been the result of a drastic decline in the cruise passenger index while the total number of tourists of the Guilin region increased sharply after 2006. Since 2006, the State Ministry of Culture and Tourism has implemented the Guiding Opinions on Promoting the Development of Rural Tourism. This includes the creation of a series of activities to promote rural tourism development, such as the China Rural Tourism Year 2006 [45]. As a result of this promotion of tourism development, the total number of tourists in Yangshuo, Xing'an, Longsheng, Gongcheng, Lipu, and Ziyuan Counties increased by 67%, 45%, 95%, 47%, 17%, and 130%, respectively, in 2006 compared to the previous year [46]. The number of cruise passengers increased rapidly again from 2012 to 2016.

Compared to the number of cruise passenger tourists, the total number of tourists in the Guilin region remained relatively stable prior to 1998, followed by an abrupt increase after 1998 ($p = 2.01 \times 10^{-6}$), peaking at 53.83 million in 2016.

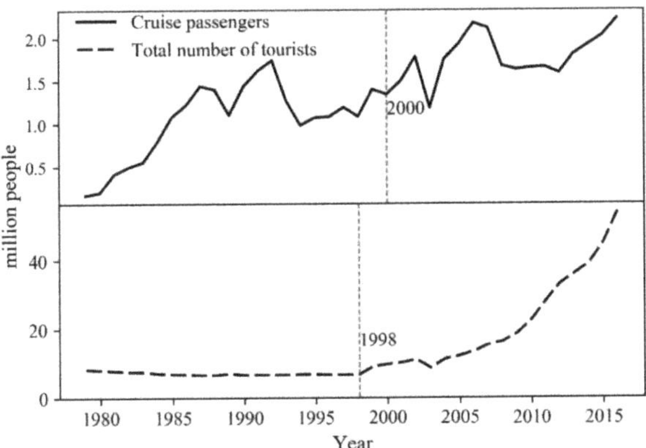

Figure 3. Temporal trends in the number of cruise passengers and total number of tourists from 1979 to 2016. The number of cruise passengers showed a tendency to increase on the whole, while the total number of tourists significantly and abruptly increased after 1998.

All cruise ticket revenue, total tourism income, and GDP for the Guilin region increased slowly before 1990 (Figure 4). From 1990 onwards, cruise ticket revenues increased rapidly until 2006, with an abrupt increase in 1997 ($p = 8.96 \times 10^{-6}$) and a decrease in 2003. This was followed by a sharp decrease in 2008, with revenues remaining relatively stable from 2008 to 2012 due to the drastic decline in cruise passenger numbers from 2006 to 2012. Cruise ticket revenue increased again from 2012 to 2016, with the highest value of 0.47 million yuan in 2016. Total tourism income increased slowly from 1990 to 2006, though it had an abrupt increase in 1994 ($p = 1.87 \times 10^{-6}$), and a peak value of 63.73 billion yuan in 2016. The GDP of the Guilin region increased gradually after 1990, with an abrupt increase in 1997 ($p = 1.86 \times 10^{-6}$), and attained a peak value of 205.48 billion yuan in 2016.

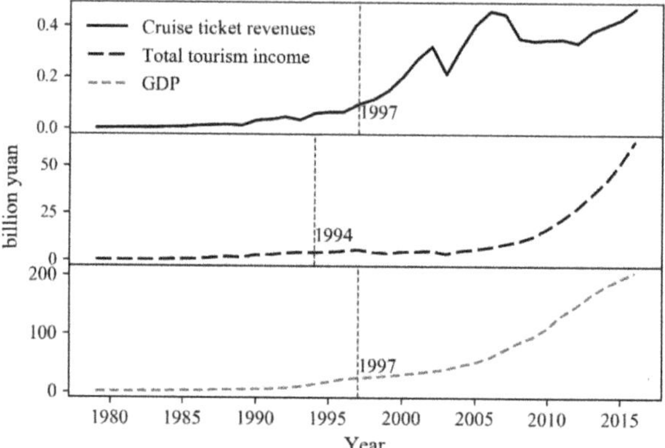

Figure 4. Temporal trends of cruise ticket revenues, total tourism income and GDP for the Guilin region, from 1979 to 2016. All cruise ticket revenues, total tourism income, and GDP increased slowly before 1990. Following 1990, cruise ticket revenues underwent a fluctuated increase until 2016. Both total tourism income and GDP for the Guilin region showed a similar gradually increasing trend prior to 2005, and then a sharp increase until 2016.

Recently, both cruise ticket revenue and cruise passenger indices have displayed a decreasing trend, particularly after 2006 (Figure 5). The cruise ticket revenue index fluctuated from 1979 to 1996, followed by a sharp increase from 1.35% in 1996 to 6.99% in 2005, with a small decrease in 2003 and 2004. A decreasing trend after 2006 was followed by an abrupt decrease in 2008 ($p = 0.02$), indicating a shrinking in the contribution of cruise ticket revenue to total tourism income. The cruise passenger index increased rapidly from 1.95% in 1979 to 26.49% in 1992, with a drastic decrease in 1989. Following 1992, there were sharp decreases in 1993 and 1994, followed by a fluctuating decrease from 1995 to 2006, and then an abrupt decrease after 2006 ($p = 0.03$).

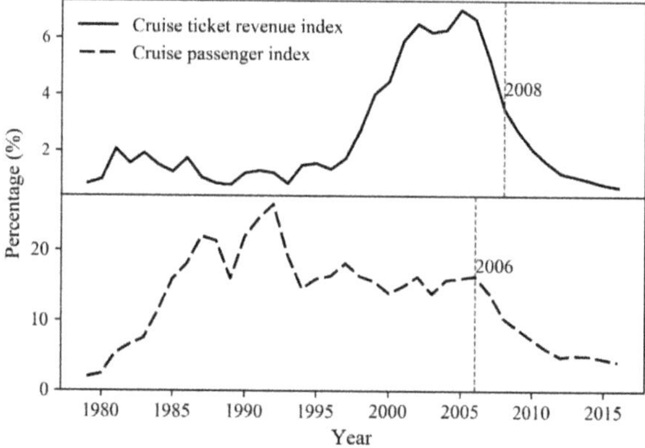

Figure 5. Temporal trends of cruise ticket revenue and cruise passenger indices from 1979 to 2016. Both cruise ticket revenue and cruise passenger indices decreased significantly after 2006.

3.2. Characteristics of Annual Stream Flow in the Lijiang River

From 1960 to 2016, the Lijiang River experienced large interannual variability in annual stream outflow (814.70 to 2219.31 mm, with a mean value of 1442.35 mm) (Figure 6). Annual stream outflow showed an overall rising trend with an increase ratio of 2.80 mm per year from 1960 to 2016. The annual dry season outflow also showed an overall upward trend. However, neither the annual stream outflow nor dry season outflow increased significantly (p value is 0.23 and 0.12, respectively) over the entire study period. The highest annual dry season outflow, in 2015, (855.62 mm) was due to heavy rain in November of that year, resulting in a November stream outflow (393.21 mm) that accounted for more than 45% of the total dry season outflow. If the annual dry season outflow for 2015 is excluded, there is no significant change during the study period ($p = 0.61$).

With the exception of April, the annual minimum flow (1 day, 3 day, 7 day, 30 day, and 90 day minimum flow), maximum flow (1 day, 3 day, 7 day, 30 day, and 90 day maximum flow), and annual monthly stream flow did not exhibit significant change. The annual April stream flow decreased significantly, with an abrupt decrease in 1982 ($p = 0.03$), suggesting that other sectors, such as agricultural irrigation, have increased water demand for spring crop growth, resulting in decreased stream flow in April.

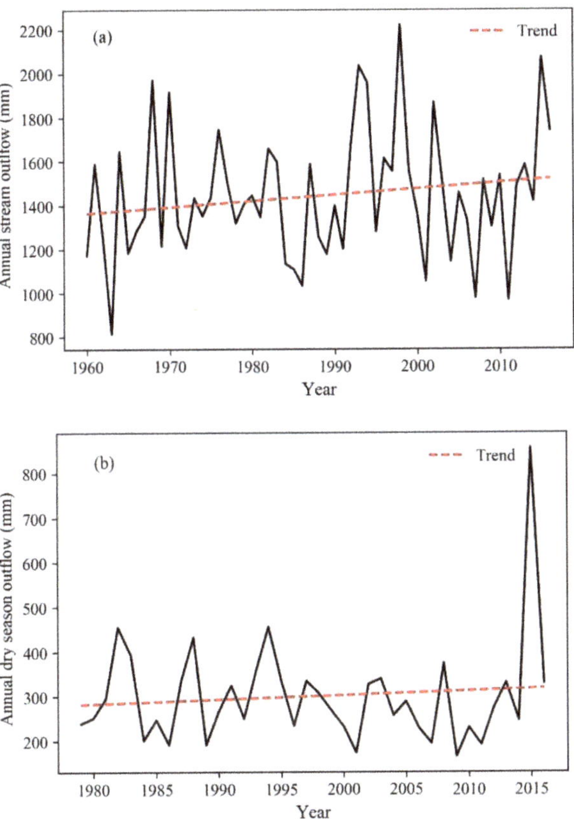

Figure 6. Trends in (**a**) annual stream outflow and (**b**) dry season outflow from 1960 to 2016. Neither annual stream outflow nor dry season outflow increased significantly, with p values of 0.23 and 0.63, respectively.

3.3. Effects of Stream Flow Changes on Cruise Tourism from Guilin to Yangshuo

Stream flow did not have a significant impact on the development of cruise tourism from 1979 to 2016, according to the Pearson's correlation matrix of cruise tourism indicators (number of cruise passengers and cruise ticket revenue) and stream flow parameters (Table 1). Stream flow in January, November, and December had a large positive correlation coefficient with the number of cruise passengers (0.18, 0.13, and 0.17, respectively) and cruise ticket revenue (0.15, 0.18, and 0.21, respectively).

Table 1. Pearson's correlation coefficients for cruise tourism indicators and stream flow.

	Jan.	Feb.	Mar.	Apr.	May	June	July	Aug.	Sept.	Oct.	Nov.	Dec.
Cruise passengers	0.18	−0.15	−0.08	−0.24	−0.10	0.04	0.10	−0.29	−0.11	−0.20	0.13	0.17
Cruise ticket revenue	0.15	−0.23	−0.23	−0.07	0.02	0.13	−0.07	−0.23	−0.13	−0.18	0.18	0.21

	1 day min [a]	3 day min	7 day min	30 day min	90 day min	1 day max [b]	3 day max	7 day max	30 day max	90 day max	base flow	annual flow	dry season flow
Cruise passengers	−0.14	0.02	0.02	0.12	−0.28	−0.07	−0.09	−0.11	−0.07	−0.01	0.05	−0.01	0.04
Cruise ticket revenue	−0.21	0.01	0.01	0.01	−0.21	−0.04	−0.07	−0.09	−0.09	0.02	−0.01	0.01	0.04

Note: [a] min, minimum flow; [b] max, maximum flow.

According to the variable selection criterion of lasso analysis, the monthly stream flow in January, February, March, April, May, June, August, and December, the 90 day minimum flow, the 1, 7, and 90 day maximum flow, annual flow, and dry season flow could be selected as the variables that have potential impacts on the number of cruise passengers (Table 2). Further, the monthly stream flow in February, March, June, August, and November, and the 7 and 30 day maximum flows could be selected as variables for exploring the effects of stream flow changes on cruise ticket revenue.

Table 2. Related coefficients of cruise tourism indicators and stream flow parameters calculated with lasso analysis.

	Jan.	Feb.	Mar.	Apr.	May	June	July	Aug.	Sept.	Oct.	Nov.	Dec.
Cruise passengers	0.0032	−0.0017	−0.0010	−0.0021	−0.0018	0.0003	0.0000	−0.0042	0.0000	0.0000	0.0000	0.0022
Cruise ticket revenue	0.0000	−0.0004	−0.0004	0.0000	0.0000	0.0005	0.0000	−0.0002	0.0000	0.0000	0.0001	0.0000

	1 day min [a]	3 day min	7 day min	30 day min	90 day min	1 day max [b]	3 day max	7 day max	30 day max	90 day max	base flow	annual flow	dry season flow
Cruise passengers	0.0000	0.0000	0.0000	0.0000	−0.0014	−0.0002	0.0002	−0.0002	0.0000	0.0012	0.0000	0.0005	0.0001
Cruise ticket revenue	0.0000	0.0000	0.0000	0.0000	0.0000	0.0000	0.0000	−0.0002	−0.0001	0.0000	0.0000	0.0000	0.0000

Note: [a] min, minimum flow; [b] max, maximum flow.

In general, stream flow changes did not have a significant impact on the development of cruise tourism on the Lijiang River from 1979 to 2016. Within the 5% confidence level, only the January stream flow had the significant p value that could Granger-cause the development of cruise tourism (Table 3); an increase in January stream flow could increase the number of cruise passengers. However, the January stream flow did not increase significantly from 1960 to 2016 ($p = 0.18$), which suggests that this could not Granger-cause the development of cruise tourism on the Lijiang River.

Table 3. Effects of stream flow changes on the development of cruise tourism using the Granger causality test with p value.

	Jan.	June	Nov.	Dec.	90-Day Max [a]	Annual Flow	Dry Season Flow
Cruise passengers	0.01	0.74		0.40	0.41	0.45	0.96
Cruise ticket revenue		0.60	0.36				

Note: [a] max, maximum flow.

3.4. Climate Variability and Socio-Economic Factors Affecting Stream Flow and Cruise Tourism

3.4.1. Climate Variability (Precipitation)

The Lijiang River experienced a large interannual variability in annual precipitation from 1960 to 2016, ranging from 1254 mm to 3012 mm, with a mean value of 1894 mm. Annual precipitation displayed an overall upward trend, with an increase ratio of 2.81 mm per year, indicating that climate change has had a positive impact on stream flow. However, annual precipitation did not increase significantly during the study period ($p = 0.51$), suggesting that changes in stream flow primarily occurred due to reservoir operation and other anthropogenic factors.

The temperature is projected to increase by more than 0.8 °C in the Lijiang River area over the next 20 years (2020–2039) compared to the 1986–2005 average. It is estimated to rise by 0.80 °C and 1.06 °C under the RCP2.6 and RCP8.5 scenarios, respectively. At the same time, precipitation is projected to decrease from 0.58 to 1.00% under RCP2.6, compared to the 1986–2005 average. Furthermore, precipitation is forecasted to decrease by approximately 0.26% in the upstream reaches of the Lijiang River, under RCP8.5. In contrast, precipitation is projected to increase from 2.32 to 3.18% under RCP4.5, and from 0.90 to 1.79% under RCP6.0, over the next 20 years. However, these increases/decreases in projected precipitation are relatively small compared to the 1986–2005 average, suggesting that future precipitation levels will not change significantly in the Lijiang River area.

3.4.2. Socio-Economic Factors

Reservoir Operation

The monthly stream flow at the Guilin water station, which is the stream flow monitoring station for cruise tourism on the Lijiang River, increased from the 1960–1986 period to the 1987–2016 period, except for the months of April, May, and August (Figure 7). The most significant increase in monthly stream flow occurred in June (277.40 to 368.20 m³/s), which resulted in the highest monthly stream flow shifting from May to June during the 1987–2016 period in contrast to the 1960–1986 period. The monthly stream flow during the dry season, from September to February, increased from the 1960–1986 period to the 1987–2016 period, especially in January (30.77 to 39.91 m³/s), with a deviation degree of 29.70%. This indicates that regulation of the downstream water supply was beneficial during the dry season.

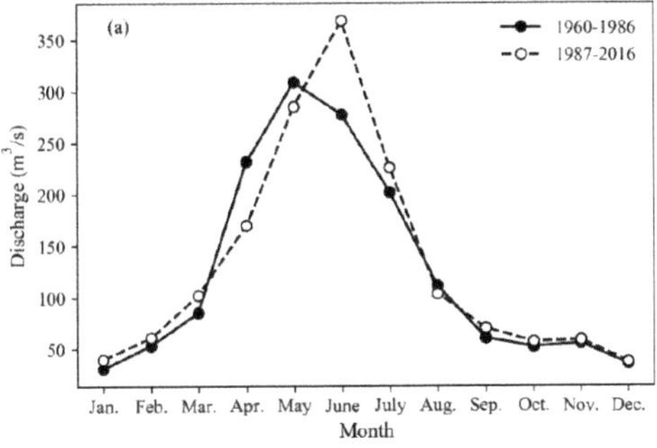

Figure 7. *Cont.*

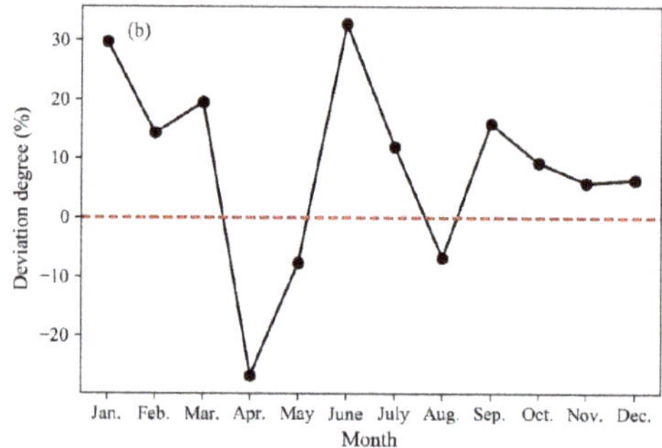

Figure 7. (a) Monthly stream flow changes and (b) their degrees of deviation during the period from 1987 to 2016 compared to the period from 1960 to 1986. Monthly stream flow in April, May, and August decreased during the period from 1987 to 2016 compared to the period from 1960 to 1986. The red line in (b) represents the zero deviation.

Land Use Change

A buffer zone of approximately 106.57 km^2 around the cruise line with a radius of 1 km was used to explore the impact of land use policy (Conversion of Cropland to Forest Program, also known as Grain for Green) on local land use changes. Land use change and its distribution patterns in the buffer zone showed that the area of forest increased from 55 km^2 in 2000 to 74 km^2 in 2015, and the water area increased from 8 km^2 in 2000 to 16 km^2 in 2015 (Figure 8). In contrast, the farmland area decreased from 10 km^2 in 2000 to 1 km^2 in 2015, indicating that the Conversion of Cropland to Forest Program implemented in the Guilin region at the beginning of the 2000s increased forest and water areas. Besides land use policy, other actions, such as sediment regulation, were adapted to improve stream flow in the Lijiang River. River bed excavation has been considerably higher in recent years compared to 1961. In 2013, sand excavation from the Lijiang River was 46,979 m^3, almost 10 times that of 1961 (4800 m^3).

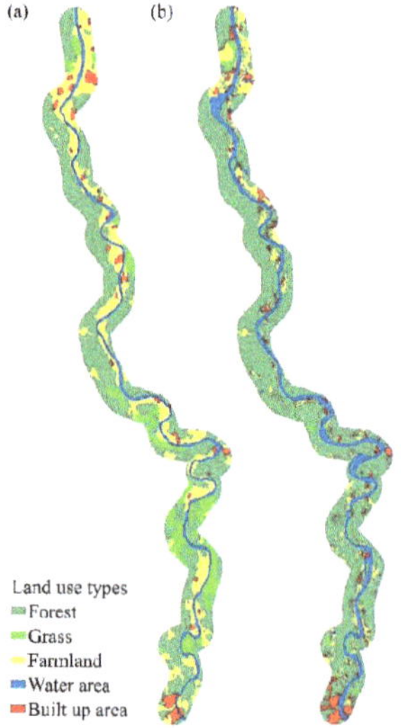

Figure 8. Spatial distribution of land use types in the buffer zone around the cruise line: (a) 2000 and (b) 2015.

4. Discussion

Cruise tourism from Guilin to Yangshuo plays a critical role in the Guilin–Lijiang River–Yangshuo tourism destination system and regional economic development. Tourism-related businesses generated almost one third of the GDP of the Guilin region in 2016. However, the cruise passenger index (ratio of the number of cruise passengers to total number of tourists in the Guilin region) and the cruise ticket revenue index (ratio of cruise ticket revenues to total tourism income for the Guilin region) displayed a decreasing trend after 2006. The cruise ticket revenue index is dependent on cruise ticket prices, number of cruise passengers, and total tourism income for the Guilin region. Since the cruise ticket price from Guilin to Yangshuo remained stable from 2006 to 2016 [46,47], a drastic decline in the number of cruise passengers resulted in a decrease in cruise ticket revenue. Furthermore, an increase in total tourism income accelerated the declining trend of the cruise ticket revenue index. The cruise passenger index strongly depends on the total number of cruise passengers, stream flow regimes of the Lijiang River, weather conditions, holiday periods, and other socioecological and political factors. Nevertheless, we found that stream flow did not decrease, particularly during the dry season from September to February. Therefore, stream flow changes did not have a significant negative impact on cruise tourism. This is likely because the total number of tourists and the total tourism income for the Guilin region increased sharply after 2006, while that of the number of cruise passengers and cruise ticket revenues remained stable or decreased. The number of cruise passengers and cruise ticket revenue have indeed shown a decreasing trend after 2006. Meanwhile, tourism opportunities in the region have been diversified, with the proliferation of new tourist destinations from 2006 to 2010 [18,48]. A series of rural tourism promoting activities created by the State Ministry of Culture and Tourism, such as the Guiding Opinions on Promoting the Sustainable Development of Rural Tourism in 2006 and 2018 [49], have effectively promoted the development of rural tourism. Rural tourism has, in turn, increased revenue and promoted employment [50]. With the development of tourist destinations within the Guilin region, the overall number of tourists and total tourism income set a new record in 2018 with, approximately 109 million people and 139 billion yuan, respectively [51].

Although we could not detect any significant impact of reduced stream flow on cruise tourism, this form of tourism is a highly important sector that is dependent on adequate water availability in the Lijiang River, particularly during the dry season, and the effects of climate change, combined with the intensification of water use, will further reduce water availability. Our projected results indicate that precipitation will not change significantly over the next 20 years (2020–2039) compared to the 1986–2005 average. However, the temperature is projected to increase from between 0.80 °C to 1.06 °C. Without a concomitant increase in precipitation, increasing temperature will likely reduce both summer and autumn stream flow [52]. Therefore, water shortages will likely worsen along the river, thus impeding regional economic development, including tourism. The effects of potential climate variability and reservoir regulation on stream flow must be carefully examined to ensure sustainable water resource management that will support viable cruise tourism.

Human activities, such as reservoir regulation, river-based tourism, and irrigation, combined with climate change, are likely to increase the stress on water availability in the Lijiang River, which will subsequently affect tourism and the aquatic ecosystem. Previous research suggests that current water discharge regulation by the Qingshitan reservoir alone cannot meet the water requirements of cruise tourism on the river as the primary function of this reservoir is the supply of water for irrigation and domestic consumption [22]. To meet the increasing water demands of economic expansion, and to support cruise tourism, particularly during the dry season, the Guilin government and the State Council have constructed several additional hydroelectric dams in the upper reaches of the Lijiang River [24]. Qingshitan reservoir, constructed in 1960, has regulated water supply downstream during the dry season since 1987. The Chuanjiang, Xiaorongjiang, and Fuzikou reservoirs were constructed in 2014, 2015, and 2018, respectively. Reservoir operation has caused significant changes in the downstream flow, improved some fish habitats on the river, especially by increasing water supply during the dry season, and also benefited human interests [22,53,54]. However, maintaining an ecological or quasi-natural flow must be ensured to conserve the river ecosystem [54]. Therefore, reservoir operation should be regulated to ensure optimal functioning of the river system, both ecologically and economically. When all of the reservoirs are fully operational and when water discharge is optimized, stream flow on the Lijiang River is expected to rise to 60 m^3/s during the dry season [55]. This will benefit both the river environment and human interests. Meanwhile, future research should consider the impact of the operation of multiple reservoirs on the Lijiang River under different climate change scenarios.

Tourism development, as well as economic growth, is associated with high water demand, and a large number of tourists may significantly exacerbate water shortage problems. The water consumption rate ranges from 84 to 2000 L per tourist per day [5]. The value of water use per tourist will increase if other amenities, such as swimming pools and golf courses, are included in the tourism area. With expanding tourism, water demands from the Lijiang River remain consistently high [21]. Meanwhile, local agricultural development may also lead to decreased stream flow. A previous study indicated that 80% of surface water (the largest quantity of regional water resources) is allocated to agriculture in the Guilin region [56]. This may explain why the monthly stream flow for the peak tourism months of April, May, and August decreased during the 1987 to 2016 period compared to the 1960 to 1986 period. Therefore, viable strategies for water regulation and irrigation planning are urgently required to avoid further adverse effects of stream flow on regional river-based tourism.

5. Conclusions

As water resources are the foundation of economic resources and critical for sustainable development, we assumed that decreases in stream flow would have negative impacts on the development of cruise tourism on the Lijiang River. However, we did not find that stream flow changes had a significant impact on the development of cruise tourism on the river from 1979 to 2016. Given that climate variability (in this case, precipitation) has experienced no significant change in previous decades, socio-economic factors, such as reservoir operation, land use policy, and river engineering, might have significant impacts on water availability. Current reservoir regulations have improved water availability in the Lijiang River, particularly during the dry season from September to February. Furthermore, additional reservoirs have been constructed to meet the increasing water demands of tourism, agriculture, and economic development. Considering that without a concomitant increase in precipitation, increases in temperature over the next 20 years (2020–2039) will likely reduce both rainy and dry season stream flow, the operation of multiple reservoirs in the Lijiang River should be conducted under conditions of (1) increasing reservoir capacities in heavy rain periods and adjusting flow regulation to support tourism during the dry season, and (2) preparing for climate change-related increases in temperature, insignificant changes in precipitation, and water shortages through adaptive measures, such as rationing water supply for cruise tourism, agriculture, domestic and municipal use.

Author Contributions: A.M. and Y.Y. conceived the idea. Y.Y. gathered and analyzed the data and wrote the first draft. A.M. provided critical feedbacks and revised the manuscript. All authors have read and agreed to the published version of the manuscript.

Funding: This study was financially supported by National Natural Science Foundation of China (grant number 41401211), National Key Research and Development Program of China (grant number 2012BAC16B01), West Light Foundation of Chinese Academy of Sciences (2014), the Foundation of the Guangxi Key Laboratory of Plant Conservation and Restoration Ecology in Karst Terrain (16-A-02-03), and the Foundation of Chinese Scholarship Council (CSC, 201808450049). A sabbatical leave awarded to A.M. from Lakehead University tenured at Guangxi Institute of Botany, Guangxi Zhuang Autonomous Region and Chinese Academy of Sciences enabled him to participate in this research.

Acknowledgments: We thank LetPub (www.letpub.com) for its linguistic assistance and scientific consultation during the preparation of this manuscript.

References

1. Li, J.; Qiao, Y.; Lei, X.; Kang, A.; Wang, M.; Liao, W.; Wang, H.; Ma, Y. A two-stage water allocation strategy for developing regional economic-environment sustainability. *J. Environ. Manag.* **2019**, *244*, 189–198. [CrossRef] [PubMed]
2. Gössling, S. Global environmental consequences of tourism. *Glob. Environ. Chang.* **2002**, *12*, 283–302. [CrossRef]
3. Valizadeh, N.; Mirzaei, M.; Allawi, M.F.; Afan, H.A.; Mohd, N.S.; Hussain, A.; El-Shafie, A. Artificial intelligence and geo-statistical models for stream-flow forecasting in ungauged stations: State of the art. *Nat. Hazards* **2017**, *86*, 1377–1392. [CrossRef]
4. Yang, W.; Liu, J. Water sustainability for China and beyond. *Science* **2012**, *337*, 649–650. [CrossRef]
5. Gössling, S.; Peeters, P.; Hall, C.M.; Ceron, J.P.; Dubois, G.; Lehmann, L.V.; Scott, D. Tourism and water use: Supply, demand, and security. An international review. *Tour. Manag.* **2012**, *33*, 1–15. [CrossRef]
6. Gössling, S. New performance indicators for water management in tourism. *Tour. Manag.* **2015**, *46*, 233–244. [CrossRef]
7. Paiano, A.; Crovella, T.; Lagioia, G. Managing sustainable practices in cruise tourism: The assessment of carbon footprint and waste of water and beverage packaging. *Tour. Manag.* **2020**, *77*, 104016. [CrossRef]
8. Papathanassis, A.; Beckmann, I. Assessing the "poverty of cruise theory" hypothesis. *Ann. Tour. Res.* **2011**, *38*, 153–174. [CrossRef]
9. Carić, H.; Mackelworth, P. Cruise tourism environmental impacts—The perspective from the Adriatic Sea. *Ocean Coast. Manag.* **2014**, *102*, 350–363. [CrossRef]
10. Lau, Y.; Sun, X. An Investigation into the Responsibility of Cruise Tourism in China. In *Maritime Transportation and Regional Sustainability*; Ng, A.K.Y., Monios, J., Jiang, C., Eds.; Elsevier Inc.: Amsterdam, The Netherlands, 2020; pp. 239–249. ISBN 9780128191347.
11. MacNeill, T.; Wozniak, D. The economic, social, and environmental impacts of cruise tourism. *Tour. Manag.* **2018**, *66*, 387–404. [CrossRef]
12. Johnson, D. Environmentally sustainable cruise tourism: A reality check. *Mar. Policy* **2002**, *26*, 261–270. [CrossRef]
13. Wang, H. *Report on the Development of Cruise Industry in China (2019)*; Wang, H., Ed.; Springer: Berlin/Heidelberg, Germany, 2020; ISBN 978-981-15-4660-0.
14. Brida, J.G.; Zapata, S. Cruise tourism: Economic, socio-cultural and environmental impacts. *Int. J. Leis. Tour. Mark.* **2010**, *1*, 205–226. [CrossRef]
15. Wang, Y. Lijiang River cruise tourism development issues and stratiges. *Econ. Reserch Guid.* **2015**, 254–256. [CrossRef]
16. Carrying Capacity for 5A Scenic Zone in China. Available online: http://www.gov.cn/xinwen/2015-07/17/content_2898961.htm (accessed on 11 August 2020).
17. Mao, X.; Meng, J.; Wang, Q. Modeling the effects of tourism and land regulation on land-use change in tourist regions: A case study of the Lijiang River Basin in Guilin, China. *Land Use Policy* **2014**, *41*, 368–377. [CrossRef]
18. Lu, L.; Bao, J.; Ling, S.; Zeng, Q.; Yu, H. The evolution progress and mechanism of Guilin-Lijiang River-Yangshuo tourism destination system. *Sci. Geogr. Sin.* **2012**, *32*, 1066–1074. [CrossRef]
19. Miao, Z. Water replenishment and tourism shipping in Lijiang River. *Site Investig. Sci. Technol.* **2000**, 36–40. [CrossRef]
20. Xu, J.; Dai, J. Variation characteristics of runoff in upper reaches of Lijiang River basin. *Yangtze River* **2018**, *49*, 41–46. [CrossRef]
21. Li, R.; Chen, Q.; Ye, F. Modelling the impacts of reservoir operations on the downstream riparian vegetation and fish habitats in the Lijiang River. *J. Hydroinformatics* **2011**, *13*, 229. [CrossRef]
22. Li, R.; Chen, Q.; Tonina, D.; Cai, D. Effects of upstream reservoir regulation on the hydrological regime and fish habitats of the Lijiang River, China. *Ecol. Eng.* **2015**, *76*, 75–83. [CrossRef]
23. Ye, F.; Chen, Q.; Li, R. Modelling the riparian vegetation evolution due to flow regulation of Lijiang River by unstructured cellular automata. *Ecol. Inform.* **2010**, *5*, 108–114. [CrossRef]
24. Yang, Y. Discussion on ecological and environmental water supplement target flow of Lijiang River. *GX Water Resour. Hydropower Eng.* **2012**, *3*, 7–10.

25. Lijiang River Eco-environmental Protection Law. 2011. Available online: http://www.gxzf.gov.cn/zwgk/flfg/dfxfg/t955429.shtml (accessed on 19 August 2020).
26. Outline Plan (2012–2020) for the Development of Guilin into an International Tourist Attraction. Available online: http://www.gov.cn/gzdt/2012-11/30/content_2278843.htm (accessed on 11 August 2020).
27. Guilin Forges Ahead in Tourism Upgrading. Available online: http://en.gxzf.gov.cn/2019-05/20/c_376274.htm (accessed on 11 August 2020).
28. Guilin Lijiang Zhi Editorial Committee. *Guilin Lijiang Zhi*; Guilin Lijiang Zhi Editorial Committee, Ed.; Guangxi People's Publishing House: Nanning, China, 2004; ISBN 7-219-05073-9/K987.
29. Guilin Economic and Social Statistical Yearbook Editorial Committee. *Guilin Economic and Social Statistical Yearbook*; Guilin Economic and Social Statistical Yearbook Editorial Committee, Ed.; China Statistics Press: Beijing, China, 2016; ISBN 9787503779817.
30. Richter, B.D.; Baumgartner, J.V.; Powell, J.; Braun, D.P. A method for assessing hydrologic alteration within ecosystems. *Conserv. Biol.* **1996**, *10*, 1163–1174. [CrossRef]
31. Guan, Q.; Yang, L.; Pan, N.; Lin, J.; Xu, C.; Wang, F.; Liu, Z. Greening and browning of the Hexi Corridor in northwest China: Spatial patterns and responses to climatic variability and anthropogenic drivers. *Remote Sens.* **2018**, *10*, 1270. [CrossRef]
32. Pohlert, T. Non-Parametric Trend Tests and Change-Point Detection. 2020. Available online: https://www.google.com.hk/url?sa=t&rct=j&q=&esrc=s&source=web&cd=&ved=2ahUKEwizspPJ1PHrAhUIMd4KHSLCDoYQFjAAegQIBRAB&url=https%3A%2F%2Fcran.r-project.org%2Fweb%2Fpackages%2Ftrend%2Fvignettes%2Ftrend.pdf&usg=AOvVaw2pWef-HxDUtvHEURSwCfMp (accessed on 20 August 2020).
33. Mann, H.B. Nonparametric tests against trend. *Econometrica* **1945**, *13*, 245–259. [CrossRef]
34. Hamed, K.H. Trend detection in hydrologic data: The Mann-Kendall trend test under the scaling hypothesis. *J. Hydrol.* **2008**, *349*, 350–363. [CrossRef]
35. Pettitt, A.N. A non-parametric approach to the change-point problem. *Appl. Stat.* **1979**, *28*, 126–135. [CrossRef]
36. Benesty, J.; Chen, J.; Huang, Y.; Cohen, I. *Noise Reduction in Speech Processing*; Springer: Berlin/Heidelberg, Germany, 2009; ISBN 9783540786115.
37. Tibshirani, R. Regression shrinkage and selection via the lasso: A retrospective. *J. R. Stat. Soc. Ser. B* **2011**, *73*, 273–282. [CrossRef]
38. Tibshirani, R. Regression shrinkage and selection via the lasso. *J. R. Stat. Soc. Ser. B* **1996**, *58*, 267–288. [CrossRef]
39. Granger, C.W.J. Testing for causality: A personal viewpoint. *J. Econ. Dyn. Control* **1980**, *2*, 329–352. [CrossRef]
40. Kamiński, M.; Ding, M.; Truccolo, W.A.; Bressler, S.L. Evaluating causal relations in neural systems: Granger causality, directed transfer function and statistical assessment of significance. *Biol. Cybern.* **2001**, *85*, 145–157. [CrossRef]
41. Mehdi, B.J.; Slim, B.Y. The role of renewable energy and agriculture in reducing CO_2 emissions: Evidence for North Africa countries. *Ecol. Indic.* **2017**, *74*, 295–301. [CrossRef]
42. Gent, P.R.; Danabasoglu, G.; Donner, L.J.; Holland, M.M.; Hunke, E.C.; Jayne, S.R.; Lawrence, D.M.; Neale, R.B.; Rasch, P.J.; Vertenstein, M.; et al. The community climate system model version 4. *J. Clim.* **2011**, *24*, 4973–4991. [CrossRef]
43. van Vuuren, D.P.; Edmonds, J.; Kainuma, M.; Riahi, K.; Thomson, A.; Hibbard, K.; Hurtt, G.C.; Kram, T.; Krey, V.; Lamarque, J.F.; et al. The representative concentration pathways: An overview. *Clim. Chang.* **2011**, *109*, 5–31. [CrossRef]
44. Xue, L.; Zhang, H.; Yang, C.; Zhang, L.; Sun, C. Quantitative assessment of hydrological alteration caused by irrigation projects in the Tarim River basin, China. *Sci. Rep.* **2017**, *7*, 1–13. [CrossRef] [PubMed]
45. Guiding Opinions on Promoting the Development of Rural Tourism. Available online: http://www.gov.cn/govweb/jrzg/2006-08/23/content_368032.htm (accessed on 20 August 2020).
46. Guilin Local Chronicies Editorial Committee. *Guilin Yearbook (2007)*; Guilin Local Chronicies Editorial Committee, Ed.; China Local Records Publishing Press: Beijing, China, 2007; ISBN 978-7-80238-175-9.
47. Guilin Local Chronicies Editorial Committee. *Guilin Nianjian (2017)*; Guilin Local Chronicies Editorial Committee, Ed.; Guangxi Normal University Press: Guilin, China, 2017; ISBN 9787559804105.

48. Feng, L.; You, H. Research on the Construction Index System of China Guilin Tourism Distribution Center. In Proceedings of the 2nd International Conference on Management, Education and Social Science (ICMESS2018), Qingdao, China, 23–24 June 2018; Atlantis Press: Beijing, China, 2018; Volume 176, pp. 1143–1146.

49. Guiding Opinions on Promoting the Sustainable Development of Rural Tourism. Available online: http://zwgk.mct.gov.cn/auto255/201812/t20181211_836468.html?keywords= (accessed on 13 August 2020).

50. Wang, L.E.; Cheng, S.K.; Zhong, L.S.; Mu, S.L.; Dhruba, B.G.C.; Ren, G.Z. Rural tourism development in China: Principles, models and the future. *J. Mt. Sci.* **2013**, *10*, 116–129. [CrossRef]

51. Guilin Local Chronicies Editorial Committee. *Guilin NianJian (2019)*; Guilin Local Chronicies Editorial Committee, Ed.; Thread-Binding Books Publishing House: Beijing, China, 2019; ISBN 9787512038479.

52. Barnett, T.P.; Adam, J.C.; Lettenmaier, D.P. Potential impacts of a warming climate on water availability in snow-dominated regions. *Nature* **2005**, *438*, 303–309. [CrossRef]

53. Li, W.; Han, R.; Chen, Q.; Qu, S.; Cheng, Z. Individual-based modelling of fish population dynamics in the river downstream under flow regulation. *Ecol. Inform.* **2010**, *5*, 115–123. [CrossRef]

54. Chen, Q.; Chen, D.; Han, R.; Li, R.; Ma, J.; Blanckaert, K. Optimizing the operation of the Qingshitan Reservoir in the Lijiang River for multiple human interests and quasi-natural flow maintenance. *J. Environ. Sci.* **2012**, *24*, 1923–1928. [CrossRef]

55. Chen, Q.; Yang, Q.; Li, R.; Ma, J. Spring micro-distribution of macroinvertebrate in relation to hydro-environmental factors in the Lijiang River, China. *J. Hydro-Environ. Res.* **2013**, *7*, 103–212. [CrossRef]

56. Jiang, Y.; Chen, Y.; Yan, Z. Water Utilization and Water Problems in Guilin City. Proceeding of the International Symposium on Water Resources and the Urban Environment, Wuhan, China, 9–10 November 2003; China Environmental Science Press: Wuhan, China, 2003; pp. 390–393.

 *sustainability*

Article

Analysis of Problems Related to the Calculation of Flood Frequency Using Rainfall-Runoff Models: A Case Study in Poland

Dariusz Młyński

Department of Sanitary Engineering and Water Management, University of Agriculture in Krakow, Mickiewicza 24–28 Street, 30-059 Krakow, Poland; dariusz.mlynski@urk.edu.pl; Tel.: +48-12-662-4041

Received: 11 July 2020; Accepted: 31 August 2020; Published: 2 September 2020

Abstract: This work aimed to quantify how the different parameters of the Snyder model influence the errors in design flows. The study was conducted for the Kamienica Nowojowska catchment (Poland). The analysis was carried out according to the following stages: determination of design precipitation, determination of design hyetograph, sensitivity analysis of the Snyder model, and quality assessment of the Snyder model. Based on the conducted research, it was found that the Snyder model did not show high sensitivity to the assumed precipitation distribution. The parameters depending on the retention capacity of the catchment had much greater impact on the obtained flow values. The verification of the model quality showed a significant disproportion in the calculated maximum flow values with the assumed return period.

Keywords: flood frequency; rainfall-runoff model

---

1. Introduction

Design flows (Q_T) are essential measures in engineering hydrology. These values are used as assumed theoretical values in all facilities where the main task is to minimize the risk associated with the rise of floods [1,2]. When these characteristics are estimated for gauged catchments, i.e., those with long series of hydrometric observations, flows are calculated on the probability distributions of random variables. However, it should be emphasized that out of all the catchments, most do not have any hydrometric observations or their series are too short to use statistical methods or are ungauged [3]. In such cases, so-called indirect methods are used, such as empirical formulas or hydrological models based on precipitation-runoff relationships [4]. In the recent years, an increasingly frequent initiative has been observed to solve the problem of calculating flood flows in ungauged catchments. In the case of small catchments, where indirect methods are the only source of hydrological information, empirical formulas are still the only one tool for determining the size of design flows [5].

Empirical formulas for calculating Q_T flows are a generalization of information regarding the physiographic and meteorological characteristics of a given region, affecting the formation of risings there. It must be emphasized that empirical formulas should be only used in areas for which they have been calibrated. However, it is possible to apply an empirical formula to an area that has not been calibrated but is similar to an area where calibration has taken place (regionalization), if it will be accepted that the results may be poor, as certain dissimilarities may have been neglected. One of the most used formula in the world for calculating Q_T flows is the so-called rational formula. The flows are determined based on characteristics such as runoff coefficient, critical precipitation, concentration time, and catchment area [6]. However, it should be emphasized that despite its simplicity, the rational formula has some limitations. They mainly concern the method of determining the runoff coefficient and determining the duration of precipitation followed by the largest runoff from the catchment [7].

Despite its limitations, the rational formula is still recommended in some regions of the world for calculating Q_T flows [8].

In engineering design practice, besides knowledge about Q_T flows, it is often necessary to use the other characteristics of the flood waves culminated over the whole Q_T event, namely volume and duration. In the case of ungauged catchments, such characteristics are determined using rainfall-runoff models [9]. Among the many models that are commonly used, one is the type of rainfall-runoff based on the Snyder synthetic unit hydrograph where the basic parameters are the size of the culmination and the time to the culmination [10]. Experience with the Snyder model and other models have shown some limitations regarding its use. They relate to the method of estimating design rainfall, high sensitivity to the design hyetograph, overestimation or underestimation of excessive rainfall from the usually used SCS-CN method, and the subjectivity of choosing the values of some indicators for estimating model parameters [11–14]. When using the Snyder model to simulate rainfall-runoff processes, attention should also be paid to other factors that may bring into question the results obtained. This is primarily the uncertainty of the input parameters and the lack of a linear relationship between the input data and the results obtained. Therefore, a sensitivity analysis and calibration of the model are fundamental. This will help obtain results with smaller errors and facilitate a better understanding of the flooding processes [15].

In the case of ungauged catchments, due to the lack of a reference value, the calibration range is very limited. Most of the studies related to the optimization of the work of hydrological rainfall-runoff models relate to comparative analyses of the floods observed and are determined by through models. However, not many reports indicate the sensitivity of Q_T flows concerning the size of parameters that can be taken subjectively. It must be emphasized that in case of Snyder model, in addition to model's parameters, the shape of design hyetograph may be subjectively determined, assuming the parameters of determining its course. Most of the studies related to the work quality of the Snyder's model concern the uncertainty analysis only for model parameters. However, in reality there may be an error uncertainty in all the parameters simultaneously. Due to this, an additional analysis of the model uncertainty regarding all parameters was performed simultaneously. This is an additional novelty of this study. Considering the above, this study aimed to analyze the problems associated with the estimation of Q_T flows using the rainfall-runoff Snyder synthetic model.

2. Study Area

The calculations were carried out for the Kamienica Nawojowska catchment. It is located in the Carpathian part of the upper Vistula basin in Poland. The catchment area is 237.7 km^2. The length of the main watercourse up to the section closing the catchment is 34.5 km. The average inclination of the drainage basin is 54.5‰. The density of the river network is 1.9 km·km^{-2}. The catchment area is used as follows (% of the total catchment area): continuous urban fabric (0.05%), discontinuous urban fabric (4.67%), commercial or industrial units (0.36%), non-irrigated arable land (15.91%), pastures (0.12%), sophisticated crop and plot systems (17.47%), agricultural areas with a high proportion of natural vegetation (2.85%), deciduous forests (16.03%), coniferous forests (21.79%), mixed forests (20.71%), and natural grasslands (0.04%). In the soil medium, there are mainly deposits with above-average permeability (medium-deep sandy soils and loess, sandy loams) and with very low permeability (clay soils, silty soils). The average annual precipitation in the basin is 901 mm. The average annual air temperature is 7.8 °C. Figure 1 shows the location of the research catchment. Figure 2 shows the use of the catchment area and Figure 3 shows its elevation differences.

Figure 1. Location of the analysed catchment.

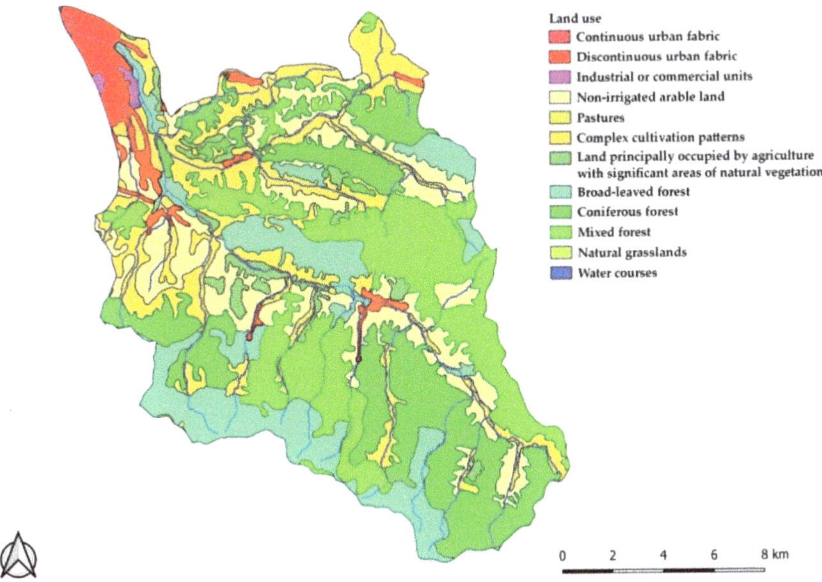

Figure 2. Land use of the Kamienica Nawojowska catchment.

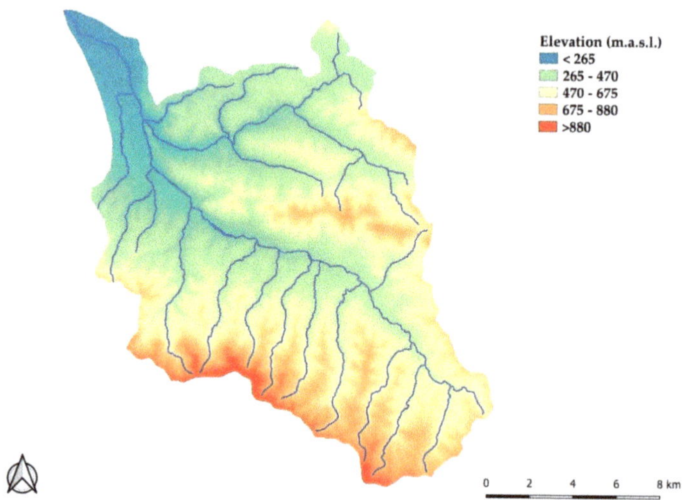

Figure 3. Elevation differences of the Kamienica Nawojowska catchment.

3. Materials and Methods

The basis for carrying out the research was the observed time series of the annual maximum daily rainfall and maximum annual flows from the 1971–2015 period for the Kamienica Nawojowska catchment. The data were obtained from the Institute of Meteorology and Water Management in Warsaw, National Research Institute. Precipitation data were obtained for four meteorological stations. Due to the rugged nature of the catchment area, they were averaged with the inverse distance method [16]. The physiographic characteristics of the catchment were established based on the Hydrographic Division Map of Poland 2010, Corine Land Cover 2018, and a digital elevation model with a grid resolution of about 25 m. The study was carried out according to the following stages: calculation of design precipitation and flows with a given period of return using statistical methods, determination of design hyetograph, calculation of excessive rainfall, determination of design flows using Snyder model, and assessment of the work quality of the Snyder model with the assumed values of its parameters.

3.1. Determination of Design Precipitation and Design Flows Using Statistical Method

In this paper, it was assumed that Q_T flows are caused by the annual maximum daily rainfall P_T with the same return period [17]. Q_T flows were determined using a statistical method to assess the quality of the Snyder model. The calculations were performed for the return periods of 1000, 100, 50, 10, 5, and 2 years. P_T precipitation and Q_T flows were calculated using the log-normal distribution, which is described as [18]:

$$f(x) = \frac{1}{(x-\varepsilon)\alpha\sqrt{2\pi}}exp\left[-\frac{1}{2}\left(\frac{ln(x-\varepsilon)-\mu}{\alpha}\right)^2\right] \tag{1}$$

$$x_p = exp\left[\mu + \frac{\alpha\sqrt{2}}{erf(2(1-p)-1)}\right] + \varepsilon \tag{2}$$

where:

x_p—quantile of the theoretical log-normal distribution;

ε—lower string limit,

$erf(2(1-p)-1)$—Gauss error function.

In practice, when Q_T is calculated, its upper limit is important. Also important is the knowledge about the degree of confidence, where the real value of Q_T will not exceed the upper limit of the confidence interval. Therefore, the probability P_β is calculated so that the real value of Q_T will not exceed the upper limit of confidence interval. This probability is known as the degree of protection. In the paper, an 84% degree of protection was assumed.

The calculated Q_T flows using the log-normal distribution were the reference level for assessing the quality of the Snyder model. After determining design precipitation, the concentration time for the catchment was determined using the Giandotti formula [19]. Then, precipitation was determined for duration equal to the concentration time using Lambor reduction curves. They are described by the following functions [20]:

$$\Psi(t) \;=\; 0.127{\cdot}t^{0.67} \text{ for } t \text{ from 5 to 120 min} \tag{3}$$

$$\Psi(t) \;=\; 0.243{\cdot}t^{0.67} \text{ for } t \text{ from 120 to 1440 min} \tag{4}$$

$$P_{T_c} \;=\; P_T(t \;=\; T_c) \tag{5}$$

where:

$\Psi(t)$—precipitation reduction factor (-),
t—duration of precipitation (min),
$P_{(T_c)}$—precipitation for a time equal to the concentration time (mm),
P_T—design precipitation with the same return period (mm),
T_c—concentration time (min).

Specified design precipitation for the assumed concentration times constituted the input signal to the Snyder model.

3.2. Determination of the Design Hyetograph

The input parameter to the Snyder model was design hyetograph which the shape was based on beta density function. This function is described by the following formula [21]:

$$f(x) \;=\; \frac{x^{\alpha-1}(1-x)^{\beta-1}}{B(\alpha,\beta)} \text{ for } 0 \le x \le t \tag{6}$$

where:

α, β—shape factors ($\alpha > 0, \beta > 0$),
B—value of the beta function,
t—duration of precipitation.

The basis for calculating the hyetograph were determined P_T precipitation with return periods of 1000, 100, 50, 20, 5 and 2 years, which were reduced for the time of concentration. 15 min were assumed as the time step of the hyetograph.

3.3. Determination of Design Flows Using the Snyder Model

The values of the Q_T flows were determined using the Snyder model. Excessive rainfall was determined using the SCS-CN method, where it depends on the soil permeability in the catchment area, the land use, and the water conditions of the catchment before precipitation causing surface runoff. Excessive rainfall was determined based on the following dependence [22–24]:

$$P_e = \begin{cases} \dfrac{(P-0.2S)^2}{P+0.8S} \text{ when } P \ge 0.2S \\ 0 \text{ when } P < 0.2S \end{cases} \tag{7}$$

where:

P_e—excessive rainfall [mm],
P—total rainfall [mm],
S—maximum potential catchment retention [mm].

The maximum potential retention of the catchment is directly related to the CN parameter and described by the following equation:

$$S = 25.4 \cdot \left(\frac{1000}{CN} - 10 \right) \tag{8}$$

The water condition of the catchment before precipitation causing the runoff are expressed by antecedent moisture condition (AMC). This parameter is considered to transform the rainfall depth in excessive rainfall. The AMC parameter includes three conditions: dry (AMC I), average (AMC II), and wet (AMC III). In this work, the CN parameter was determined for the wet condition (AMC III). The CN parameter was calculated for the catchment as a weighted average, according to the guidelines presented in the paper [25].

The Snyder model is based on the assumptions of the unit hydrograph, which is described by two parameters: the culmination of the hydrograph and the time of reaching culmination. These parameters are estimated according to the following dependence [26,27]:

$$T_L = C_t \cdot (L \cdot L_c)^{0.3} \tag{9}$$

where:

T_L—delay time [h],
C_t—coefficient related to the catchment retention range from 1.8 to 2.2 [-],
L—distance along the watercourse from the closing cross-section to the intersection of the dry valley with the watershed [km],
L_c—distance along the main watercourse from the mouth section to the center of gravity of the catchment area [km].

$$Q_p = \frac{2.78 \cdot C_p \cdot A}{T_L} \tag{10}$$

where:

Q_p—peak flow of unit hydrograph [m^3·s^{-1}·mm],
C_p—empirical coefficient resulting from the simplification of the hydrograph to triangle, taking values from 0.4 to 0.8 [-],
A—catchment area [km^2].

This study evaluates the sensitivity of the Snyder model to the assumed values of the input parameters of the Snyder model. Factors such as the influence of the shape of the hydrograph of precipitation and changes in the values of C_t and C_p parameters were analyzed. At the beginning of the calculations, it was assumed that the maximum precipitation intensity occurs in the middle of the hyetograph, hence, the parameters α and β were assumed as 5.0 and 5.0, respectively. The impact of changes of the shape of the precipitation hyetograph on the Q_T values calculated with the Snyder model was determined by changing the values of the parameters of the hydrograph by ±0.5, assuming its values from 1.0 to 9.0. By changing the values of the one parameter, the other was considered as a constant (5.0). The values of the C_t and C_p parameters were initially set at 2.0 and 0.6, respectively. The impact of the assumed values of these parameters on Q_T values was determined by changing them by ±0.025, where the C_t parameter took values from 1.800 to 2.200 and the C_p parameter from

0.400 to 0.800. The analysis of the sensitivity of the Snyder model to changes in the values of C_t and C_p parameters was conducted for the precipitation hyetograph with the maximum precipitation intensity in the middle of its duration ($\alpha = 5.0$ and $\beta = 5.0$).

3.4. Evaluation of the Quality of Work of the Snyder Model

The assessment of the work quality of Snyder models in relation to the assumed values of input parameters was made using mean absolute percentage error (*MAPE*), which is described by the following dependence [28]:

$$MAPE = \frac{Q_T - Q_{T_{Snyder}}}{Q_T} \cdot 100 [\%] \tag{11}$$

where:

Q_T—maximum flow with a given frequency of occurrence, calculated using the log-normal distribution [m^3·s^{-1}],

$Q_{T_{Snyder}}$—maximum flow with a given occurrence frequency, calculated using the Snyder model [m^3·s^{-1}].

4. Results and Discussion

The study was carried out in the following stages: determination of the P_T precipitation and Q_T flows using log-normal distribution, determination of the height precipitation for concentration time, determination of the course of precipitation hyetograph, assessment of the sensitivity of the Snyder model to the input parameters, and assessment of the quality of the Snyder model.

4.1. Determination of Design Precipitation and Flows

Precipitation P_T and Q_T flows were determined using log-normal distribution. The results of the calculations are presented in Figures 4 and 5. Based on the results summarized in Figure 4, the following precipitation values with a return period of 1000, 100, 50, 20, 10, 5, and 2 years were found: 152.4, 114.2, 103.1, 88.5; 77.3, 65.6, and 48.2 mm, respectively. The values of Q_T flows with the same return periods are 1311.5, 709.9, 570.2, 410.5, 306.6, 215.3, and 109.5 m^3·s^{-1}. Determination of P_T precipitation and Q_T flows should be conducted based on observed time series that are homogeneous and independent, i.e., that do not show significant trends in their values. Research presented by Młyński et al. [29] showed that in the upper Vistula basin, the maximum annual daily precipitation is characterized by a lack of trends. Kundzewicz et al. [30] showed that the maximum annual flows in the upper Vistula basin are also characterized by a lack of trends in their course. Therefore, it proves that the factors determining peak flows were stationary over the considered time period. The essential element of P_T precipitation and Q_T flows estimation is the selection of the best-fitted probability distribution functions. According to Węglarczyk [31], hydrometeorological data usually support hypotheses about the compatibility of their empirical distributions with more than one theoretical distribution. Then the best-fitted functions should be sought from the group of the distribution candidates. Research by Młyński et al. [32] showed that in the upper Vistula basin the best distribution for calculating design precipitation are log-normal and GEV. Młyński et al. [33] showed that in the same area the best-fitted distributions for calculating Q_T flows are Pearson type III and log-normal.

In the next stage of the study, the assumed precipitation was determined for the duration equal to the concentration time. Lambor reduction curves were used. The results are summarized in Figure 6. The determined value of concentration time for the Kamienica Nawojowska catchment was 8 h. Analyzing the course of precipitation reduction curves, its amount for this time was for return periods of 1000, 100, 50, 20, 10, 5, and 2 years: 119.7, 89.7, 81.0, 69.5, 60.7, 51.6, 37.9 mm, respectively. The time of concentration is one of the basic parameters of hydrological models that determines the culmination in the peak flows [34]. It should be emphasized that this indicator may be vary depending on the methodology used. Research by Grimaldi et al. [35] showed a difference in the obtained values as much as 500%, depending on the method used. These differences could be due to differences in the definition of this phenomenon and variable calculation assumptions of this parameter. Therefore, further observations should be made to understand the phenomenon of the concentration time better.

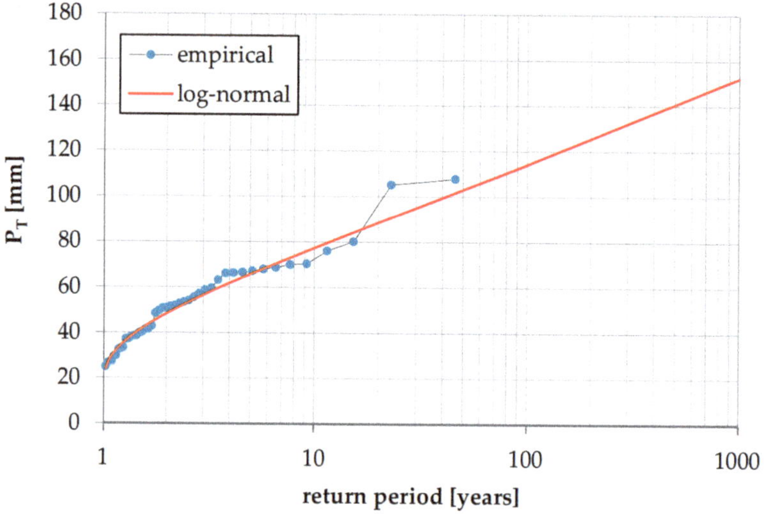

Figure 4. P_T precipitation for the Kamienica Nawojowska catchment determined with the log-normal distribution.

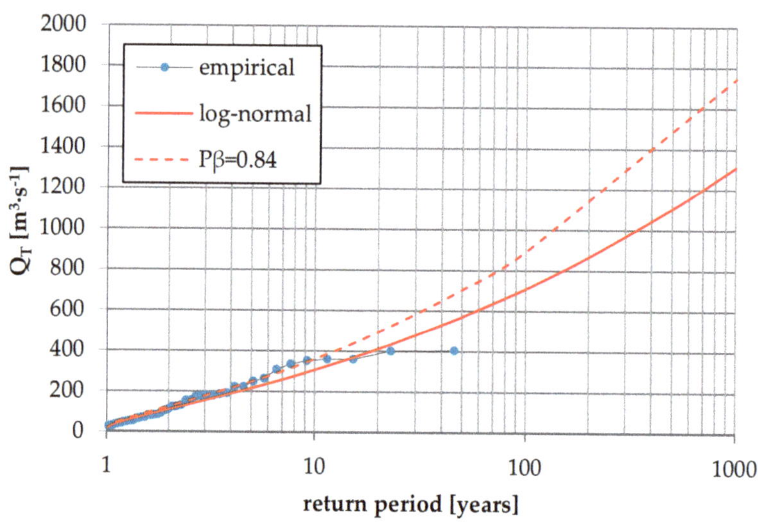

Figure 5. Q_T flow values for the Kamienica Nawojowska catchment determined with the log-normal distribution.

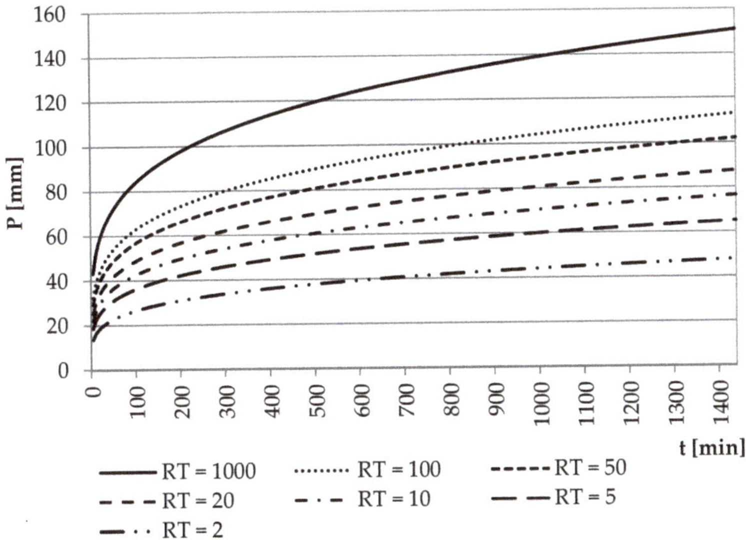

Figure 6. Precipitation reduction curves.

4.2. Determination of Precipitation Hyetograph

The design precipitation hyetographs were determined using beta distribution. Figure 7 presents the impact of changes in the values of the hyetograph parameters on time to reach maximum precipitation intensity. Based on the results summarized in Figure 7, it was found that assuming a constant value of the parameter $\beta = 5.0$, increasing the value of the parameter α causes a longer time to reach the maximum precipitation intensity. For $\alpha = 1.0$, the maximum precipitation was achieved after 3.1% of its total duration. For $\alpha = 9.0$, the maximum precipitation was achieved after 68.8% of its total duration. Assuming a constant value of $\alpha = 5.0$, the value of the parameter β has an inverse effect on the time to reach the maximum precipitation intensity. For $\beta = 1.0$, the maximum intensity was reached for 100% of the total duration of precipitation. For $\beta = 9.0$, the maximum intensity was achieved for 34.4% of the total duration of the event.

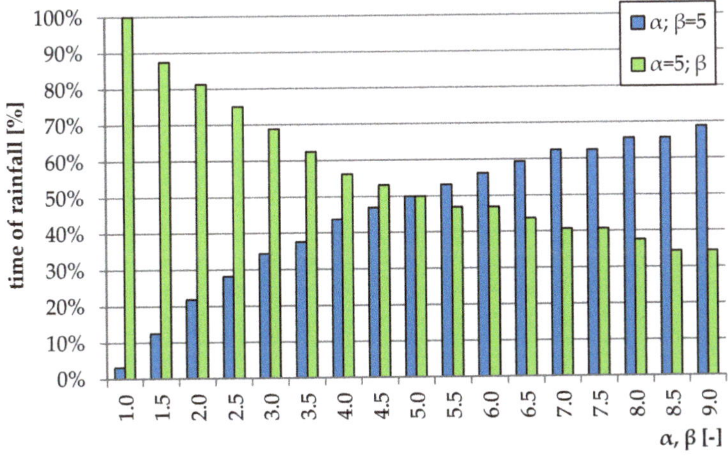

Figure 7. Impact of α and β parameter values on achieving the maximum precipitation time.

4.3. Determining the Value of Design Flows Using the Snyder Model

In order to determine Q_T flows with the Snyder model, excessive rainfall was determined by the SCS-CN method for assumed return periods. The results of the calculations are presented in Table 1.

Table 1. Effective precipitation compared to the total value for the Kamienica Nawojowska catchment.

Return Period	CN [-]	P [mm]	P_{net}
1000		119.7	75.9
100		89.7	49.5
50		81.0	42.1
20	83.8	69.5	32.7
10		60.7	25.9
5		51.6	19.2
2		37.9	10.2

In order to determine excessive rainfall, the CN parameter for AMC III was determined for the assumed return periods. It should be emphasized that when estimating Q_T flows, the assumption of an appropriate level of moisture is a significant problem. In the case of mountain catchments, the initial moisture content is determined not only by atmospheric precipitation, but also by the high groundwater level. Also, such catchments usually have soils with reduced permeability, which makes it difficult to infiltrate rainfall. Hence, it is assumed that too low of a level of moisture can lead to an underestimation of the culmination. As reported by De Paulo et al. [36], when Q_T flows calculated with the use of precipitation models are used as design values, it is recommended to adopt AMC III in order to minimize the risk of their underestimation. It should also be emphasized that the SCS-CN method itself has some limitations. As demonstrated by the studies conducted by Grimaldi et al. [37], due to the assumption of constant infiltration, this method causes an underestimation of effective precipitation at the beginning of its occurrence and an overestimation at the end of the episode.

After determining the excessive rainfall, the Q_T flows were determined for the analyzed return periods. Table 2 summarizes the values of the basic flood characteristics: culmination Q_T, volume V, and time to reach culmination. These characteristics were initially determined assuming the maximum intensity of precipitation in the middle of its duration ($\alpha = 5.0$, $\beta = 5.0$) and assuming $C_t = 2.000$ and $C_p = 0.600$.

Table 2. Characteristics of high hydrographs for assumed return periods.

Characteristic	Return Period						
	1000	100	50	20	10	5	2
Q_T [m^3·s^{-1}]	304.1	198.6	169.2	131.7	104.3	77.4	41.2
V [mln m^3]	18.233	11.877	10.118	7.869	6.157	4.617	2.462
t [h]	14.50	14.50	14.50	14.50	14.50	14.50	14.75

The essential element to use the Snyder model is the determination of the shape of the precipitation hyetograph. Figure 8 presents an analysis of the sensitivity of the Snyder model to the change in the shape of the precipitation hyetograph. The percentage change in the values of Q_T flows, depending on the values of parameters α and β, was determined in relation to the Q_T flows determined for the hyetograph with the maximum intensity in the middle of the duration of precipitation and assuming $C_t = 2.000$ and $C_p = 0.600$.

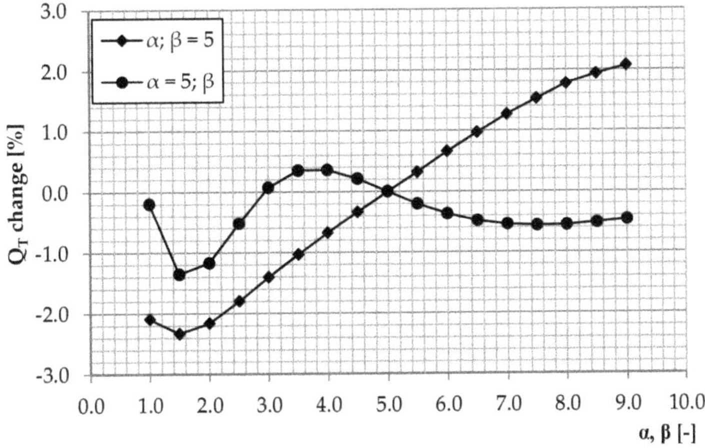

Figure 8. Analysis of the sensitivity of the Snyder model to changes in the shape of the hyetograph of precipitation.

The obtained percentage changes in the value of Q_T flows in relation to the assumed shape of the precipitation hyetograph were the same, regardless of the return period. The results presented in Figure 8 indicate that with the constant size of parameter $\beta = 5.0$, the maximum difference between Q_T obtained for variable values of α is 4.4%. The lowest Q_T value, with constant β was obtained for $\alpha = 1.5$ and the highest for $\alpha = 9.0$. The Q_T flow rates increase linearly for parameters α from 1.5 to 9.0. Also, they are lower than the culmination specified for the hyetograph with the maximum precipitation intensity in the middle of its duration. When the values of α parameters were greater than 5.0, increasing flow rates were noticed. Showing slightly different sensitivity, the Snyder model showed the change in the β parameter, with a constant value of $\alpha = 5.0$. The maximum difference between Q_T values at constant α was 1.7%. However, it should be emphasized that in this case there is no functional relationship between the β parameter and Q_T flows. The lowest value of Q_T with the variable β was obtained when its value was 1.5, while the highest for the sizes 3.5 and 4.0. In the case when the value of the parameter β was lower than 5.0 then it was noticed that for the range from 1.0 to 4.5 Q_T flows were lower than those obtained for the hyetograph with the maximum precipitation intensity in the middle of duration. Also, the values of these flows increased systematically for β from 1.5 to 4.0. In the case when the values of this parameter increased from 4.5 and higher, decreasing Q_T flows were found. When analyzing the impact of the precipitation hyetograph on Q_T values, it can be concluded that the Snyder model is more sensitive to changes in the α parameter. As the calculations showed, this sensitivity is not very high. It should be emphasized, however, that the sensitivity analysis was carried out assuming a constant value of one of the parameters describing the shape of the precipitation hyetograph. Only assuming α and β values from 1.0 to 9.0 with their change every 0.5, 289 different combinations of these parameters can be obtained. It should be emphasized that their values can take any numbers greater than 0. Therefore, in the absence of information about the distribution of precipitation, using beta distribution should be carefully set its parameters, because they can affect the results of the simulation. This indicates the need to optimize parameters α and β in such a way that they can be used in ungauged catchments [38]. Studies related to the impact of precipitation hyetograph on culmination values obtained from simulations were also conducted by Sigaroodi and Chen [39] and by Petroselli et al. [40]. The authors also unambiguously indicated the differences in the size of the culmination concerning changes in the shape of the precipitation hyetograph.

In the next stage of the study, the sensitivity of the Snyder model to changes in C_t and C_p parameters was analyzed. By changing the size of one parameter, the other was considered constant. The results of the analysis are summarized in Figures 9 and 10.

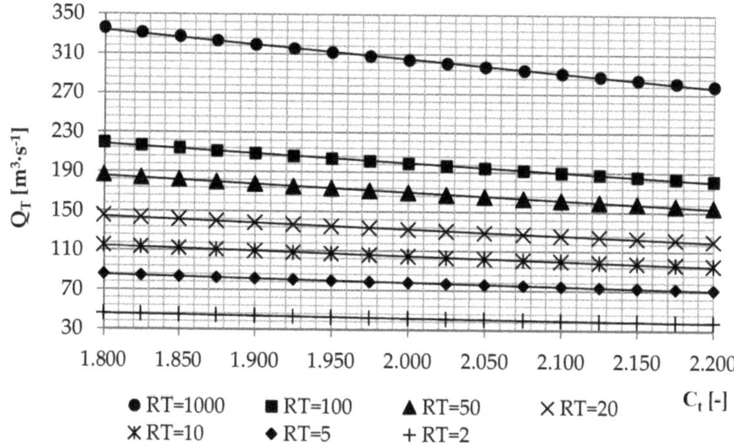

Figure 9. Analysis of sensitivity of the Snyder model to changes in the C_t parameter.

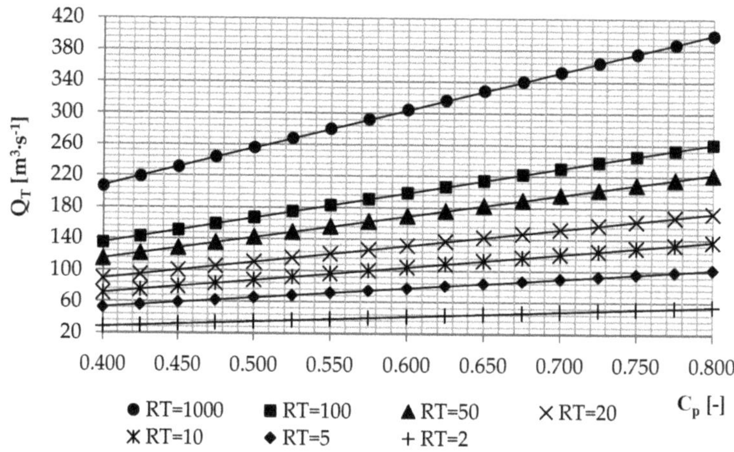

Figure 10. Analysis of sensitivity of the Snyder model to changes in the C_p parameter.

When analyzing the values presented in Figure 9, it was found that at a constant value $C_p = 0.6$, the value of the parameter C_t is inversely proportional to the value of the flow Q_T. The C_t parameter is the main component of the formula for the lag time in the Snyder model (formula 9). As the lag time increases, the number of culmination decreases. The differences between the minimum and maximum Q_T for the assumed values of C_t parameters were significant and reached almost 18% for all return periods. In the case of the C_p parameter values, its changing at constant $C_t = 2.000$, caused even greater disproportions in the obtained Q_T values. The size of this parameter is directly proportional to the flow with the assumed return period. The difference between the minimum and maximum Q_T for the assumed C_p values with a constant C_t was almost 50%. This indicates that the Snyder model is more sensitive to changing this parameter. Both the C_t and C_p parameters in the Snyder model are values describing the retention capacity of the catchment. However, they were developed for specific local conditions in other climate zones, as in Poland. Hence, the assumed values of these parameters may not fully reflect the real conditions affecting the flooding in the other climate zone. Therefore, in order to minimize this problem, it is necessary to work out the values of these parameters in the form of a function depending on the catchment characteristics. Such analyses were carried out by Wałega [41] who defined equations describing the C_t and C_p parameters. According to his research, the C_t parameter depends on CN parameter and average slope of catchment. In the case of the C_p parameter, the lag time and the average slope of the catchment have the greatest impact on its values. However, the values of its coefficients of determination pointed to unsatisfactory model quality. This could be due to the lack of consideration of the rest of the critical characteristics affecting the retention capacity of the catchment.

4.4. Determining Relative Error

Complementing the research was determining the values of the relative error in Q_T flow estimation with the Snyder model, at various values of its parameters. The results of the analysis are presented in Table 3.

Table 3. Relative error estimation for Q_T flows using the Snyder model.

Return Period	$\alpha; \beta = 5$			$\alpha = 5; \beta$			$C_t; C_p = 0.600$			$C_t = 2.000; C_p$		
	Min	Mean	Max	Min	Mean	Max	Min	Mean	Max	Min	Mean	Max
1000	77.3	77.3	77.3	74.4	76.7	78.8	74.4	76.7	78.8	69.6	76.9	84.3
100	71.3	72.0	72.7	69.1	71.9	74.4	69.1	71.9	74.4	63.3	72.1	81.0
50	69.6	70.3	71.0	67.3	70.2	72.9	67.3	70.2	72.9	61.1	70.4	79.9
20	67.1	67.9	68.7	64.6	67.8	70.7	64.6	67.8	70.7	57.9	68.0	78.3
10	65.1	66.0	66.9	62.5	65.9	68.9	62.5	65.9	68.9	55.3	66.0	77.0
5	63.1	64.0	65.0	60.3	63.9	67.1	60.3	63.9	67.1	52.7	64.1	75.7
2	61.3	62.2	63.3	58.2	62.1	65.5	58.2	62.1	65.5	50.2	62.2	74.4

Based on the results presented in Table 3, it was found that the average relative error in Q_T flow estimation, regardless of the assumed parameter values, is about 69%. According the scale presented in work [42], the performance rating is unsatisfactory. In the case of changes in the shape of the hyetograph of precipitation, it can be seen that the assumed values of parameters α and β have a low effect on the quality of the results obtained, regardless of the return periods. The C_t and C_p parameters have a much greater impact on the quality of the model's work. In the case of the former, the relative error rates ranged from 65 to 72% on average. For the C_p parameter, these disparities were even larger and ranged from 59% to 79% on average.

It should be emphasized that α and β parameters may take any values above zero. In case of the C_t and C_p parameters, the values should be in the range of 1.8 to 2.2 and 0.4 to 0.8 respectively. Due to this, the error or uncertainty in all four parameters must be simultaneously established. Therefore, the additional calculations were made, consisting on randomly sampling value for each parameter repeating this 1000 times, and presented the results as a histogram of errors. The results are presented in Figure 11. Analyzing the results presented in Figure 11, it was found that the error of Q_T calculated by Snyder model decreases with respect to the return period. For Q_{1000} the highest count of values was for error for the range 80–85% (332). For Q_{100} it was for the range of 85–90% (503 observations). For Q_{50} and Q_{20} it was for the range of 65–70% (295 and 263 observations, respectively). For Q_{10} it was for the range of 70–75% (244 observations). For Q_5 and Q_2 the highest count of values was for errors in the range 60–65% (238 and 218 observations, respectively).

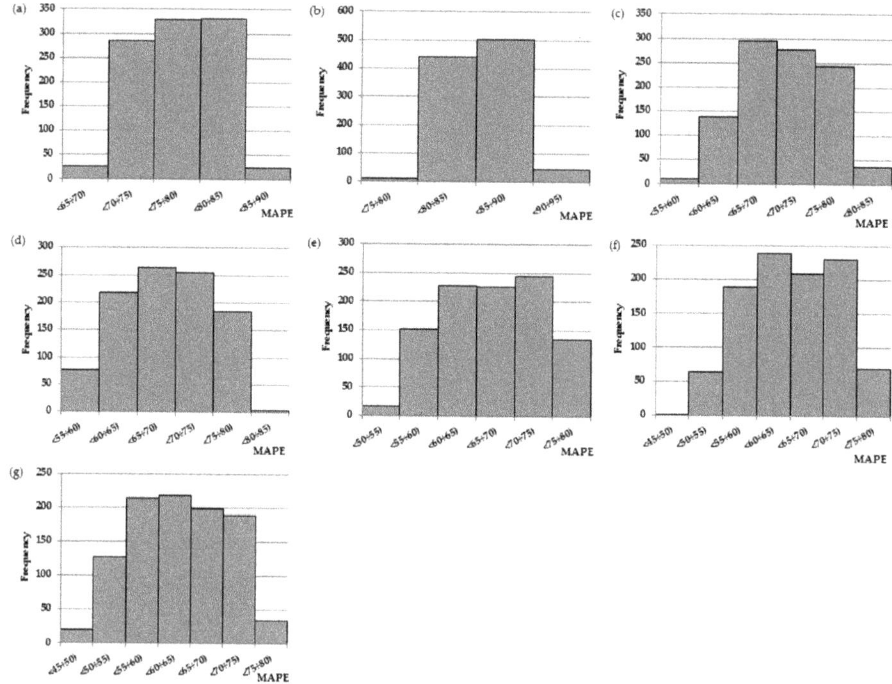

Figure 11. The histograms of MAPE for Q_T calculated by Snyder model with randomly sampled parameters values: (**a**) Q_{1000}; (**b**) Q_{100}; (**c**) Q_{50}; (**d**) Q_{20}; (**e**) Q_{10}; (**f**) Q_5; (**g**) Q_2.

It must be emphasized that an important factor affecting the model results is the quality of input data, mainly the precipitation data. When the model is used for the Q_T calculation, the precipitation data are in the form of precipitation with the same return period. Bormann [43] indicated that high quality model results require also high-quality precipitation data. However, this is not always highly resolved. The studies conducted by Bárdossy and Das [44] showed that the number and spatial distribution of the rain station affect at the model's results. Antcil et al. [45] showed that simulation performance was reduced when the mean rainfall was calculated using a number of rainfall stations lower than a certain number. Spatial distribution and the accuracy of the rainfall input to a rainfall-runoff model considerably influence the volume of rainfall runoff, peak runoff, and time-to-peak. It needs to be highlighted that flood modeling requires knowledge for local conditions related to flood shaping [46]. To accurately estimate floods using hydrological models, their parameters and the initial state variables must be known. Good estimations of parameters and initial state variables are required to enable the models to make accurate estimations [47]. It should be emphasized that trustworthiness is important in hydrologic modeling. Hence, the uncertainty analyses are very important. A variety of methods have been developed to deal with parameter uncertainty. Among these methods, the generalized likelihood uncertainty estimation method, the formal Bayesian method using the Metropolis–Hastings (MH) algorithm and a Markov Chain Monte Carlo (MCMC) methodology are extensively used. However, most studies were carried out in the catchments where detailed hydrological and meteorological data were available. Smaller catchments (as in the case study) often feature just one or two gauges and rainfall stations, and sometimes this infrastructure is completely absent. The implementation of flood protection plans requires also hydrological analyses to be conducted, often with the use of hydrological models, in the catchments with under-developed measurement network. Hence, designers usually base their calculations on basic hydrological models, mostly those of lumped parameters, due to their simplicity and ease of obtaining and setting the parameters [14].

The conducted research allowed to indicate significant disproportions between Q_T determined by the Snyder model and the statistical method. They clearly emphasize the problems associated with the use of rainfall-runoff models to determine these flows. The first is the assumption of equal probability for design rainfall and flows. Research by Pilgrim and Cordery [48] has shown that this is simplifying things too much. The actual relationship between the probability of precipitation and flow is not clear, because each characteristic of assumed precipitation introduces some changes in the assumed probability. Therefore, it is also necessary to consider other characteristics describing design precipitation, such as duration, distribution over time, and intensity. Studies by Viglione and Blöschl [17] have shown that equal precipitation and flow probability can be assumed, provided that the precipitation episodes causing the surges are of the same duration. In this work, the flow value with a return period of 1000 years, determined using the Snyder model is 304.1 m^3·s^{-1}. This corresponds to the frequency of occurrence every ten years, determined by the statistical method. As demonstrated by Hirabayashi et al. [49], due to climate change, the incidence of significant floods will gradually increase. The calculations also allowed to indicate the necessity associated with the calibration of rainfall-runoff models. The values of the input parameters to the models, determined from certain numerical intervals, should be determined by optimization. In the case of gauged catchments, where the flow values are known, calibration is not a major obstacle. It should be emphasized, however, that these models are usually used in ungauged catchments with no hydrometric measurements. A designer with a specific range of input parameter values is unable to determine their optimal value. Hence, continuous research is needed on methods to determine Q_T in ungauged catchments. In Poland, empirical formulas are the most used. However, due to the fact that they were developed in the last century, bearing in mind the ongoing climate change and the use of catchment areas, their use may raise justified doubts. Research conducted by Młyński et al. [50] showed that the differences between Q_T determined by empirical formulas and statistical methods could reach 70% and above in the upper Vistula drainage basins. Given the above, Młyński et al. [51] implemented the EBA4SUB model for estimating Q_T in the upper Vistula basin. The conducted analyses allowed the authors to state that this model gives a lower value of Q_T errors than empirical formulas. This is due to climate change and the land use of the catchment area over the years. In the paper of Młyński et al. [20], a new methodology for calculating Q_T in ungauged catchments has been proposed. It is also based on the EBA4SUB model, however, this idea boils down not to directly determining the Q_T value, but to determine it by simulating a series of observational maximum annual flows and then determining the Q_T value with statistical distributions. The main advantage of the methodology is the minimization of the problem of equal precipitation and flow probability. Also, the EBA4SUB model does not require calibration, which results in unambiguous results. It should also be emphasized that for design purposes, it is often necessary to know the other parameters of the floods, such as volume or duration time. Research conducted by Młyński et al. [52] has also shown that the EBA4SUB model is a handy tool when it comes to the modeling rainfall-runoff events. The analyses made by the authors for mountain catchments clearly confirmed that the obtained flow hydrographs are similar to the real ones as determined by the Snyder and SCS-UH models. Because more and more urbanization of the catchment area is affecting the water cycle, further research is focused on developing a methodology for determining Q_T in urban catchments using the EBA4SUB model.

5. Conclusions

The study aimed to analyze problems related to the determination of Q_T flows in ungauged catchments of the upper Vistula river basin in Poland. The studies were carried out according to the following stages: determination of design precipitation P_T and Q_T flows using log-normal distribution, determination the height of precipitation for concentration time, determination the course of precipitation hyetograph, assessment of the sensitivity of the Snyder model to the input parameters, and assessment of the quality of the Snyder model. The calculations were made using the Snyder model. The calculations made it possible to state that the shape of the precipitation hyetograph did not significantly affect the magnitude of the culmination determined using the analyzed model. However, it should be emphasized that in the case of the beta distribution used, the model was more sensitive to changes in the α parameter. In the case of the analysis of parameters related to the retention of the catchment area: C_t and C_p, the model was significantly more sensitive to their change. Based on the calculations made, it was found that for the assumed return periods, the average error in Q_T estimation was 65%. The conducted studies clearly emphasized the importance of calibration of the Snyder model and the problems of use in ungauged catchments.

Funding: This research received no external funding.

Acknowledgments: Special thanks to Andrzej Wałęga, from University of Agriculture in Krakow for consultations and critical comments to the article. Thank you for all your support.

Conflicts of Interest: The author declares no conflict of interest.

References

1. Karabová, B.; Sikorska-Senoener, A.E.; Banasik, K.; Kohnová, S. Rainfallrunoff model for a small catchment in Carpathians. *Ann. Warsaw Univ. Life Sci. SGGW. Land Reclam.* **2012**, *44*, 155–162. [CrossRef]
2. Vaššová, D. Comparison of rainfall-runoff models for design discharge assessment in a small ungauged catchment. *Soil Water Res.* **2013**, *8*, 26–33. [CrossRef]
3. Khaleghi, M.R.; Ghodusi, J.; Ahmadi, H. Regional analysis using the Geomorphologic Instantaneous Unit Hydrograph (GIUH) method. *Soil Water Res.* **2014**, *9*, 25–30. [CrossRef]
4. Jena, J.; Nath, S. An empirical formula for design flood estimation of un-gauged catchments in Brahmani Basin, Odisha. *J. Inst. Eng. India Ser. A* **2020**, *101*, 1–6. [CrossRef]
5. Hrachowitz, M.; Savenije, H.H.G.; Blöschl, G.; McDonnell, J.J.; Sivapalan, M.; Pomeroy, J.W.; Arheimer, B.; Blume, T.; Clark, M.P.; Ehret, U.; et al. A decade of Predictions in Ungauged Basins (PUB)—A review. *Hydrol. Sci. J.* **2013**, *58*, 1198–1255. [CrossRef]
6. Dhakal, N.; Fang, X.; Asquith, W.H.; Cleveland, T.G.; Thompson, D.B. Return period adjustment for runoff coefficients based on analysis in undeveloped Texas watersheds. *J. Irrig. Drain. Eng. ASCE* **2013**, *139*, 476–482. [CrossRef]
7. Piscopia, R.; Petroselli, A.; Grimaldi, S. A software package for the prediction of design flood hydrograph in small and ungauged basins. *J. Agric. Eng.* **2015**, *432*, 74–84.
8. Petroselli, A.; Grimaldi, S. Design hydrograph estimation in small and fully ungauged basin: A preliminary assessment of the EBA4SUB framework. *J. Flood Risk Manag.* **2018**, *11*, 197–2010. [CrossRef]
9. Salami, A.W.; Bilewu, S.O.; Ibitoye, A.B.; Ayanshola, A.M. Runoff hydrographs using Snyder and SCS unit hydrograph methods: A case study of selected rivers in south west Nigeria. *J. Ecol. Eng.* **2017**, *18*, 25–34. [CrossRef]
10. Federova, D.; Kovář, P.; Gregar, J.; Jelínková, A.; Novotná, J. The use of Snyder synthetic hydrograph for simulation of overland flow in small ungauged and gauged catchments. *Soil Water Res.* **2018**, *13*, 185–192.
11. Blair, A.; Sanger, D.; White, D.; Holland, A.F.; Vandiver, L.; Bowker, C.; White, S. Quantifying and simulating stormwater runoff in watersheds. *Hydrol. Process.* **2012**, *28*, 559–569. [CrossRef]
12. Banasik, K.; Rutkowska, A.; Kohnová, S. Retention and Curve Number variability in a small agricultural catchment: The probabilistic approach. *Water* **2014**, *6*, 1118–1133. [CrossRef]
13. Kowalik, T.; Wałęga, A. Estimation of CN Parameter for small agricultural watersheds using asymptotic functions. *Water* **2015**, *7*, 939–955. [CrossRef]
14. Wałęga, A.; Książek, L. Influence of rainfall data on the uncertainty of flood simulation. *Soil Water Res.* **2016**, *11*, 277–284. [CrossRef]
15. Wałęga, A. The importance of calibration parameters on the accuracy of the floods description in the Snyder's model. *J. Water Land Dev.* **2016**, *28*, 19–25. [CrossRef]
16. Chen, F.W.; Liu, C.W. Estimation of the spatial rainfall distribution using inverse distance weighting (IDW) in the middle of Taiwan. *Paddy Water Environ.* **2012**, *10*, 209–222. [CrossRef]

17. Viglione, A.; Blöschl, G. On the role of storm duration in the mapping of rainfall to flood return period. *Hydrol. Earth Syst. Sci. Discuss.* **2008**, *5*, 3419–3447. [CrossRef]
18. Hogg, R.V.; Craig, A.T. *Introduction to Mathematical Statistics*; Macmilan Publishing Co.: New York, NY, USA, 1978.
19. Petroselli, A.; Vojtek, M.; Vojteková, J. Flood mapping in small ungauged basins: A comparisonof different approaches for two case studies in Slovakia. *Hydrol. Res.* **2019**, *50*, 379–392. [CrossRef]
20. Młyński, D.; Wałęga, A.; Ozga-Zieliński, B.; Ciupak, M.; Petroselli, A. New approach for determining the quantiles of maximum annual flows in ungauged catchments using the EBA4SUB model. *J. Hydrol.* **2020**, *589*, 1–12. [CrossRef]
21. Gądek, W.; Bodziony, M. The hydrological model and formula for determining the hypothetical flood wave volume in non-gauged basins. *Meteorol. Hydrol. Water Manag.* **2015**, *3*, 1–7. [CrossRef]
22. Xiao, B.; Wang, Q.; Fan, J.; Han, F.; Dai, Q. Application of the SCS-CN Model to Runoff Estimation in a Small Watershed with High Spatial Heterogeneity. *Pedosphere* **2011**, *26*, 738–749. [CrossRef]
23. Soulis, K.X.; Valiantzas, J.D. SCS-CN parameter determination using rainfall-runoff data in heterogeneous watersheds—The two-CN system approach. *Hydrol. Earth Syst. Sci.* **2012**, *16*, 1001–1015. [CrossRef]
24. Soulis, K.X. Estimation of SCS Curve Number variation following forest fires. *Hydrol. Sci. J.* **2018**, *63*, 1332–1346. [CrossRef]
25. Wałęga, A.; Salata, T. Influence of land cover data sources on estimation of direct runoff according to SCS-CN and modified SME methods. *Catena* **2019**, *172*, 232–242. [CrossRef]
26. Singh, P.K.; Mishra, S.K.; Jain, M.K. A review of the synthetic unit hydrograph: From the empirical UH to advanced geomorphological methods. *Hydrol. Sci. J.* **2014**, *59*, 239–261. [CrossRef]
27. Sudhakar, B.S.; Anupam, K.S.; Akshay, A.J. Snyder unit hydrograph and GIS for estimation of flood for un-gauged catchments in lower Tapi basin, India. *Hydrol. Curr. Res.* **2015**, *6*, 1–10.
28. Kim, S.; Kim, H. A new metric of absolute percentage error for intermittent demand forecast. *Int. J. Forecast.* **2016**, *32*, 669–679. [CrossRef]
29. Młyński, D.; Cebulska, M.; Wałęga, A. Trends, variability, and seasonality of Maximum annual daily precipitation in the upper Vistula basin, Poland. *Atmosphere* **2018**, *9*, 313. [CrossRef]
30. Kundzewicz, Z.W.; Stoffel, M.; Kaczka, R.J.; Wyżga, B.; Niedźwiedź, T.; Pińskwar, I.; Ruiz-Villanueva, V.; Łupikasza, E.; Czajka, B.; Ballesteros-Canovas, J.A. Floods at the Northern Foothills of the Tatra Mountains—A Polish–Swiss Research Project. *Acta Geophys.* **2014**, *62*, 620–641. [CrossRef]
31. Węglarczyk, S. Eight reasons to revise the formulas used in calculation of the maximum annual flows with a set exceedance probability in Poland. *Gospod. Wodna* **2015**, *11*, 323–328. (In Polish)
32. Młyński, D.; Wałęga, A.; Petroselli, A.; Tauro, F.; Cebulska, M. Estimating maximum daily precipitation in the upper Vistula basin, Poland. *Atmosphere* **2019**, *10*, 43. [CrossRef]
33. Młyński, D.; Wałęga, A.; Stachura, T.; Kaczor, G. A new empirical approach to calculating flood frequency in ungauged catchments: A case study of the upper Vistula basin, Poland. *Water* **2019**, *11*, 601. [CrossRef]
34. Wachulec, K.; Wałęga, A.; Młyński, D. The effect of time of concentration and rainfall characteristics on runoff hydrograph in small ungauged catchment. *Sci. Rev. Eng. Environ. Sci.* **2016**, *71*, 72–82. (In Polish)
35. Grimaldi, S.; Petroselli, A.; Tauro, F.; Porfiri, M. Time of concentration: A paradox in modern hydrology. *Hydrol. Sci. J.* **2012**, *57*, 217–228. [CrossRef]
36. De Paola, F.; Ranucci, A.; Feo, A. Antecedent moisture condition (SCS) frequency assessment: A case study in southern Italy. *Irrig. Drain.* **2013**, *62*, 61–71. [CrossRef]
37. Grimaldi, S.; Petroselli, A.; Romano, N. Curve-Number/Green–Ampt mixed procedure for streamflow predictions in ungauged basins: Parameter sensitivity analysis. *Hydrol. Process.* **2013**, *27*, 1265–1275. [CrossRef]
38. Wałęga, A.; Cupak, A.; Miernik, W. Influence of entrance parameters on maximum flow quantity receive from NRCS-UH model. *Infrastruct. Ecol. Rural Areas* **2011**, *7*, 85–95.
39. Sigaroodi, S.K.; Chen, Q. Effects and consideration of storm movement in rainfall–runoff modelling at the basin scale. *Hydrol. Earth Syst. Sci.* **2016**, *20*, 5063–5071. [CrossRef]
40. Petroselli, A.; Grimaldi, S.; Piscopia, R.; Tauro, F. Design hydrograph estimation in small and ungauged basins: A comparative assessment of event based (EBA4SUB) and continuous (COSMO4SUB) modeling approaches. *Acta Sci. Pol. Form. Circumiectus* **2019**, *18*, 113–124.

41. Wałęga, A. An attempt to establish regional dependencies for the parameter calculation of the Snyder's synthetic unit hydrograph. *Infrastruct. Ecol. Rural Areas* **2012**, *2*, 5–16.
42. Ajmal, M.; Waseem, M.; Kim, D.; Kim, T.-W. A pragmatic slope-adjusted curve number model to reduce uncertainty in predicting flood runoff from steep watersheds. *Water* **2020**, *12*, 1469. [CrossRef]
43. Bormann, H. Impact of spatial data resolution on simulated catchments water balances and model performance of the multi scale TOPLAST model. *Hydrol. Earth Syst. Sci.* **2006**, *10*, 165–179. [CrossRef]
44. Bárdossy, A.; Das, T. Rainfall network on model calibration and application. *Hydrol. Earth Syst. Sci.* **2008**, *12*, 77–89. [CrossRef]
45. Anctil, F.; Lauzon, N.; Andreassian, V.; Oudin, L.; Perrin, C. Improvement of rainfall-runoff forecasts through mean areal rainfall optimization. *J. Hydrol.* **2006**, *328*, 717–725. [CrossRef]
46. Kovář, P.; Hrabalíková, M.; Neruda, M.; Neruda, R.; Šrejber, J.; Jelínková, A.; Bačinová, H. Choosing an appropriate hydrological model for rainfall-runoff extremes in small catchments. *Soil Water Res.* **2015**, *10*, 137–146. [CrossRef]
47. Lü, H.; Hou, T.; Horton, R.; Zhu, Y.; Chen, X.; Jia, Y.; Wang, W.; Fu, X. The streamflow estimation using the Xinanjiang rainfall runoff model and dual state-parameter estimation method. *J. Hydrol.* **2013**, *480*, 102–114. [CrossRef]
48. Pilgrim, D.H.; Cordery, I. Rainfall temporal patterns for design floods. *J. Hydrol. Div. ASCE* **1975**, *101*, 81–95.
49. Hirabayashi, Y.; Kanae, S.; Emori, S.; Oki, T.; Kimoto, M. Global projections of changing risks of floods and droughts in a changing climate. *Hydrol. Sci. J.* **2008**, *53*, 754–773. [CrossRef]
50. Młyński, D.; Wałęga, A.; Petroselli, A. Verification of empirical formulas for calculating annual peak flows with specific return period in the upper Vistula basin. *Acta Sci. Pol. Form. Circumiectus* **2018**, *17*, 145–154.
51. Młyński, D.; Petroselli, A.; Wałęga, A. Flood frequency analysis by an event-based rainfall-runoff model in selected catchments of southern Poland. *Soil Water Res.* **2018**, *13*, 170–176.
52. Młyński, D.; Wałęga, A.; Książek, L.; Florek, J.; Petroselli, A. Possibility of Using Selected Rainfall-Runoff Models for Determining the Design Hydrograph in Mountainous Catchments: A Case Study in Poland. *Water* **2020**, *12*, 1450. [CrossRef]

MDPI
St. Alban-Anlage 66
4052 Basel
Switzerland
Tel. +41 61 683 77 34
Fax +41 61 302 89 18
www.mdpi.com

Sustainability Editorial Office
E-mail: sustainability@mdpi.com
www.mdpi.com/journal/sustainability